Veterinary Bacteriology

Veterinary Bacteriology

Edited by:

Patricia Marques

Delve PUBLISHING

www.delvepublishing.com

Veterinary Bacteriology

Patricia Marques

Delve Publishing
224 Shoreacres Road
Burlington, ON L7L 2H2
Canada
www.delvepublishing.com

Email: orders@arclereducation.com

ISBN: 978-1-77407-784-9 (Hardcover)

DECLARATION

ABOUT THE EDITOR

Patricia obtained her PhD form University College Dublin in 2010. Her interests are on Microbiology and Parasitology. She is currently working as a Postdoc at University of Maryland Baltimore, USA on Chlamydia infections.

TABLE OF CONTENTS

Section 3: Swine

Section 4: Canines

Section 5: Fish

Section 6: Avian

LIST OF CONTRIBUTORS

Tormod Mørk
Department of Animal Health, National Veterinary Institute, N-0033 Oslo, Norway

Steinar Waage
Department of Production Animal Clinical Sciences, Norwegian School of Veterinary Science, N-0033 Oslo, Norway

Tore Tollersrud
Department of Animal Health, National Veterinary Institute, N-0033 Oslo, Norway

Bjørg Kvitle
Department of Animal Health, National Veterinary Institute, N-0033 Oslo, Norway

Ståle Sviland
Department of Animal Health, National Veterinary Institute, N-0033 Oslo, Norway

Gabriela Herna´ndez-Mora
Servicio Nacional de Salud Animal (SENASA), Ministerio de Agricultura y Ganaderı´a, Heredia, Costa Rica

Roberto Bonilla-Montoya
Servicio Nacional de Salud Animal (SENASA), Ministerio de Agricultura y Ganaderı´a, Heredia, Costa Rica

Osvaldo Barrantes-Granados
Servicio Nacional de Salud Animal (SENASA), Ministerio de Agricultura y Ganaderı´a, Heredia, Costa Rica

Andrea Esquivel-Sua´rez
Servicio Nacional de Salud Animal (SENASA), Ministerio de Agricultura y Ganaderı´a, Heredia, Costa Rica

Danilo Montero-Caballero
Instituto Nacional de Aprendizaje, San Jose´, Costa Rica

Rocı´o Gonza´lez-Barrientos
Servicio Nacional de Salud Animal (SENASA), Ministerio de Agricultura y Ganaderı´a, Heredia, Costa Rica

Zeanne Fallas-Monge
Servicio Nacional de Salud Animal (SENASA), Ministerio de Agricultura y Ganaderı´a, Heredia, Costa Rica

Jose´ David Palacios-Alfaro
Fundacio´n Keto, San Jose´, Costa Rica

Mario Baldi
Programa de Investigacio´n en Enfermedades Tropicales (PIET), Escuela de Medicina Veterinaria, Universidad Nacional, Heredia, Costa Rica

Elena Campos
Centro Nacional de Referencia en Bacteriolog´ıa, Instituto Costarricense de Investigacio´n y Enseñanza en Nutricio´n y Salud (INCIENSA), Cartago, CR

Grettel Chanto
Centro Nacional de Referencia en Bacteriolog´ıa, Instituto Costarricense de Investigacio´n y Enseñanza en Nutricio´n y Salud (INCIENSA), Cartago, CR

Elı´as Barquero-Calvo
Programa de Investigacio´n en Enfermedades Tropicales (PIET), Escuela de Medicina Veterinaria, Universidad Nacional, Heredia, Costa Rica

Carlos Chaco´n-Dı´az
Facultad de Microbiologı´a, Centro de Investigacio´n en Enfermedades Tropicales (CIET), Universidad de Costa Rica, San Jose´, Costa Rica

Esteban Chaves-Olarte
Facultad de Microbiologı´a, Centro de Investigacio´n en Enfermedades Tropicales (CIET), Universidad de Costa Rica, San Jose´, Costa Rica

Caterina Guzma´n Verri
Programa de Investigacio´n en Enfermedades Tropicales (PIET), Escuela de Medicina Veterinaria, Universidad Nacional, Heredia, Costa Rica

Juan-Jose´ Romero-Zu´ñiga
Programa de Investigacio´n en Medicina Poblacional, Escuela de Medicina Veterinaria, Universidad Nacional, Heredia, Costa Rica

Edgardo Moreno
Programa de Investigacio´n en Enfermedades Tropicales (PIET), Escuela de Medicina Veterinaria, Universidad Nacional, Heredia, Costa Rica
Instituto Clodomiro Picado, Universidad de Costa Rica, San Jose´, Costa Rica

Miia Kontturi
Department of Production Animal Medicine, Faculty of Veterinary Medicine, University of Helsinki, Paroninkuja 20, 04920 Saarentaus, Finland

Reijo Junni
Department of Production Animal Medicine, Faculty of Veterinary Medicine, University of Helsinki, Paroninkuja 20, 04920 Saarentaus, Finland

Heli Simojoki
Department of Production Animal Medicine, Faculty of Veterinary Medicine, University of Helsinki, Paroninkuja 20, 04920 Saarentaus, Finland

Erja Malinen
Veterinary Bacteriology and Pathology unit, Finnish Food Authority, Mustialankatu 3, 00790 Helsinki, Finland

Eija Seuna
Veterinary Bacteriology and Pathology unit, Finnish Food Authority, Mustialankatu 3, 00790 Helsinki, Finland

Kirstine Klitgaard
National Veterinary Institute, Technical University of Denmark, Kemitorvet, Building 202, 2800 Kgs. Lyngby, Denmark

Minna Kujala-Wirth
Department of Production Animal Medicine, Faculty of Veterinary Medicine, University of Helsinki, Paroninkuja 20, 04920 Saarentaus, Finland

Timo Soveri
Department of Production Animal Medicine, Faculty of Veterinary Medicine, University of Helsinki, Paroninkuja 20, 04920 Saarentaus, Finland

Sinikka Pelkonen
Veterinary Bacteriology and Pathology unit, Finnish Food Authority, Mustialankatu 3, 00790 Helsinki, Finland

Ana Sofia Ferreira
Instituto Nacional de Investigação Agrária e Veterinária, I.P. (INIAV, I.P.), Unidade Estratégica de Investigação e Serviços em Produção e Saúde Animal, Lisboa, Portugal, School of Pharmacy
Instituto de Cie^ncias Biome´dicas de Abel Salazar, Universidade do Porto, Porto, Portugal

Pedro Costa
Instituto Nacional de Investigação Agrária e Veterinária, I.P. (INIAV, I.P.), Unidade Estratégica de Investigação e Serviços em Produção e Saúde Animal, Lisboa, Portugal, School of Pharmacy
Unidade de Microbiologia Me´dica, Instituto de Higiene e Medicina Tropical, Universidade Nova de Lisboa, Lisboa, Portugal

Teresa Rocha
Instituto Nacional de Investigação Agrária e Veterinária, I.P. (INIAV, I.P.), Unidade Estratégica de Investigação e Serviços em Produção e Saúde Animal, Lisboa, Portugal, School of Pharmacy

Ana Amaro
Instituto Nacional de Investigação Agrária e Veterinária, I.P. (INIAV, I.P.), Unidade Estratégica de Investigação e Serviços em Produção e Saúde Animal, Lisboa, Portugal, School of Pharmacy

Maria Luı´sa Vieira
Unidade de Microbiologia Me´dica, Instituto de Higiene e Medicina Tropical, Universidade Nova de Lisboa, Lisboa, Portugal

Ahmed Ahmed
WHO/ FAO/OIE and National Collaborating Centre for Reference and Research on Leptospirosis, KIT Biomedical Research, Amsterdam, The Netherlands

Gertrude Thompson
Instituto de Cie^ncias Biome´dicas de Abel Salazar, Universidade do Porto, Porto, Portugal
Research Center in Biodiversity and Genetic Resources (CIBIO-ICETA), Universidade do Porto, Porto, Portugal

Rudy A. Hartskeerl
WHO/ FAO/OIE and National Collaborating Centre for Reference and Research on Leptospirosis, KIT Biomedical Research, Amsterdam, The Netherlands

João Inácio
Instituto Nacional de Investigação Agrária e Veterinária, I.P. (INIAV, I.P.), Unidade Estratégica de Investigação e Serviços em Produção e Saúde Animal, Lisboa, Portugal, School of Pharmacy
School of Pharmacy and Biomolecular Sciences, University of Brighton, Brighton, United Kingdom

Åsa Lundberg
Department of Animal Health and Antimicrobial Strategies, National Veterinary Institute, SE-751 89 Uppsala, Sweden
Department of Clinical Sciences, Swedish University of Agricultural Sciences, SE-750 07 Uppsala, Sweden

Ann Nyman
Department of Animal Health and Antimicrobial Strategies, National Veterinary Institute, SE-751 89 Uppsala, Sweden

Helle Ericsson Unnerstad
Department of Animal Health and Antimicrobial Strategies, National Veterinary Institute, SE-751 89 Uppsala, Sweden

Karin Persson Waller
Department of Animal Health and Antimicrobial Strategies, National Veterinary Institute, SE-751 89 Uppsala, Sweden
Department of Clinical Sciences, Swedish University of Agricultural Sciences, SE-750 07 Uppsala, Sweden

Aure´lie Courcoul
Epidemiology Unit, Paris-Est University, Anses, Laboratory for Animal Health, Maisons-Alfort, France

Jean-Louis Moyen
Regional Analysis and Research Laboratory of Dordogne (LDAR24), Coulounieix-Chamiers, France

Laure Bruge` re
Regional Analysis and Research Laboratory of Dordogne (LDAR24), Coulounieix-Chamiers, France

Sandy Faye
Regional Analysis and Research Laboratory of Dordogne (LDAR24), Coulounieix-Chamiers, France

Sylvie He´nault
Bovine Tuberculosis Reference Laboratory, Paris-Est University, Anses, Laboratory for Animal Health, Bacterial Zoonoses Unit, MaisonsAlfort, France

He´le`ne Gares
Regional Analysis and Research Laboratory of Dordogne (LDAR24), Coulounieix-Chamiers, France

Maria-Laura Boschiroli
Bovine Tuberculosis Reference Laboratory, Paris-Est University, Anses, Laboratory for Animal Health, Bacterial Zoonoses Unit, MaisonsAlfort, France

Gerard M. Murray
Sligo Regional Veterinary Laboratory, Department of Agriculture, Food and Marine, Doonally, Sligo, Ireland

Ro´nan G. O'Neill
Central Veterinary Research Laboratory, Department of Agriculture, Food and Marine, Backweston, Celbridge, County Kildare, Ireland

Alison M. Lee
School of Veterinary Medicine, University College Dublin, Belfield, Dublin, Ireland

Ma´ire C. McElroy
Central Veterinary Research Laboratory, Department of Agriculture, Food and Marine, Backweston, Celbridge, County Kildare, Ireland

Simon J. More
Centre for Veterinary Epidemiology and Risk Analysis, UCD School of Veterinary Medicine, University College Dublin, Belfield, Dublin, Ireland

Aisling Monagle
School of Veterinary Medicine, University College Dublin, Belfield, Dublin, Ireland

Bernadette Earley
Animal and Bioscience Research Department, Animal & Grassland Research and Innovation Centre, Teagasc, Grange, Dunsany, County Meath, Ireland

Joseph P. Cassidy
School of Veterinary Medicine, University College Dublin, Belfield, Dublin, Ireland

Stefanie Gobeli Brawand
Institute of Veterinary Bacteriology, University of Bern, Bern, Switzerland

Sonja Kittl
Institute of Veterinary Bacteriology, University of Bern, Bern, Switzerland

Martina Dettwiler
Institute of Animal Pathology, University of Bern, Bern, Switzerland

Andreas Thomann
Institute of Veterinary Bacteriology, University of Bern, Bern, Switzerland

Simon Feyer
Institute of Veterinary Bacteriology, University of Bern, Bern, Switzerland

José Cachim
Animal Health, General Management of Agriculture, Viticulture and Veterinary Affairs (DGAV), Canton of Vaud, Switzerland

Grégoire Theubet
Vétérinaires Mont-Terri Sàrl, Courgenay, Switzerland

Nicole Liechti
Spiez Laboratory, Federal Office for Civil Protection, Spiez, Switzerland
Interfaculty Bioinformatics Unit, University of Bern, Bern, Switzerland

Matthias Wittwer
Spiez Laboratory, Federal Office for Civil Protection, Spiez, Switzerland

Nadia Schürch
Spiez Laboratory, Federal Office for Civil Protection, Spiez, Switzerland

Simone Oberhänsli
Interfaculty Bioinformatics Unit, University of Bern, Bern, Switzerland

Andreas Heinimann
Institute of Geography and Centre for Development and Environment, University of Bern, Bern, Switzerland

Jörg Jores
Institute of Veterinary Bacteriology, University of Bern, Bern, Switzerland

Niels Dekker
Faculty of Veterinary Medicine, Department of Farm Animal Health, Utrecht University, Utrecht, The Netherlands

Faculty of Veterinary Medicine, Department of Infectious Diseases and Immunology, Utrecht University, Utrecht, The Netherlands

Annemarie Bouma
Faculty of Veterinary Medicine, Department of Farm Animal Health, Utrecht University, Utrecht, The Netherlands

Ineke Daemen
Faculty of Veterinary Medicine, Department of Farm Animal Health, Utrecht University, Utrecht, The Netherlands

Don Klinkenberg
Faculty of Veterinary Medicine, Department of Farm Animal Health, Utrecht University, Utrecht, The Netherlands

Leo van Leengoed
Faculty of Veterinary Medicine, Department of Farm Animal Health, Utrecht University, Utrecht, The Netherlands

Jaap A. Wagenaar
Faculty of Veterinary Medicine, Department of Infectious Diseases and Immunology, Utrecht University, Utrecht, The Netherlands

Arjan Stegeman
Faculty of Veterinary Medicine, Department of Farm Animal Health, Utrecht University, Utrecht, The Netherlands

Tijs J Tobias
Department of Farm Animal Health, Faculty of Veterinary Medicine, Utrecht University, Yalelaan 7, Utrecht 3584 CL, The Netherlands

Annemarie Bouma
Department of Farm Animal Health, Faculty of Veterinary Medicine, Utrecht University, Yalelaan 7, Utrecht 3584 CL, The Netherlands

Angeline JJM Daemen
Department of Farm Animal Health, Faculty of Veterinary Medicine, Utrecht University, Yalelaan 7, Utrecht 3584 CL, The Netherlands

Jaap A Wagenaar
Department of Infectious Diseases and Immunology, Faculty of Veterinary Medicine, Utrecht University, Yalelaan 1, Utrecht 3584 CL, The Netherlands
Central Veterinary Institute of Wageningen UR, Lelystad 8200 AB, The Netherlands

Arjan Stegeman
Department of Farm Animal Health, Faculty of Veterinary Medicine, Utrecht University, Yalelaan 7, Utrecht 3584 CL, The Netherlands

Don Klinkenberg
Department of Farm Animal Health, Faculty of Veterinary Medicine, Utrecht University, Yalelaan 7, Utrecht 3584 CL, The Netherlands

Kristien De Puysseleyr
Department of Molecular Biotechnology, Faculty of Bioscience Engineering, University of Ghent, Ghent, Belgium

Leentje De Puysseleyr
Department of Molecular Biotechnology, Faculty of Bioscience Engineering, University of Ghent, Ghent, Belgium

Julie Geldhof
Department of Molecular Biotechnology, Faculty of Bioscience Engineering, University of Ghent, Ghent, Belgium

Eric Cox
Department of Virology, Parasitology and Immunology, Faculty of Veterinary Medicine, University of Ghent, Ghent, Belgium

Daisy Vanrompay
Department of Molecular Biotechnology, Faculty of Bioscience Engineering, University of Ghent, Ghent, Belgium

Manuela Oliveira
CIISA/Faculdade de Medicina Veterinária, Universidade de Lisboa, Avenida da Universidade Técnica, 1300-477 Lisboa, Portugal

Cynthia Barroco
CIISA/Faculdade de Medicina Veterinária, Universidade de Lisboa, Avenida da Universidade Técnica, 1300-477 Lisboa, Portugal

Carla Mottola
CIISA/Faculdade de Medicina Veterinária, Universidade de Lisboa, Avenida da Universidade Técnica, 1300-477 Lisboa, Portugal

Raquel Santos
CIISA/Faculdade de Medicina Veterinária, Universidade de Lisboa, Avenida da Universidade Técnica, 1300-477 Lisboa, Portugal

Abdelhak Lemsaddek
BioFIG, Centro para a Biodiversidade, Genómica Integrativa e Funcional, Faculdade de Ciências, Universidade de Lisboa, 1749-016 Lisboa, Portugal

Luis Tavares
CIISA/Faculdade de Medicina Veterinária, Universidade de Lisboa, Avenida da Universidade Técnica, 1300-477 Lisboa, Portugal

Teresa Semedo-Lemsaddek
CIISA/Faculdade de Medicina Veterinária, Universidade de Lisboa, Avenida da Universidade Técnica, 1300-477 Lisboa, Portugal

Jisun Sun
Department of Veterinary Population Medicine, University of Minnesota, 385 ASVM, 1988 Fitch Ave, St. Paul, MN 55108, USA

My Yang
Department of Veterinary Population Medicine, University of Minnesota, 385 ASVM, 1988 Fitch Ave, St. Paul, MN 55108, USA

Srinand Sreevatsan
Department of Veterinary Population Medicine, University of Minnesota, 385 ASVM, 1988 Fitch Ave, St. Paul, MN 55108, USA

Jeffrey B. Bender
Department of Veterinary Population Medicine, University of Minnesota, 385 ASVM, 1988 Fitch Ave, St. Paul, MN 55108, USA

Randall S. Singer
Department of Veterinary and Biomedical Sciences, University of Minnesota, St. Paul, MN 55108, USA

Todd P. Knutson
Department of Veterinary Population Medicine, University of Minnesota, 385 ASVM, 1988 Fitch Ave, St. Paul, MN 55108, USA

Douglas G. Marthaler
Department of Veterinary Population Medicine, University of Minnesota, 385 ASVM, 1988 Fitch Ave, St. Paul, MN 55108, USA

Peter R. Davies
Department of Veterinary Population Medicine, University of Minnesota, 385 ASVM, 1988 Fitch Ave, St. Paul, MN 55108, USA

Floyd E. Dewhirst
Department of Molecular Genetics, The Forsyth Institute, Cambridge, Massachusetts, United States of America
Department of Oral Medicine, Infection and Immunity, Harvard School of Dental Medicine, Boston, Massachusetts, United States of America

Erin A. Klein
Department of Molecular Genetics, The Forsyth Institute, Cambridge, Massachusetts, United States of America

Emily C. Thompson
Department of Molecular Genetics, The Forsyth Institute, Cambridge, Massachusetts, United States of America

Jessica M. Blanton
Department of Molecular Genetics, The Forsyth Institute, Cambridge, Massachusetts, United States of America

Tsute Chen
Department of Molecular Genetics, The Forsyth Institute, Cambridge, Massachusetts, United States of America

Lisa Milella
The Veterinary Dental Surgery, Byfleet, United Kingdom

Catherine M. F. Buckley
WALTHAM Centre for Pet Nutrition, Melton Mowbray, United Kingdom

Ian J. Davis
WALTHAM Centre for Pet Nutrition, Melton Mowbray, United Kingdom

Marie-Lousie Bennett
Mars Pet Care Europe, Birstall, United Kingdom

Zoe V. Marshall-Jones
WALTHAM Centre for Pet Nutrition, Melton Mowbray, United Kingdom

Mia Holmberg
Department of Epidemiology and Disease Control, National Veterinary Institute, SVA, SE-751 89 Uppsala, Sweden

Thomas Rosendal
Department of Epidemiology and Disease Control, National Veterinary Institute, SVA, SE-751 89 Uppsala, Sweden

Eva O Engvall
Department of Bacteriology, National Veterinary Institute, SVA, SE-751 89 Uppsala, Sweden

Anna Ohlson
Växa Sverige, SE-104 25 Stockholm, Sweden

Ann Lindberg
Department of Epidemiology and Disease Control, National Veterinary Institute, SVA, SE-751 89 Uppsala, Sweden

Brian Austin
Institute of Aquaculture, Pathfoot Building, University of Stirling, Stirling FK9 4LA, Scotland, UK

Leydis Zamora
Centro de Vigilancia Sanitaria Veterinaria (VISAVET), Universidad Complutense, Madrid, Spain

Jose´ F. Ferna´ndez-Garayza´bal
Centro de Vigilancia Sanitaria Veterinaria (VISAVET), Universidad Complutense, Madrid, Spain
Departamento de Sanidad Animal, Facultad de Veterinaria, Universidad Complutense, Madrid, Spain

Cristina Sa´nchez-Porro
Departamento de Microbiologı´a y Parasitologı´a, Facultad de Farmacia, Universidad de Sevilla, Sevilla, Spain

Mari Angel Palacios
Piszolla, S.L., Alba de Tormes, Salamanca, Spain

Edward R. B. Moore
Culture Collection University of Gothenburg (CCUG) and Department of Infectious Disease, Sahlgrenska Academy of the University of Gothenburg, Go¨teborg, Sweden

Lucas Domı´nguez
Centro de Vigilancia Sanitaria Veterinaria (VISAVET), Universidad Complutense, Madrid, Spain

Antonio Ventosa
Departamento de Microbiologı´a y Parasitologı´a, Facultad de Farmacia, Universidad de Sevilla, Sevilla, Spain

Ana I. Vela
Centro de Vigilancia Sanitaria Veterinaria (VISAVET), Universidad Complutense, Madrid, Spain
Departamento de Sanidad Animal, Facultad de Veterinaria, Universidad Complutense, Madrid, Spain

Rocio Crespo
Avian Health and Food Safety Laboratory, Washington Animal Disease Diagnostic Laboratory, Washington State University, 2607 West Pioneer, Puyallup, WA 98371, USA

Michael M Garner
Northwest ZooPath, 654 W. Main St., Monroe, WA 98272, USA

Sharon G Hopkins
Public Health–Seattle & King County, 401 5th Avenue, Ste 1100, Seattle, WA 98104, USA

Devendra H Shah
Department of Veterinary Microbiology, Washington State University, Pullman, WA 99164-7040, USA
Pathology and Paul Allen School for Global Animal Health, Washington State University, Pullman, WA 99164-7040, USA

Shaohui Wang
Shanghai Veterinary Research Institute, Chinese Academy of Agricultural Sciences, Shanghai, China

Qingmei Meng
Shanghai Veterinary Research Institute, Chinese Academy of Agricultural Sciences, Shanghai, China
Key Lab of Animal Bacteriology, Ministry of Agriculture, Nanjing Agricultural University, Nanjing, Jiangsu, China

Jianjun Dai
Key Lab of Animal Bacteriology, Ministry of Agriculture, Nanjing Agricultural University, Nanjing, Jiangsu, China

Xiangan Han
Shanghai Veterinary Research Institute, Chinese Academy of Agricultural Sciences, Shanghai, China

Yue Han
Shanghai Veterinary Research Institute, Chinese Academy of Agricultural Sciences, Shanghai, China

Chan Ding
Shanghai Veterinary Research Institute, Chinese Academy of Agricultural Sciences, Shanghai, China

Haiwen Liu
Shanghai Veterinary Research Institute, Chinese Academy of Agricultural Sciences, Shanghai, China

Shengqing Yu
Shanghai Veterinary Research Institute, Chinese Academy of Agricultural Sciences, Shanghai, China

LIST OF ABBREVIATIONS

AIC	Akaike's Information Criterion
AASV	American Association of Swine Veterinarians
AFLP	amplified fragment-length polymorphism PCR
AT	Array tube
AHFSL	Avian Health and Food Safety Laboratory
BRD	Bovine respiratory disease
BRSV	bovine respiratory syncytial virus
bTB	Bovine tuberculosis
CMT	California Mastitis Test
CLA	caseous lymphadenitis
CVRL	Central Veterinary Research Laboratory
CAMP	Christie–Atkins–Munch–Petersen
CM	Clinical mastitis
CS	clinical score
CC	clonal complex
CNS	coagulase-negative staphylococci
CFU	colony forming units
CFT	Complement Fixation Test
CR	Costa Rica
CR-NAHS	Costa Rican National Animal Health Service
CI	credible intervals
CAAT	Cross Agglutinin Absorption Test
DPI	days post inoculation
DAP	diaminopimelic acid
DD	digital dermatitis
DD	Disk Diffusion
EMJH	Ellinghausen-McCullough-JohnsonHarris
GE	genome equivalents

HEX	hexachloro-fluoresceine
HOMD	Human Oral Microbiome Database
IHC	immunohistochemistry
iFAT	indirect fluorescent antibody test
iNOS	inducible nitric oxide synthase
ITPA	Institute of Animal Pathology
ID	interdigital dermatitis
IP	interdigital phlegmon
IC	intermittent carriers
IAC	internal amplification control
ITS 1	internal transcribed spacer 1
IMI	intramammary infection
LSM	Least square means
MEGA	Molecular Evolutionary Genetics Analysis
MAT	Microscopic Agglutination Test
MST	minimal spanning tree
mCCDA	modified charcoal, cefoperazone, desoxycholate agar
MLSA	multi locus sequence analysis
MLST	multi-locus sequence typing
NGS	next generation sequence
NO	nitric oxide
NTC	No-template controls
OD	optical density
PP	Percent positivity
PC	persistent carriers
PCR	polymerase chain reaction
PME	post-mortem examination
PFGE	Pulsed Field Gel Electrophoresis
RAPD	random amplified polymorphic DNA
RT-PCR	real time polymerase chain reaction
RVL	Regional Veterinary Laboratory
RBT	Rose Bengal test
SBE	selective bacteriologic examination
SKDM	selective kidney disease medium

SNPs	single nucleotide polymorphisms
SRVL	Sligo Regional Veterinary Laboratory
SCC	somatic cell count
SADRS	Swedish Animal Disease Recording System
DIC	Deviance Information Criterion
TPC	true persistent carriers
VTCM	veterinary-treated cases of clinical mastitis

PREFACE

Bacteriology is the branch and specialty of biology that studies the morphology, ecology, genetics and biochemistry of bacteria as well as many other aspects related to them. This subdivision of microbiology involves the identification, classification, and characterization of bacterial species. Because of the similarity of thinking and working with microorganisms other than bacteria, such as protozoa, fungi, and viruses, there has been a tendency for the field of bacteriology to extend as microbiology. The terms were formerly often used interchangeably. However, bacteriology can be classified as a distinct science.

Veterinary Bacteriology is the study of bacteria that cause diseases on animals.

In the current book, is separated by groups of animals, with section 1 dedicated to Ovines and Caprines, focusing in bacteria that cause Mastitis, Brucellosis and Interdigital phlegmon.

Section 2 is dedicated to Bovine bacterial infections, Leptospira, Streptococcus, Tuberculosis, Antrax and respiratory diseases.

Section 3 is dedicated to Swine, Streptococcus, Actinobacillus, Chlamydia, Corynebacterium and Staphylococcus.

Studying the canine Microbiome is part of section 4. The second chapter of section 4 cover campylobacter in canines.

Section 5 studies the taxonomy of bacteria in fish.

Section 6 is dedicated to Avian bacteriology, focusing in Listeria and Escherichia coli.

SECTION 1:
OVINES AND CAPRINES

CHAPTER 1

Clinical Mastitis in Ewes; Bacteriology, Epidemiology and Clinical Features

Tormod Mørk[1], Steinar Waage[2], Tore Tollersrud[1], Bjørg Kvitle[1] and Ståle Sviland[1]

[1]Department of Animal Health, National Veterinary Institute, N-0033 Oslo, Norway
[2]Department of Production Animal Clinical Sciences, Norwegian School of Veterinary Science, N-0033 Oslo, Norway

ABSTRACT

Background

Clinical mastitis is an important disease in sheep. The objective of this work was to identify causal bacteria and study certain epidemiological and clinical features of clinical mastitis in ewes kept for meat and wool production.

Citation: Mørk, T., Waage, S., Tollersrud, T. et al. Clinical mastitis in ewes; bacteriology, epidemiology and clinical features. Acta Vet Scand 49, 23 (2007). https://doi.org/10.1186/1751-0147-49-23.

METHODS

The study included 509 ewes with clinical mastitis from 353 flocks located in 14 of the 19 counties in Norway. Clinical examination and collection of udder secretions were carried out by veterinarians. Pulsed-field gel electrophoresis (PFGE) was performed on 92 *Staphylococcus aureus* isolates from 64 ewes.

Results and conclusion

S. aureus was recovered from 65.3% of 547 clinically affected mammary glands, coagulase-negative staphylococci from 2.9%, enterobacteria, mainly *Escherichia coli*, from 7.3%, *Streptococcus* spp. from 4.6%, *Mannheimia haemolytica* from 1.8% and various other bacteria from 4.9%, while no bacteria were cultured from 13.2% of the samples. Forty percent of the ewes with unilateral clinical *S. aureus* mastitis also had a subclinical *S. aureus* infection in the other mammary gland. Twenty-four of 28 (86%) pairs of *S. aureus* isolates obtained from clinically and subclinically affected mammary glands of the same ewe were indistinguishable by PFGE. The number of identical pairs was significantly greater than expected, based on the distribution of different *S. aureus* types within the flocks. One-third of the cases occurred during the first week after lambing, while a second peak was observed in the third week of lactation. Gangrene was present in 8.8% of the clinically affected glands; *S. aureus* was recovered from 72.9%, *Clostridium perfringens* from 6.3% and *E. coli* from 6.3% of the secretions from such glands. This study shows that *S. aureus* predominates as a cause of clinical ovine mastitis in Norway, also in very severe cases. Results also indicate that *S. aureus* is frequently spread between udder halves of infected ewes.

BACKGROUND

Mastitis is an important disease in sheep. Clinical cases are often severe; systemic signs are present and the condition is obviously painful. Clinically affected glands frequently suffer partial or complete damage and do not resume normal function. Reduced milk yield leads to decreased growth of the lambs [1-3]. Additional losses associated with clinical mastitis are costs of treatment and culling of ewes due to permanent udder damage [3-7]. In very severe cases, gangrene may develop in the mammary gland and the ewe may die. Thus, mastitis has a major impact on both economy and animal welfare in sheep production.

Although a wide range of microorganisms may cause ovine mastitis, most cases are reported to be due to staphylococci [8]. Several reports indicate that coagulase-negative staphylococci (CNS) are the most common cause of subclinical mastitis in dairy ewes [9-14], while both CNS and *Staphylococcus aureus* are frequent causes in meat sheep [5,15,16]. With regard to organisms associated with clinical mastitis, there are fewer reports published. *S. aureus* has been reported to be the most common causal organism in both meat [5,15,17-19] and dairy ewes [13,20,21]. *Mannheimia haemolytica* [5,18,19,22], *Escherichia coli* [13,18,19] and various streptococci [15,18,19] are other important causative organisms.

Differences in climate, production forms and management practices may give rise to differences both in the epidemiology, bacteriology and clinical manifestations of mastitis. In Norway, sheep are kept exclusively for meat and wool production. They are housed during the winter and early spring, including the lambing season.

The objective of this study was to identify bacteria associated with clinical ovine mastitis in Norway. In addition, certain epidemiological and clinical features of the disease were studied.

METHODS

Animals and Clinical Data

Udder secretions were collected and clinical data recorded from 509 ewes with clinical mastitis. The ewes belonged to 353 flocks located in 14 counties in Norway (Figure 1). The geographical distribution of the cases is shown in Table 1. Clinical mastitis was present in one gland in 471 ewes and in both glands in 38 ewes. The study was carried out in 2002, 2003 and 2004. Only cases that occurred between 1 week prepartum and 8 weeks postpartum were included. In Norway, lambing generally takes place in April and May.

Table 1: Distribution by region and county of 547 milk samples obtained from ovine mammary glands with clinical mastitis, and of the 509[a] ewes and 353 flocks from which the samples originated

Region	County	No. of flocks	No. of ewes	No. of glands
East	Akershus	13	27	31
	Hedmark	68	121	128

	Oppland	60	85	86
	Buskerud	4	5	6
South	Aust-Agder	34	56	57
	Vest-Agder	4	6	7
	Rogaland	40	48	53
West	Hordaland	22	26	28
	Sogn og Fjordane	25	25	29
	Møre og Romsdal	32	39	40
North	Sør-Trøndelag	33	46	50
	Nord-Trøndelag	6	10	13
	Nordland	3	3	4
	Troms	9	12	15

[a] Four hundred and seventy-one ewes with unilateral and 38 with bilateral intramammary infection.

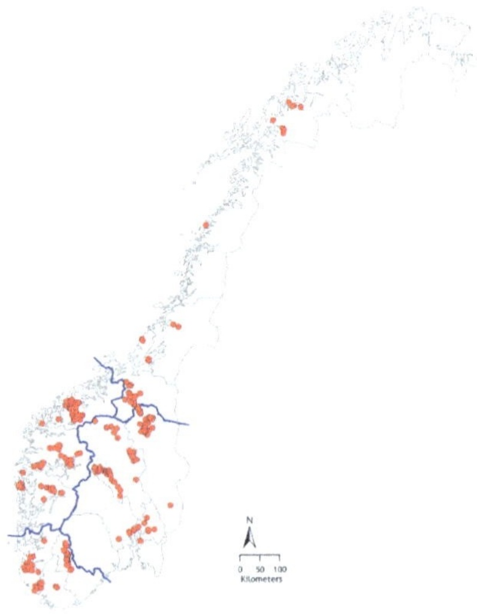

Figure 1: Map of Norway showing the location of the sheep flocks from which cases of clinical mastitis were obtained. Thin lines show county boundaries and thick lines region boundaries.

Data and Sample Collection

Thirty-two veterinary practitioners contributed to the study. When called to a case of clinical mastitis, the veterinarian was to examine the ewe, collect udder secretions and record information regarding the identity, age, date of parturition, the number of lambs and the clinical condition of the ewe on a standardized form. Clinical data included the rectal temperature, an assessment of the severity of systemic signs (graded as none, weak, moderate or severe) and local clinical signs of the affected gland, including whether or not gangrene was present (i.e., cold and blue udder and teat skin).

Prior to treatment, samples were collected aseptically from the clinically affected glands in 10-ml sterile plastic vials by the veterinary practitioner according to the International Dairy Federation's standards [23]. Additionally, samples were taken from the clinically unaffected gland of 252 of the ewes with unilateral clinical mastitis. The samples were sent by mail to the laboratory as soon as possible after sampling, or frozen and stored at -20°C until submission.

If a ewe experienced more than one episode of mastitis during the observation period, only the first episode was included in the study.

Microbiological Methods

The samples were examined at the National Veterinary Institute or at the TINE Mastitis Laboratory in Molde, Norway, and bacteria were identified according to the recommendations of the International Dairy Federation [23] with additions. The National Veterinary Institute and the TINE Mastitis Laboratory are both quality assured in accordance with NS-EN ISO/IEC 17025. Briefly, the secretions were brought to room temperature, assessed visually and characterized by appearance before they were mechanically shaken and 10 μl plated on Bacto Blood Agar Base No 2 (Difco Laboratories, Detroit, MI, USA) containing 5% washed bovine erythrocytes and incubated for 48 hours in a 5% CO_2 atmosphere at 37°C. Cultures were read at 24 and 48 hours. If growth was not detected after incubation for 24 hours, the original sample was incubated for 4 hours at 37°C and 50 μl aliquots plated and incubated for 24 hours under aerobic (5% CO_2 atmosphere) and anaerobic conditions.

Bacterial species were identified tentatively by their gross colony morphology and by Gram staining, and further confirmatory tests were used as necessary. All suspected staphylococcal colonies were tested using the tube coagulase test (Becton Dickinson Microbiology Systems, Bedford,

MA, USA). Coagulase-positive staphylococci were streaked on peptone agar (p-agar) (Difco, Sparks, MD) supplemented with 7 mg/l of acriflavin (Sigma-Aldrich Chemie, Steinheim, Germany) [24], and incubated at 37°C for 24 hours. Bacterial growth in the full length of the streak on p-agar was considered confirmative of *S. aureus*. Isolates identified as *S. aureus* were stored at -70°C in Bacto Heart Infusion Broth (Difco) with 15% glycerol. *E. coli* was identified by the lactose and indole tests, and other enterobacteria were identified to the species or genus level by using a microtube identification system (API 20 E®; bioMérieux S.A., Marcy-l'Etoile, France). *Streptococcus uberis*, *Streptococcus dysgalactiae*, *Streptococcus agalactiae*, *Streptococcus* spp. and *Enterococcus* spp. were distinguished by the CAMP reaction, the aesculin and inulin tests and by culture on the bromthymolblue lactose-sucrose agar. Bacteria within the family *Pasteurellaceae* were identified to the species level by the CAMP reaction, the indole, mannitol, sorbitol, trehalose, dulcitol, oxidase and beta-galactosidase tests and the haemolysis patterns. *Clostridium perfringens* was differentiated from other *Clostridium* spp. by colony morphology, immobility and the presence of a zone of partial haemolysis and a zone of complete haemolysis. *Arcanobacterium pyogenes* was identified by Gram staining and the presence of pinpoint colonies surrounded by a narrow zone of clear haemolysis at 48 hours. None of the samples were from ewes with arthritis, conjunctivitis or pneumonia; therefore, the mammary secretions were not checked for the presence of mycoplasms.

Total DNA was prepared and pulsed-field gel electrophoresis (PFGE) performed as described previously [25] on 92 *S. aureus* isolates from 21 flocks in which at least one ewe had bilateral *S. aureus* intramammary infection (IMI) and where at least two ewes experienced clinical *S. aureus* mastitis. The band patterns were compared visually. Isolates with indistinguishable patterns were considered identical PFGE types while those with at least one band difference were considered to be different types.

Statistical Methods

The chi-square test was used to compare the frequencies of cases within different time intervals in relation to parturition and the relative proportions of clinical *S. aureus* cases and gangrenous mastitis cases in ewes of different parity and with different number of lambs.

The distribution of pairs of *S. aureus* PFGE types within flocks (equal vs. unequal) in ewes with bilateral IMI was compared with the corresponding distribution that would be expected if all *S. aureus* isolates found within each flock were paired randomly. All isolates from the flocks that supplied two or more cases of clinical *S. aureus* mastitis, of which at least one ewe had bilateral IMI, were included, and Fisher's exact test was used to test the probability of identical distributions of the observed and expected pairs.

$P < 0.05$ was considered statistically significant.

RESULTS

Bacteriological and Epidemiological Findings

The distribution of bacteria cultured from secretions from the glands with clinical mastitis is shown in Table Table2.2. *S. aureus* was the predominant pathogen and was found in 65.3% of the samples from affected glands. In the samples from the southern, eastern, western and northern regions, *S. aureus* was found in 76.0%, 59.0%, 63.9% and 69.5%, respectively.

Table 2: Results of culture of secretions recovered from 547 mammary glands with clinical mastitis

Bacteriological finding	n	%
Staphylococcus aureus	357	65.3
Coagulase-negative staphylococci	16	2.9
Streptococcus uberis	9	1.6
Streptococcus dysgalactiae[a]	8	1.5
Streptococcus spp.[b]	8	1.5
Enterococcus spp.	4	0.7
Escherichia coli	35	6.4
Klebsiella pneumoniae	2	0.4
Enterobacter spp.	3	0.5
Mannheimia haemolytica	10	1.8
Arcanobacterium pyogenes	4	0.7
Clostridium perfringens	7	1.3
Pasteurella spp.[c]	4	0.7
No growth	72	13.2
Contaminated samples[d]	8	1.5

[a] Subsp. *dysgalactiae*.

[b] Other than *Str. dysgalactiae* and *Str. uberis*.
[c] *Pasteurella mairii* (2 samples), *Pasteurella multocida* (2 samples).
[d] If blood agar plates contained more than two different types of colonies.

Information about the date of parturition was received from 318 of the 471 cases of unilateral clinical mastitis. The distribution of the observed clinical mastitis cases in relation to the time of parturition is shown in Figure 2. The relative proportion of cases was greatest during the first week after lambing. Sixty-four (20.1%) of the 318 ewes for which the times of lambing and treatment were recorded, were treated for clinical mastitis during the first two days after parturition. The proportion of cases was significantly greater in the first ($P < 0.005$) and the third ($P < 0.05$) week postpartum as compared with the second week postpartum.

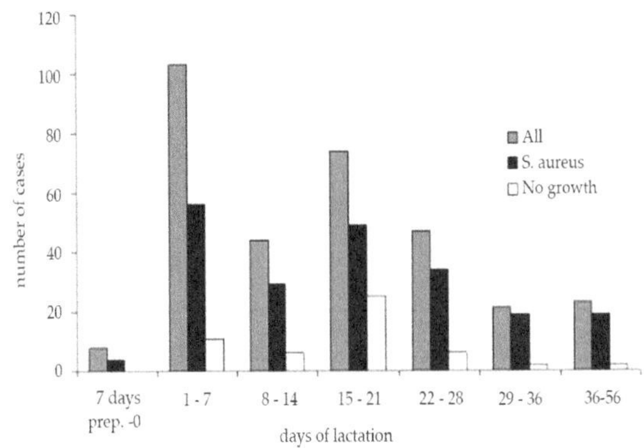

Figure 2: Distribution of 318 cases of clinical mastitis in relation to weeks of lactation.

The proportion of the clinical *S. aureus* cases among all the clinical cases did not differ significantly between weeks of lactation, between ewes of different parities or between ewes with different number of lambs (data not shown).

In 180 ewes with unilateral clinical *S. aureus* mastitis, from which samples from both mammary glands were examined, *S. aureus* was found in secretions from 72 (40.0%) of the glands without clinical signs. In 72 ewes with unilateral clinical mastitis not caused by *S. aureus*, a subclinical *S. aureus* infection was found to be present in 10 (13.9%) of the glands without clinical signs.

S. aureus PFGE types in Ewes with Bilateral IMI

A total of 22 different PFGE types were found among 92 *S. aureus* isolates from 21 farms. Comparisons of *S. aureus* pairs revealed that 24 (86.0%) of 28 pairs had indistinguishable band patterns (Table (Table3).3). Given the PFGE types of all *S. aureus* isolates within each flock, expected flock-specific distributions of PFGE type pairs based on all possible random pairwise combinations of isolates were arranged. The number of pairs with identical types observed in the flocks was significantly greater than that expected, when assuming a random distribution of isolates ($P < 0.0001$).

Table 3: Distribution within flocks of 22 different PFGE types of 92 *S. aureus* isolates from ewes with intramammary infection (IMI). Only flocks in which at least one ewe experienced bilateral *S. aureus* IMI and two ewes experienced clinical mastitis are included. The observed pairs of PFGE types from the ewes with bilateral *S. aureus* IMI are shown. Also shown is the expected distribution of pairs with equal and unequal PFGE type combinations when assuming random pairwise distribution of the observed isolates within each flock

				Expected pairs (n)	
Flock	Ewes with IMI (n)	*S. auerus* types (n) present in flock	Bilateral PFGE type combinations	Equal	Unequal
F1	6	K, P, Q (4), R	QQ	6	15
F2	9	I, J, L (7), M	LL	21	24
F3	2	R (3)	RR	3	0
F4	3	R (4)	RR	6	0
F5	2	H (2), O (2)	HH, OO	2	4
F6	2	D (2), H	DD	1	2
F7	2	H, I, U	IU	0	3
F8	2	H (3), X	HH, HX	3	3
F9	2	D, H (3)	DH, HH	3	3
F10	2	I, M, V	IV	0	3
F11	6	H (5), I, W	HH	10	11
F12	3	H, I (3)	II	3	3
F13	3	H (4), I (2)	HH, HH, II	7	8
F14	3	F, J (3)	JJ	3	3
F15	2	H, S (2)	SS	1	2

F16	2	D (3)	DD	3	0
F17	4	H (2), I (2), N, G (2)	HH, II, GG	3	18
F18	2	C (2), E	CC	1	2
F19	2	C (2), G	CC	1	2
F20	3	I (2), R, T	II	1	5
F21	2	A (2), B	AA	1	2

Clinical Signs

Of the 471 cases of unilateral clinical mastitis systemic signs were recorded in 325 (Table (Table4)4) and the rectal temperature was measured in 342. Moderate or severe systemic signs were present in 159 (48.9%) ewes and 193 (56.4%) ewes had a rectal temperature above 40.0°C. Thirty-seven (11.8%) out of 313 ewes showed no systemic signs and had a rectal temperature below 40.1°C.

Table 4: Distribution of 325 of the 471 ewes with unilateral clinical mastitis by causal organism and the systemic signs[a]

Pathogen	1 No. (%)	2 No. (%)	3 No. (%)	4 No. (%)	Total
All	51 (15.7)	115 (35.4)	119 (36.6)	40 (12.3)	325
S. aureus	30 (13.5)	77 (34.5)	89 (39.9)	27 (12.1)	223
Enterobacteria	2 (7.7)	9 (34.6)	11 (42.3)	4 (15.4)	26
No growth	11 (34.4)	12 (37.5)	4 (12.5)	5 (15.6)	32

[a] 1 = no systemic signs, 2 = weak systemic signs, 3 = moderate systemic signs, 4 = severe systemic signs.

Gangrene was present in 48 (8.8%) of 547 clinically affected udder halves and *S. aureus* was found in 35 (72.9%), *C. perfringens* in 3 (6.3%), *E. coli* in 3 (6.3%) and *A. pyogenes* in 1 (2.1%). Bacteria were not found in 4 (8.3%) samples from such cases and two samples were contaminated. The degree of systemic influence was recorded in 33 ewes with gangrenous mastitis. Thirty (90.9%) of these cases exhibited moderate or severe signs. The proportion of the gangrenous mastitis cases among all the clinical cases did not differ significantly between weeks of lactation, between ewes of different parities or between ewes with different number of lambs (data not shown).

DISCUSSION

A random and representative selection of sheep flocks was not deemed feasible for this type of study. However, in order to have a reasonably representative geographical spread of the mastitis cases ewes from 14 of the 19 counties in Norway were included. Thus, the present flocks were housed and pastured under various conditions, and the variations in climatic conditions, flock size and management routines of Norwegian sheep production were reasonably well represented. During the lambing season and subsequent weeks the ewes are paid close attention. Later, most flocks are moved to pastures in the forests or mountains. For this reason, the study was restricted to cases that occurred between 1 week prepartum and 8 weeks postpartum in order to obtain clinical cases of recent origin.

S. aureus was found in 65% of the samples from clinically affected glands. The dominance of *S. aureus* as a cause of clinical ovine mastitis has also been shown in regional studies in Norway, where *S. aureus* was isolated from udder secretions of between 64 and 87% of the ewes [26,27]. In the present study, the largest proportion of *S. aureus* was found in the southern region (76%). A similar proportion of such cases (75%) was previously observed in a study including cases from one of the municipalities of this region [28]. Studies of clinical mastitis in meat sheep in other countries have found varying, though mostly relatively great, proportions of *S. aureus* infections (20–60%) [5,17-20,22]. In a study of dairy sheep in Jordan, Lafi et al. [13] found that 22% of the clinical mastitis cases were caused by *S. aureus*.

The main *S. aureus* reservoirs in sheep are suggested to be infected mammary glands and teat lesions [29]. However, *S. aureus* can also be cultured from intact teat skin and other body sites [10,30,31]. In dairy flocks, transfer during milking is considered an important mechanism for the spread of this organism [29]. In flocks of meat sheep, transmission of *S. aureus* between ewes could be a result of the herdsman transmitting *S. aureus* between ewes during manual udder control, or the udder being exposed to bedding material contaminated from infected ewes [6,7]. Some lambs occasionally suck other ewes than their dam, which might be a mechanism for the spread of *S. aureus* [32]. In Norway, routine examination of teats and udders are performed after weaning, and ewes with palpable abnormalities or which have experienced clinical mastitis are usually slaughtered before the breeding season. This contributes to decreasing the reservoir of *S. aureus*, but it obviously does not eliminate it.

Studies in other countries have reported prevalences of subclinical *S. aureus* IMI to be between 1 and 6% [12,13,15,33]. This indicates that subclinically infected glands are an important reservoir of *S. aureus* that can only be detected through bacteriological examination. In this study, 14% of the ewes with unilateral clinical mastitis caused by other pathogens than *S. aureus* had a subclinical *S. aureus* infection in the other mammary gland, while 40% of the ewes with clinical *S. aureus* infection in one gland had a subclinical *S. aureus* infection in the other. PFGE typing showed that 86% of the pairs of isolates from ewes with bilateral *S. aureus* IMI were indistinguishable. This percentage was much higher than what would be expected if the isolates from each of the flocks were paired at random, thus demonstrating a considerably greater tendency for spread of *S. aureus* between the udder halves of a ewe than between ewes within a flock.

The very low percentage of *M. haemolytica* in the clinically affected glands in this study (1.8%) contrasts results of clinical mastitis surveys in meat sheep in the UK and Ireland, where the proportions of cases caused by this organism were found to be approximately 50% and 21%, respectively [5,18]. Enterobacteria, mainly *E. coli*, were obtained from 7.3% of the clinically affected udder halves, which is similar to the proportions found in other studies on meat sheep [5,18,22]. However, the number of clinical cases caused by Gram negative bacteria and *A. pyogenes* may be underestimated because the samples were frozen before bacteriological analysis [34,35].

It is noteworthy that nearly 85% of the ewes exhibited systemic signs and that gangrene was present in as much as 9% of the clinically affected udder halves, clearly showing that ovine clinical mastitis must be considered a serious animal welfare problem. However, mild cases of clinical mastitis are most likely underrepresented in this study. According to Norwegian legislation, antibiotic treatment of animals must be initiated by a veterinarian and, for economical reasons, farmers might be reluctant to call for a veterinary surgeon to treat mild clinical cases.

Most mastitis cases occurred close to lambing. One-third of the ewes developed clinical mastitis during the first week after lambing, and a second peak, although somewhat smaller, was observed in the third week postpartum. This is in accordance with data from the Norwegian Sheep Recording System [36]. Likewise, studies in the UK and Ireland found that cases of acute clinical mastitis occurred most frequently during the first week of lactation, while a second peak occurred between the third and fourth week [5] or the fourth and seventh week [18] after lambing. Clinical cases around parturition might be newly acquired IMI or aggravations of existing subclinical infections [37]. The proportion of very severe cases, in which gangrene had developed, was not greater among cases occurring close to lambing as compared with those occurring later. The second peak could be explained by increased milk demand from the lambs and the eruption of incisors, which increases the risk of teat lesions. It has been reported that teat lesions frequently are present in ewes with clinical mastitis three to four weeks after lambing [38].

CONCLUSION

This study shows that *S. aureus* is the most common cause of clinical mastitis in sheep in Norway and that this organism is frequently spread between glands in infected ewes. Further studies identifying predisposing factors, including reservoirs, transmission routes and factors facilitating *S. aureus* infection of the ovine mammary gland, are needed in order to improve strategies to reduce the occurrence of mastitis in sheep.

AUTHORS' CONTRIBUTIONS

TM, SW, TT and SS have been involved in the design of the study and the protocols. TM has been responsible for the field project. BK, TM and SS have performed most of the microbiological work in the laboratory. BK has performed the PFGE. SW and SS have been responsible for data analysis in cooperation with TM and TT. TM drafted the manuscript, but all the authors have contributed substantially to the final manuscript. All authors have read and approved the final manuscript.

ACKNOWLEDGEMENTS

Of the veterinary surgeons who kindly contributed to the study, the authors would particularly like to thank Gunvald Mosdøl, Olav Koltveit, Grethe Steihaug, Dag Aasland, Jan I. Holøymoen, Anne Aaby, Anders Avdem, Berit C. Brændvang, Aud Solberg, Odd S. Skregelid and Knut I. Dragset.

REFERENCES

1. Fthenakis GC, Jones JET. The effect of experimentally induced subclinical mastitis on milk yield of ewes and on the growth of lambs. Br Vet J. 1990;146:43–49.
2. Johnston WS, Maclachlan GK, Murray IS. A survey of sheep losses and their causes on commercial farms in the north of Scotland. Vet Rec. 1980;106:238–240.
3. Larsgard AG, Vaabenoe A. Genetic and environmental causes of variation in mastitis in sheep. Small Rumin Res. 1993;12:339–347. doi: 10.1016/0921-4488(93)90069-T.
4. Gibson IR, Hendy PG. Mastitis in dry ewes. Vet Rec. 1976;98:511–512.
5. Jones JET. Mastitis in sheep. In: Owen JB, Axfor RFE, editor. Breeding for Disease Resistance in Farm Animals. Bangor: Tucson, AZ, CAB International; 1991. pp. 412–423.
6. Kirk JH, Glenn JS. Mastitis in ewes. Compend Cont Educ Pract Vet. 1996;18:582–591.
7. Watson DJ, Buswell JF. Modern aspects of sheep mastitis. Br Vet J. 1984;140:529–534.
8. Bergonier D, Berthelot X. New advances in epizootiology and control of ewe mastitis. Livest Prod Sci. 2003;79:1–16. doi: 10.1016/S0301-6226(02)00145-8.
9. Ariznabarreta A, Gonzalo C, San Primitivo F. Microbiological quality and somatic cell count of ewe milk with special reference to staphylococci. J Dairy Sci. 2002;85:1370–1375.
10. Burriel AR. Dynamics of intramammary infection in the sheep caused by coagulase-negative staphylococci and its influence on udder tissue and milk composition. Vet Rec. 1997;140:419–423.
11. Fthenakis GC. Prevalence and aetiology of subclinical mastitis in ewes of Southern Greece. Small Rumin Res. 1994;13:293–300. doi: 10.1016/0921-4488(94)90078-7.
12. Gonzalo C, Ariznabarreta A, Carriedo JA, San Primitivo F. Mammary pathogens and their relationship to somatic cell count and milk yield losses in dairy ewes. J Dairy Sci. 2002;85:1460–1467.
13. Lafi SQ, Al-Majali AM, Rousan MD, Alawneh JM. Epidemiological studies of clinical and subclinical ovine mastitis in Awassi sheep in

northern Jordan. Prev Vet Med. 1998;33:171–181. doi: 10.1016/S0167-5877(97)00048-2.

14. Pengov A. The role of coagulase-negative *Staphylococcus* spp. and associated somatic cell counts in the ovine mammary gland. J Dairy Sci. 2001;84:572–574.
15. Watson DL, Franklin NA, Davies HI, Kettlewell P, Frost AJ. Survey of intramammary infections in ewes on the New England Tableland of New South Wales. Aust Vet J. 1990;67:6–8.
16. Hariharan H, Donachie W, Macaldowie C, Keefe G. Bacteriology and somatic cell counts in milk samples from ewes on a Scottish farm. Can J Vet Res. 2004;68:188–192. [PMC free article]
17. Kryzanowski J, Wawron W, Malinowski V, Gluszak J, Orlik S. Bacterial flora isolated from the secretion of mastitic udders of ewes and their sensitivity to antibiotics. Med Vet. 1983;39:462–464.
18. Onnasch H, Healy AM, Brophy PO, Kinsella A, Doherty ML. A study of mastitis in sheep. Res Vet Sci. 2002;72:42. doi: 10.1016/S0034-5288(02)90120-7.
19. Quinlivan TD. Survey observations on ovine mastitis in New Zealand stud Romney flocks. 2. The bacteriology of ovine mastitis. N Z Vet J. 1968;16:153–160.
20. Al-Samarrae SAG, Sharma VK, Yousif AA. Mastitis in sheep in Iraq. Vet Rec. 1985;116:323.
21. Kirk JH, Glenn JS, Maas JP. Mastitis in a flock of milking sheep. Small Rumin Res. 1996;22:187–191. doi: 10.1016/S0921-4488(96)00881-4.
22. Bocklisch H, Wetzstein D. Clinical, diagnostic laboratory and therapeutic studies of mastitis in a large sheep breeding flock (article in German) Tierärztl Prax. 1994;22(6):524–528.
23. International Dairy Federation . Laboratory methods for use in mastitis work. Brussels: IDF, Document 132; 1981.
24. Capurro A, Concha C, Nilsson L, Östensson K. Identification of coagulase-positive staphylococci isolated from bovine milk. Acta Vet Scand. 1999;40:315–321.
25. Mørk T, Tollersrud T, Kvitle B, Jørgensen HJ, Waage S. Genetic diversity of *Staphylococcus aureus* isolated from ovine intramammary infections in Norway. Vet Microbiol. 2005;106:265–273. doi: 10.1016/j.vetmic.2004.12.020.
26. Kvitrud A, Lysne I. Examinations of mastitis in sheep (article in

Norwegian) Nord Vet Med. 1959;11:129–140.

27. Sæter EA, Eieland E. Ovine mastitis in mountain districts in Norway (article in Norwegian) Nord Vet Med. 1961;13:32–44.
28. Indrebø A, Mosdøl G. Mastitis in the ewe. Microbiological findings and sensitivity to antibiotics (article in Norwegian) Nor Vet Tidsskr. 1991;103:107–113.
29. Vautor E, Abadie G, Guibert JM, Huard C, Pepin M. Genotyping of *Staphylococcus aureus* isolated from various sites on farms with dairy sheep using pulsed-field gel electrophoresis. Vet Microbiol. 2003;96:69–79. doi: 10.1016/S0378-1135(03)00207-4.
30. Scott MJ, Jones JET. The carriage of *Pasteurella haemolytica* in sheep and its transfer between ewes and lambs in relation to mastitis. J Comp Pathol. 1998;118:359–363. doi: 10.1016/S0021-9975(07)80011-9.
31. Vautor E, Abadie G, Guibert JM, Chevalier N, Pepin M. Nasal carriage of *Staphylococcus aureus* in dairy sheep. Vet Microbiol. 2005;106:235–239. doi: 10.1016/j.vetmic.2004.11.019.
32. Bergonier D, de Cremoux R, Rupp R, Lagriffoul G, Berthelot X. Mastitis of dairy small ruminants. Vet Res. 2003;34:689–716. doi: 10.1051/vetres:2003030.
33. Deutz A, Pernthaner A, Schlerka G, Baumgartner W. Investigations on cell count of milk and distribution of bacterial mastitis in Lower Austrian sheep and goat flocks (article in German) Wien Tierärztl Mschr. 1990;77:70–77.
34. Sanchez A, Contreras A, Jimenez J, Luengo C, Corrales JC, Fernandez C. Effect of freezing goat milk samples on recovery of intramammary bacterial pathogens. Vet Microbiol. 2003;94:71–77. doi: 10.1016/S0378-1135(03)00066-X.
35. Schukken YH, Grommers FJ, Smit JA, Vandegeer D, Brand A. Effect of freezing on bacteriologic culturing of mastitis milk samples. J Dairy Sci. 1989;72:1900–1906.
36. Vatn S. Epidemiology of mastitis in Norwegian sheep. In: Fthenakis GC, McKaller FA, editor. Proceedings of the 6th International Sheep Veterinary Congress:17–21 June 2005; Hersonissos. 6th International Sheep Veterinary Congress; 2005. pp. 318–319.
37. Watkins GH, Burriel AR, Jones JET. A field investigation of subclinical mastitis in sheep in southern England. Br Vet J. 1991;147:413–420.

38. Indrebø A. Mastitis and teat injuries in the ewe in relation to age, partus and number of lambs (article in Norwegian) Nor Vet Tidsskr. 1991;103:197–204.

CHAPTER 2

Brucellosis in mammals of Costa Rica: An Epidemiological Survey

Gabriela Herna´ndez-Mora[1] , Roberto Bonilla-Montoya[1] , Osvaldo Barrantes-Granados[1] , Andrea Esquivel-Sua´rez[1] , Danilo Montero-Caballero[2] , Rocı´o Gonza´lez-Barrientos[1] , Zeanne Fallas-Monge[1] , Jose´ David Palacios-Alfaro[3] , Mario Baldi[4] , Elena Campos[5] , Grettel Chanto[5] , Elı´as Barquero-Calvo[4] , Carlos Chaco´n-Dı´az[6] , Esteban Chaves-Olarte[6] , Caterina Guzma´n Verri[4] , Juan-Jose´ Romero-Zu´ñiga[7] , Edgardo Moreno[4,8]

[1] Servicio Nacional de Salud Animal (SENASA), Ministerio de Agricultura y Ganaderı´a, Heredia, Costa Rica

[2] Instituto Nacional de Aprendizaje, San Jose´, Costa Rica

[3] Fundacio´n Keto, San Jose´, Costa Rica

[4] Programa de Investigacio´n en Enfermedades Tropicales (PIET), Escuela de Medicina Veterinaria, Universidad Nacional, Heredia, Costa Rica

[5] Centro Nacional de Referencia en Bacteriolog´ıa, Instituto Costarricense de Investigacio´n y Enseñanza en Nutricio´n y Salud (INCIENSA), Cartago, CR

[6] Facultad de Microbiologı´a, Centro de Investigacio´n en Enfermedades Tropicales (CIET), Universidad de Costa Rica, San Jose´, Costa Rica

[7] Programa de Investigacio´n en Medicina Poblacional, Escuela de Medicina Veterinaria, Universidad Nacional, Heredia, Costa Rica

[8] Instituto Clodomiro Picado, Universidad de Costa Rica, San Jose´, Costa Rica

---

Citation: Hernández-Mora G, Bonilla-Montoya R, Barrantes-Granados O, Esquivel-Suárez A, Montero-Caballero D, González-Barrientos R, et al. (2017) Brucellosis in mammals of Costa Rica: An epidemiological survey. PLoS ONE 12(8): e0182644. https://doi.org/10.1371/journal.pone.0182644.

ABSTRACT

Brucellosis has been an endemic disease of cattle and humans in Costa Rica since the beginning of XX century. However, brucellosis in sheep, goats, pigs, water buffaloes, horses and cetaceans, has not been reported in the country. We have performed a brucellosis survey in these host mammal species, from 1999–2016. In addition, we have documented the number of human brucellosis reported cases, from 2003–2016. The brucellosis seroprevalence in goat and sheep herds was 0.98% and 0.7% respectively, with no *Brucella* isolation. Antibodies against *Brucella* were not detected in feral or domestic pigs. Likewise, brucellosis seroprevalence in horse and water buffalo farms was estimated in 6.5% and 21.7%, respectively, with no *Brucella* isolation. Six cetacean species showed positive reactions against *Brucella* antigens, and *B. ceti* was isolated in 70% (n = 29) of striped dolphins (*Stenella coeruleoalba*). A steady increase in the diagnosis of human brucellosis cases was observed. Taking into account the prevalence of brucellosis in the various host mammals of Costa Rica, different measures are recommended.

INTRODUCTION

Costa Rica (CR) is a Central American country with a surface area of 51100 Km^2 and a human population close to five million. Most of the inhabitants are located in the Central Valley, flanked by the volcanic chain and the mountain range. The country is divided in six administrative areas: Chorotega, Central Pacific, Brunca, Central, Northern Huetar and Caribbean Huetar. CR has two ocean fronts: the Pacific Ocean and the Caribbean Sea. In addition, there is the Cocos Island located in the Pacific Ocean [1].

Bovine brucellosis is a significant problem in CR [2] and human brucellosis has been endemic since the beginning of last century [3,4]. However, the presence of *Brucella* organisms in sheep, goats, pigs, water buffaloes, horses and cetaceans and the impact that brucellosis has in these animals has been barely explored in CR [5]. Moreover, very little information in the number of human cases arriving to the CR health centers has been recorded.

Up to now, five species of *Brucella* have been isolated in CR: *Brucella abortus* (biotypes 1, 2 and 3) in cattle and humans, *Brucella suis* (biotype 1) in domestic swine, *Brucella canis* in dogs, *Brucella neotomae* in humans and *Brucella ceti* (dolphin type) in dolphins [2,5–7]. *B. melitensis* and *B. ovis* have not been reported in CR.

In this work we describe the distribution and the prevalence of brucellosis in different mammal species and the cumulative number of human brucellosis cases in CR. We discuss our findings in concordance to the conditions and measures carried out in the country and the zoonotic potential. Brucellosis in cattle is not reported here, since it has been thoroughly described in the accompanying manuscript [2].

MATERIALS AND METHODS

Serum Samples

Sheep and goats

The total number of sheep and goats in CR is close to 12358 and 4626, distributed in about 164 and 271 herds, respectively (Table 1). For sampling purposes CR was divided in six administrative areas by the Costa Rican National Animal Health Service (CR-NAHS) of the Ministry of Agriculture and Livestock Management: Chorotega, Central Pacific, Brunca, Central, Northern Huetar and Caribbean Huetar. Herds from each species were divided in three sections. For sheep the first section "A" included 6200 animals in 22 herds of broodstock farms with ≥150 individuals; section "B" were 3577 animals in 37 herds from farms with eventual broodstock activities, with populations ranging from 149–60 animals; and section "C" were 2691 in 105 herds for productive farms with population ≤59 animals. For goats, we used the same criteria used for sheep. Section "A" included 1406 goats in 13 herds; section "B" were 1603 distributed in 14 herds; and section "C" were 1617 from 137 farms. Seventy-eight caprine and 139 ovine herds, corresponding to 2013 and 1668 animals respectively, were sampled nationwide as part of the surveillance program, during 2014–2016.

Table 1: Numbers of ovine and caprine herds and numbers of animals by geographical region in Costa Rica (2015)

Region	Ovine		Caprine	
	Herd	Animals	Herd	Animals
1. Northern Huetar	36	2440	39	2077

2. Central	59	4295	117	1973
3. Brunca	21	1246	41	128
4. Chorotega	28	2792	22	312
5. Caribbean Huetar	9	637	28	79
6. Central Pacific	11	948	24	57
Total	**164**	**12358**	**271**	**4626**

Water Buffalos

The estimated water buffalo population in the country corresponds to 13000 animals, distributed in about 100 herds. About 70% of the water buffalo farms are devoted to mozzarella cheese production. The rest, are dedicated to meat production, leather industry or as wild fauna in zoological parks [8,9]. A total of 2586 animal blood samples, corresponding to 46 herds located in the six administrative areas were taken during 2014–2016.

Pigs

The estimated number of domestic swine in continental CR is close to 435500, most of them under intensive management farms, located in the Northern Huetar and Central Pacific regions [10]. A total of 2256 pigs from eight herds were sampled from 2014–2016. In addition, 160 blood samples collected at the slaughter house in the Central region were also studied. As part of the control of Wildlife Service of National Parks of CR, 58 feral pigs were sampled in the East side of Cocos Island National Park (23.85 km^2) located in the West Pacific Ocean (5°31′08″N 87°04′18″O), during 1998–2000. This region included close to half of the area. The sampling spots were chosen randomly and their location estimated on the basis of recognized pathways and reference points already established in maps used by the National Park rangers. Ages were estimated on the basis of size, weight, secondary sexual organ development, hair distribution, hoof size and dentition. Samples were analyzed at the CR-NAHS Laboratory or at the Veterinary Medicine School, National University, Heredia, CR.

Horses

The estimated population of horses in CR is close to 67000 in about 20000 farms [10]. In CR there is little tradition for eating horse meat. Therefore, most of the equines are devoted to sports, recreation and work. A total of 1270 horse blood samples from 215 farms located in the six administrative areas were taken during 2014–2016.

Cetaceans

Thirty cetacean species have been documented in Costa Rican waters, representing about 36% of the 83 species known worldwide [11]. From 2004–2016, 115 individuals from sixteen species were reported stranded in the Costa Rican shorelines (Table 2). Cetacean blood samples were taken at the stranding sites. After death, the animals were transported to the Veterinary School of the National University of CR, for necropsy and bacteriological studies.

Table 2: Number of cetaceans stranded in Costa Rica from January 2004 to September 2016

Common name	**Specie**	**Number of animals**
Striped dolphin	*Stenella coeruleoalba*	51
Bottlenose dolphin	*Tursiops truncatus*	10
Spotted dolphin	*Stenella attenuata*	8
Humpback whale	*Megaptera novaengliae*	8
False killer whale	*Pseudorca crassidens*	6
Spinner dolphin	*Stenella longirostris*	4
Rough tooth dolphin	*Steno bredanensis*	4
Dwarf sperm whale	*Kogia sima*	4
Cuvier beaked whale	*Ziphius cavirostris*	3
Risso's dolphin	*Grampus griseus*	2
Pilot whale	*Globicephala macrorhynchus*	2
Sperm whale	*Physeter machrocephalus*	2
Common dolphin	*Delphinus delphis*	1
Beaked whale	*Mesoplodon* spp.	1
Beaked whale	*Mesoplodon* spp.	1
Sei Whale	*Balaenoptera borealis*	1
Unknown species*	Unknown	7
Total		**115**

Humans

Brucellosis in humans has been documented in CR since 1915 [3,4]. A survey for human brucellosis from 2003–2016 was carried out at the laboratories of Public Health Services (CCSS) of CR. In addition, a total of 250 abattoir workers were monitored for antibodies against *Brucella* antigens, from 2015–

2016. All human case reports and bacteriology were received at the National Reference Bacteriology Laboratory at the Costa Rican Institute for Research and Training in Nutrition and Health (INCIENSA), for confirmation.

Information Collected and Blood Animal Samples

Relevant data concerning geographical localization, size of the farm, management and characteristics of the herds or individual animals were collected. The information also included veterinary services, reproductive parameters, history of abortion/stillbirth and the presence of other domestic and wildlife species in the farms. Breeds and identifications were registered.

Blood samples were collected with syringes or a sterile vacutainers with Z serum clot activator (Vacutainer System, Greiner Bio-one), transported under refrigeration, and sera obtained by centrifugation. Each sample received a consecutive number. Analyses of the sera were performed within 24–72 hours after collection at the CR-NAHS Brucellosis Serology Laboratory or at the Immunology Laboratory at the School of Veterinary Medicine, National University, Heredia, CR. Humans blood samples were sent to the National Reference Bacteriology Laboratory (INCIENSA) for confirmation.

Serological Tests

Rose Bengal test (RBT) (ID-VET, France), indirect protein A/G ELISA (iELISA) (ID-VET, France) and competitive ELISA (cELISA) (Svanovir, SVANOVA, Sweden) and fluorescent polarization assay (FPA) (Sentry 100 instrument, Diachemix, United States) were used as diagnostic tools, as described elsewhere [12–14]. For the standardization of small ruminant brucellosis diagnostic tests, positive and negative sera from sheep and goats were obtained from Spain and Mexico respectively. Twenty sera from *B. melitensis* biotype 1 culture positive sheep, twenty sera from *B. melitensis* biotype 1 culture positive goats, twenty- one sera from non-vaccinated negative sheep and twenty-one sera from non-vaccinated negative goats were obtained and used for validation as previously described [14,15]. In Costa Rica sheep and goats are not vaccinated. Therefore, the specificity of RBT in the absence of vaccination has been estimated to be ~100%; likewise, under these conditions the sensitivity has also been estimated in ~100% [14]. The cut off values for iELISA, cELISA and FPA in sheep and goats were 120% S/P, 30% positivity and 20 milipolarization units, respectively. Since standardized diagnostic tests for water buffalo brucellosis are not available, RBT, iELISA and cELISA, including the cut-off values, were used as reported for cattle [16]. Dolphin sera were collected and tested in RBT, iELISA and cELISA as described before [17]. For swine, modified RBT, iELISA and cELISA was used as described elsewhere [18]. Likewise, for horses, background levels for the same tests were estimated with sera from 20 healthy horses with no signs of brucellosis and with no contact with cattle or small ruminants. All animal sera samples were initially screened by RBT and then by iELISA, cELISA and FPA, following the procedures described elsewhere [13,15,17]. For humans, RBT and microagglutination in 96/well round bottom plates were used for screening, as described before [19].

Culture Conditions and *Brucella* Identification

Bacteriological cultures and identification of *Brucella* isolates were performed as described in the accompanying paper [2]. Briefly, various reference *Brucella* species were used as positive controls for genetic and bacteriological identification of samples [2]. According to the National Brucellosis Control Program of the CR-NAHS, seropositive sheep, goats, buffalos or pigs are selected for obligatory culling and pathological examination [20]. Necropsies were carried out at the Pathology Department in the Veterinary School of the National University, CR. Animal samples, included milk and other secretions such as vaginal swabs, semen and cerebrospinal fluid. Tissues samples included reproductive organs lymph nodes, spleen, kidney, liver and brain. In some cases aborted fetuses were also collected and sampled. Cultures were performed at the CR-NAHS or at the Bacteriology Laboratory of the Veterinary School. Non-selective and selective media, including blood agar and Columbia agar, supplemented with 5% of dextrose and sheep blood as well as Modified *Brucella* Selective Supplement Oxoid® (SR0209) and CITA medium, under 10% CO2 atmosphere, were used [21]. The selected bacterial colonies were subjected to Gram staining, agglutination with acriflavine and acridine orange dyes, tested for urease and oxidase activity, citrate utilization, nitrate reduction, H_2S production, growth in the presence of CO_2, thionin (20 μg/mL) and basic fuchsin (20 μg/mL) and uptake of crystal violet, according to described procedures [12].

Brucella DNA samples from each isolate and control strains were extracted with DNeasy Blood & Tissue kit from QIAGEN, and stored at -80°C until used. Identification of *Brucella* species was performed by bruce-ladder, single-nucleotide polymorphisms and MLVA16 analysis following standard procedures [22–25]. *Brucella* control strains were used for validation. The profiles were analyzed following standardized procedures (http://mlva.u-psud.fr/brucella/) and thereafter entered in the database MLVA-NET (http://microbesgenotyping.i2bc.paris-saclay.fr/).

Ethical Considerations

Sampling of domestic and wildlife animals is part of the National Brucellosis Control Program of the CR-NAHS [20] and the Law of Reportable Infectious Diseases of the Ministry of Health of CR [26]. Dolphin serum samples were taken from stranded dolphins following the procedures described before [27]. Protocols for the use of animal serum samples were revised and approved by the "Comité Institucional para el Cuido y Uso de los Animales de la Universidad de CR" (CICUA 057–16366), and "Comité Institucional para el Cuido y Uso de los Animales" of the National University, Heredia, CR (SIA 0545–15), and in agreement with the corresponding law "Ley de Bienestar de los Animales", CR (Ley 7451 on Animal Welfare), and according to the "International Convention for the Protection of Animals" endorsed by Costa Rican Veterinary General Law on the CR-NAHS (Ley 8495).

Human samples were handled by the authorities of the Public Health Service of CR (Social Security Services CCSS and Ministry of Health) and then submitted to National Reference Bacteriology Laboratory at INCIENSA for diagnostic confirmation. In this institution the samples were handled according to the INCIENSA ethical committee specifications and the agreement between INCIENSA and SENASA (Oficio 16-06-2013). Upon registration to the Medical Health Centre, all patients were informed regarding the purpose of the work and provided the corresponding written consents according to the respective Law (Ley 9234, La Gaceta 79). All samples were taken following the procedures dictated by the Costa Rican National Health system (Ley 9234, La Gaceta 79), and the World Medical Association Declaration of Helsinki (Ethical Principles for Medical Research Involving Human Subjects, General Assembly, Seoul, October 2008), regarding blood samples.

Statistics

For sheep and goats, the sample sizes were determined according to Cannon and Roe [28] using Win Episcope 2.0 software [29], with an expected brucellosis prevalence of 0.6% for sheep and 0.7% for goats, with a confidence level of 95%. This estimation included 500 sheep and 413 goats to be sampled, distributed in 10 and 13 herds respectively, sorted by region as described above. Herd selection was chosen assuming that the management and biosecurity actions, regarding these two ruminants, are similar in CR. Herds were chosen randomly from sections "A" and "B", which are the broodstock herds, and largely reflected the sanitary conditions of section "C". From each herd selected, a proportional sample population was calculated based on the clinical signs compatible with brucellosis, with a confident level of 95% and an expected prevalence of 5%, according with Cannon and Roe [28]. In addition to the random sampling, and in order to increase the probability of positive results, a biased priority was given to females with a history of abortions, weak or stillborn births, placenta retention, or with conditions that rendered individuals more susceptible to any infection, such as low body condition and pale mucous membranes. If the total number of animals defined for the herd was not covered with these specifications, random adult females were selected. Breeding rams in each farm were also examined for the detection of orchitis, epididymitis and reproductive problems. For feral pigs, the size of the sample was selected for an expected maximum population of 500 pigs distributed in the entire island, with a 95% confidence level and a tentative prevalence of 5%. The rest of the animal species sampled corresponded to the surveillance performed as part of the National Brucellosis Control Program of the CR-NAHS and according to the OIE specifications [13].

RESULTS

Sheep and Goats

Most of the ovine and caprine herds are located in the lowlands of CR (below 1000 m) and are mainly devoted to dairy (caprine) and meat (ovine) production (Table 1). The sampling procedure was carried out at the indicated regions, from 2015–2016 (Fig 1). From a total of 510 sheep sampled, corresponding to 10 herds, eleven animals (five herds) were RBT positive and five cELISA positive. None of the RBT positive animals were positive in iELISA, cELISA or FPA. Likewise, from a total of 424 goats, covering close to 10% of the Costa Rican population, only five animals demonstrated positive reactions in RBT. However, none of these RBT positive samples resulted positive in iELISA, cELISA or FPA. According to these results, the estimated brucellosis RBT prevalence values for goat and sheep herds were 0.98% and 0.7%, respectively. The RBT positive animals were culled and tested for the presence of *Brucella* spp. in lymph nodes, spleen, liver, placenta, mammary gland, milk and fetus organs. All cultured samples tested negative for *Brucella* spp. Epidemiological and clinical surveys of the sheep and goat populations and the corresponding farms did not demonstrate clinical brucellosis.

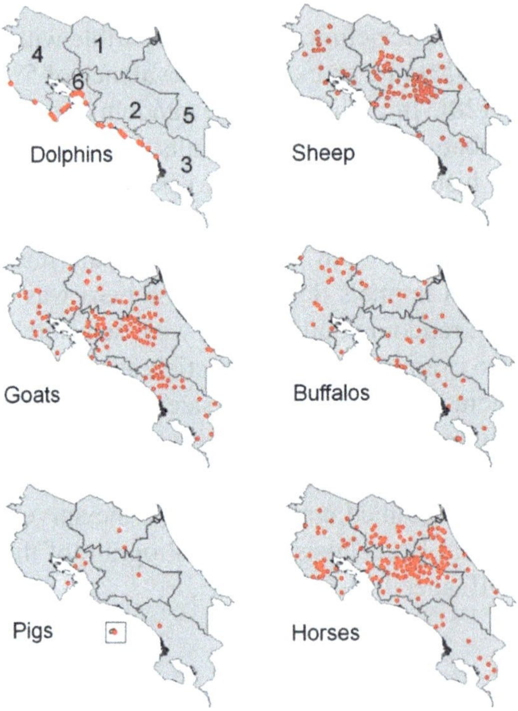

Figure 1: Sampling of animal stocks, in the six regions of CR.

The epidemiological regions are as follow: 1, Northern Huetar; 2, Central; 3, Brunca; 4, Chorotega; 5, Caribbean Huetar; 6, Central Pacific. Each red dot represents an animal stock facility.

From the 3681 ovine and caprine routinely sampled at the CR-NAHS laboratories for regular diagnosis, only one caprine was classified as positive in RBT and iELISA. The animal was slaughter and their various organs tested for the presence of *Brucella*, with negative results. Clinical disease compatible with *B. ovis* infection was not detected in rams. Likewise, this bacterium was not isolated from semen samples. Taken together these data, the "positive" RBT reactions were estimated as unspecific and the presence of brucellosis in ovine and caprine herds ruled out.

Water Buffalos

Most water buffalos are located in the low lands, since they require fresh water habitats for subsistence. From a total of 2586 samples distributed in 46 herds, collected from 2014–2016 (Fig 1), 17 animals tested positive in RBT, 38 in cELISA and 77 in the iELISA. The total number of herds positive in these three assays was ten. All RBT positive samples were also positive in iELISA and cELISA; and all samples positive for cELISA were also positive in iELISA. FPA was not performed. In spite of the efforts, *Brucella* organisms were not isolated from vaginal swabs, dairy products, placental tissues, fetuses, testes, lymph nodes, mammary gland, blood, spleen or liver of the culled seropositive animals. However, due to the reported clinical characteristics and the testimonies of persistent abortions and positive serological reactions, *Brucella* infections were suspected. Moreover, it is likely that *B. abortus* constitutes an infection source for water buffaloes, since bovine brucellosis caused by this *Brucella* specie is highly prevalent in CR [2].

Pigs

From the number of herds studied and the samples obtained at the slaughter house (Fig 1), only two pigs of one herd were RBT positive. From these, only one pig was also positive in iELISA and cELISA. The FPA assay was not performed. Positive animals were culled and different tissues were cultured for the presence of *Brucella*, with negative results. In addition, tissues of aborted fetuses in some farms were also tested for the presence of *Brucella*, all with negative results. Likewise, positive serological reactions were not detected in the feral pig population in the Cocos Island. Histopathological examination of the liver in the feral swine sample showed chronic inflammation in 84% of the cases, while 20% had multifocal granulomatous inflammation with eosinophilic infiltration, probably related to the presence of parasite nematode *Stephanurus dentatus*, but not *Brucella*. Taken together these data, the positive serological reactions were estimated as non-specific.

Horses

Most horses are located in North Huetar, Chorotega and the northern part of the Caribbean Huetar regions of CR. Therefore, most of the samples are from these areas (Fig 1). From the total number of farms studied 14 (6.5%) had seropositive animals, including 18 horses positive in RBT;

from these, only four were also positive in both iELISA and cELISA. In spite of the efforts, *Brucella* was not isolated from horses. However, it is likely that *B. abortus* is a source of infection in horses, since many of these animals are in close contact with infected bovines in CR. In addition, some clinical features such as fistulous withers and nonspecific lameness due to joint infection, have occasionally been observed in horses.

Cetaceans

Cetacean brucellosis in Costa Rican was investigated from 2004–2016. RBT and iELISA, designed for cetacean diagnosis, were positive in 54 (46.9%) individuals from six different species. They included 38 striped dolphins (*Stenella coeruleoalba*), one bottlenose dolphins (*Tursiops truncatus*), one spotted dolphins (*S. attenuata*), one common dolphin (*Delphinus delphis*), one rough toothed dolphin (*Steno bredanensis*), and one Cuvier beaked whale (*Ziphius cavirostris*). However, striped dolphin (*S. coeruleoalba)* remains as the only cetacean specie from which *B. ceti* has been isolated from different organs in CR.

Strong positive RBT and iELISA reactions were obtained in sera from 37 out of 38 striped dolphins stranded at the Pacific coast of CR (Fig 1). Thirty-seven out of 38 striped dolphins, stranded alive. At the time of stranding, all live animals presented neurological symptoms such as tremors, buoyancy difficulties, weakness, seizures and locomotion problems. With exception of two dolphins (one seropositive and one seronegative), all other *S. coeruleoalba* dolphins displayed neurobrucellosis, following previous diagnosis [27]. All of them died at the stranding site within hours after the event. Necropsy was performed in all cases and *B. ceti* was isolated from the cerebrospinal fluid of 29 individuals (70%). In addition, *B. ceti* was also present in placenta, umbilical cord, amniotic and allantoic fluids, multiple fetal organs, milk, cardiac valve, atlanto occipital joint fluid, lung and lung nematodes (*Halocercus spp.*) [6,27,30,31]. All *B. ceti* isolates belonged to the MLVA16 type P [32], corresponding to the Pacific Ocean (data accessible at: http://microbesgenotyping.i2bc.paris-saclay.fr/ [and the following entries: public databases, Brucella v4_1, bmarCR+number, years 2006–2014]).

Humans

According with the Costa Rican National Reference Bacteriology Laboratory (INCIENSA), the number of human cases reported by the health centers over

12 year (2003–2015) period corresponded to 124 patients (Fig 2A): fifty one were from the Central region 37 from the Caribbean Huetar region and 36 cases from all other regions. Male and female patients represented 79 and 41 cases (Fig 2B), respectively, with ages ranging between 8–76 year-old, with a large proportion of veterinarians, farmers and slaughter plant workers (Fig 2C). From a total of the 250 abattoir workers only three presented high antibody titers (>1/160) compatible with an active brucellosis. With the exception of two *B. neotomae* isolates [7], all other human brucellosis cases corresponded to *B. abortus*.

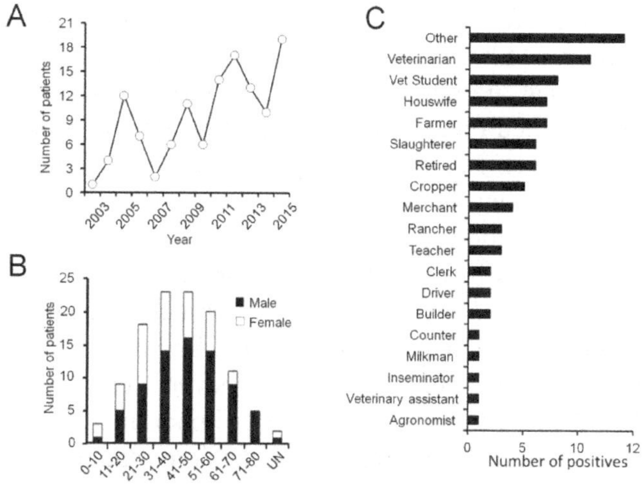

Figure 2: Occurrence human brucellosis cases in CR from 2003–2015.

(A) Number of human brucellosis cases diagnosed per year in CR for the period. All cases recorded were due to *B. abortus*. (B) Distribution per age and proportion of male and female brucellosis cases in CR, diagnosed for the period. (C) Proportion of 250 seropositive abattoir workers from 2015 to 2016.

DISCUSSION

For most of the history of CR, sheep and goats have been raised in very low numbers and the dairy products and meat of these animals barely consumed [33]. Until 1975 the number of goats and sheep in the country were close to 1000 animals, all together [33]. However, in the nineties the population of these small ruminants started to increase. Already, in the first decade of the XXI century, the numbers of goats and sheep were close to 5000 and

3000, respectively [34]. With the enhanced acceptance of ovine and caprine dairy and meat products, the emergent industries for small ruminants have increased. Indeed, the numbers of goats and sheep have augmented almost three fold (12852) and twelve fold (35800) [10], respectively.

An epidemiological survey for caprine and ovine brucellosis was performed from 2015–2016. Although we detected a minor number of RBT positive reactions in small ruminants, they were regarded as false positives. In spite of the high specificity and sensitivity displayed by the RBT under controlled conditions with a limited number of known sera, this assay is not perfect and some non-specific reactions are expected to occur under field conditions. Therefore, exhaustive clinical, pathological and epidemiological investigations in the serologically positive sheep and goats were carried out, all rendering negative results for the presence of *Brucella* infections. Bacteria displaying similar antigenic determinants as smooth brucellae may be the source of false positive reactions [35]. In addition, positive serological reactions due to *B. abortus* infections cannot be ruled out, since this bacterium is highly prevalent in CR [2]. However, we did not isolate *B, abortus* or any other brucellae from the tissues of goats and sheep. Although *B. melitensis* may be present in some Central American countries [36], this bacterium has never been isolated in animals or humans in CR [5, 36]. Following this, it is important to keep these small ruminants free of brucellosis, restricting the importation of animals and semen from *B. melitensis* free countries.

Similar to goats and sheep, the number of water buffalos has steadily increased in CR during the last ten years. In 2006 the number of water buffalos in CR was close to 615 animals [37]; in ten years the population has increased twenty fold, most of them devoted to the production of dairy products. Taking into account the persistent positive serological reactions, their close association of water buffalo with *B. abortus* infected cattle and the reported cases of abortions compatible with clinical disease; we believe that some water buffalo populations are infected with *Brucella* in CR. Moreover, a significant number of the CR water buffalo population originates from Trinidad-Tobago, country endemic for water buffalo brucellosis [8, 38]. The fact that we did not isolate *Brucella* from water buffalos may be related to the natural resistance of these animals to brucellosis in relation to other bovines [39].

B. suis was isolated from a domestic pig in the Central region of CR in 1984 [5]. Since then, the bacterium has not been isolated from boars, in spite of the efforts. In CR pigs seldom roam freely around the houses and most animals are confined to intensive management facilities, under good health conditions. Moreover, with the exception of Cocos Island, no feral pigs are present in the CR territory. Since no clinical or epidemiological surveys indicate swine brucellosis, it is unlikely that *B. suis* is currently infecting pigs in the country.

Horses are not primary *Brucella* hosts and commonly they do not have the ability to transmit the bacterium to other animals or humans. Therefore, horses are not of epidemiological relevance in keeping the bacterium life cycle; however, these animals are sentinels for the presence of *Brucella* in other animals, mainly in cattle. Like humans, they become infected by contact with abortions or with infected cattle, and display a wide range of clinical manifestations including articular swelling and general weakness [40]. The fact that close to 18 horses displayed recurrent positive reactions against *Brucella*, may be an indication of the high seroprevalence of *Brucella* infections in cattle [2], including water buffalo.

B. ceti infections in dolphins stranded in the CR Pacific coast were detected for the first time in 2004 [6]. A total of 115 stranding events from at least 16 different species of cetaceans have been recorded in CR seashores from 2004–2016 (Table 2). From these, six species displayed positive serological reactions. However, *B. ceti* active infections have been only documented in striped dolphins from the Pacific Ocean of CR. All *B. ceti* isolates belong to the same MLVA16 type P. This bacterial group corresponds to a particular cluster distinct from other *B. ceti* strains isolated in various oceanic latitudes, and it is a hallmark for *S. coeruleoalba* infections in the Eastern Tropical Pacific [32]. Moreover, all the 29 dolphin cases in which *B. ceti* organisms were isolated suffered from neurobrucellosis [27]. It seems, therefore, that this dolphin specie is highly susceptible to *B. ceti* and that many of the stranding events were due to brain infections, as recorded in other latitudes [41]. The surveillance of cetacean brucellosis in Central American littorals requires attention. This is mandatory to understand the impact that brucellosis has in the Eastern Tropical Pacific marine mammal populations and to ensure prevention measures for potential human and animal infections [42].

In a previous study in the Central region (Cartago, CR), in which 71% of the human population consumed unpasteurized dairy products; an overall seroprevalence of 0.87% was detected [19]. However, no statistically significant association was found between unpasteurized milk consumption and the presence of antibodies against *Brucella* organisms. Here, we reported a steady increase in the number of human brucellosis cases during a lapse of 12 years. Whether the steady increase of human brucellosis reports corresponded to improved diagnosis or to intensification in the number of cases, is not known. The number of human brucellosis cases due to *B. abortus* is consistent with the high prevalence of bovine brucellosis in CR, and the absence of *B. melitensis* in sheep and goats, and *B. suis* in pigs, two *Brucella* species that display a higher zoonotic potential than former bacteria [43]. In CR there are other zoonotic brucellae such as *B. neotomae* [7] and *B. canis* [44], which were not considered in this study. Nevertheless, a careful identification of strains is required, even with those *Brucella* species that are considered of low zoonotic risk.

From the epidemiological perspective, it seems that the population of sheep, goats and pigs in CR are free of *B. melitensis* infections. This seems to be also the case for *B. ovis* in rams and *B. suis* for pigs. Consequently, humans are also free of these bacterial species. However, with the increasing number of small ruminant species in the country the risk of *Brucella* infections arriving from other latitudes requires permanent surveillance, improved management and sensitive and specific diagnostic tools.

CONCLUSIONS

- Domestic ovine, caprine and swine herds are free of brucellosis in CR.
- The presence of *Brucella* infections in water buffaloes is highly suspected in CR.
- The presence of *B. abortus* infections in horses is highly suspected in CR.
- Striped dolphins from the Pacific Ocean of CR are the main host of *B. ceti* cluster type P.
- The main clinical symptom found in striped dolphins corresponded to neurobrucellosis.
- Detection of human infections, due to *B. abortus*, has steadily increased since 2005 in CR.
- Estimating the presence of *Brucella* infections in different hosts inhabiting CR is relevant for understanding the impact that brucellosis has in the country and for prevention measures.

ACKNOWLEDGMENTS

The authors thank the field veterinarians and personnel from the laboratories of CR-NAHS, SENASA-CR, PIET, CIET and the Pathology Department of the Veterinary School of the National University. Likewise, the authors thank the assistance provided by the farmers, and owners of the different establishments. We thank William Rossiter from Cetacean Society International, Tourist Police, Life guards, Fire Department, Emergency Service (911), and personnel from Parque Marino del Pacífico, National System of Conservation Areas.

REFERENCES

1. Vargas-Ulate T. Geografía de Costa Rica. Editorial Tecnológica, Cartago, Costa Rica. 2004;124.
2. Hernández-Mora G, Ruiz-Villalobos N, Bonilla-Montoya R, Romero-Zúñiga JJ, Jiménez-Arias J, Barquero-Calvo E, et al. Epidemiology of bovine brucellosis in Costa Rica: lessons learned from failures in the control of the disease. Submitted to Plos One. 2017. (accompanying manuscript).
3. Quirós D. El primer caso de fiebre de Bruce en Costa Rica. Gaceta Médica de Costa Rica. 1915;11:241–244.
4. Zeledón-Alvarado J. Primera historia clínica de brucelosis humana en Costa Rica. Rev. Med. 1940;72:153–167.
5. Sequeira A, Campos E, Mendoza L, San-Román MA, Moreno E. Identificación de especies y biotipos de *Brucella* aisladas en Costa Rica. Turrialb. 1984;34:525–526.
6. Hernández-Mora G, González-Barrientos R, Morales JA, Chaves-Olarte E, Guzmán-Verri C, Baquero-Calvo E, et al. Neurobrucellosis in stranded dolphins, Costa Rica. Emerg Infect Dis. 2008;14:1429–1433.
7. Suárez-Esquivel M, Ruiz-Villalobos N, Jiménez-Rojas C, Barquero-Calvo E, Chacón-Díaz C, Víquez-Ruiz E, et al. Human neurobrucellosis caused by *Brucella neotomae*. Emerg. Infect. Dis. 2017; 23:997–1000.
8. Rosales-Rodríguez R. Situación del búfalo de agua en Costa Rica. Tecnología en Marcha (Costa Rica). 2011;24:19–24.
9. Rosales-Rodríguez R. Costa Rica. In: Crudeli GA, Konrad JL, Patiño EM, editors. Reproducción en búfalas. 1st ed. Moglia.2016;251.
10. Instituto Nacional de Estadísticas- INEC. VI Censo Nacional Agropecuario. San José, Costa Rica. 2014;72–74.
11. May Collado L. Marine Mammals. In: Wehrtmann IS, Cortes J, editors. Marine biodiversity of Costa Rica, Central America. Monographiae Biologicae. 1st ed. Springer. 2009;479–496.
12. Alton GG, Jones LM, Angus RD, Verger JM. Techniques for brucellosis laboratory. INRA: Paris;1988.
13. OIE-World Organization for Animal Health. Bovine brucellosis: In manual of diagnostic tests and vaccines for terrestrial animals. 12 rue de Prony 75017. Paris; 2009.

14. Marín CM, Moreno E, Moriyón I, Díaz R, Blasco JM. Performance of competitive and indirect enzyme-linked immunosorbent assays, gel immunoprecipitation with native hapten polysaccharide, and standard serological tests in diagnosis of sheep brucellosis. Clin Diagn Lab Immunol. 1999;6:269–272.
15. OIE-World Organization for Animal Health. Caprine and ovine brucellosis: In Manual of Diagnostic Tests and Vaccines. 12 rue de Prony 75017. Paris; 2009.
16. Moreno E, Rojas N, Nielsen K, Gall D. Comparison of different serological assays for the differential diagnosis of brucellosis. In International Atomic Energy, Diagnosis and Epidemiology of Animal Diseases in Latin America IAEA-TECDOC 1055. Viena; 1998;153–161.
17. Hernández- Mora G, Manire CA, González-Barrientos R, Barquero-Calvo E, Guzmán-Verri C, Staggs L, et al. Serological diagnosis of *Brucella* infections in odontocetes. Clin Vaccine Immunol. 2009;16:906–915. doi: 10.1128/CVI.00413-08
18. Muñoz PM, Blasco JM, Engel B, De Miguel MJ, Marín CM, Dieste L, et al. Assessment of performance of selected serological tests for diagnosing brucellosis in pigs. Vet Immunol Immunopathol. 2012;146:150–158. doi: 10.1016/j.vetimm.2012.02.012
19. Chanto G, Rojas N, Ching A, Zúñiga R, Castro E, Chaverri S, et al. Prevalencia de anticuerpos séricos contra la bacteria *Brucella* sp. en una población humana tropical. Rev Biol Trop 2007;55:385–391.
20. Reglamento para la intervención de la brucelosis bovina. Decreto 34858-MAG-2008. Gaceta, 230, (7-11- 2008) 2008; Available from: http://www.pgrweb.go.cr/scij/Busqueda/Normativa/Normas/nrm_texto_completo.aspx?param1=NRTC&nValor1=1&nValor2=64425&nValor3=74734&strTipM=TC
21. De Miguel MJ, Marín CM, Muñoz PM, Dieste L, Grilló MJ, Blasco JM. Development of a selective culture medium for primary isolation of the main *Brucella* species. J Clin Microbiol. 2011;49:1458–63. doi: 10.1128/JCM.02301-10
22. Le Flèche P, Jacques I, Grayon M, Al Dahouk S, Bouchon P, Denoeud F, et al. Evaluation and selection of tandem repeat loci for a *Brucella* MLVA typing assay. BMC Microbiol. 2006;6:9 doi: 10.1186/1471-2180-6-9

23. García-Yoldi D, Marín CM, de Miguel MJ, Muñoz PM, Vizmanos JL, López-Goñi I. Multiplex PCR Assay for the identification and differentiation of all *Brucella* Species and the vaccine strains *Brucella abortus* S19 and RB51 and *Brucella melitensis* Rev1. Clin Chem. 2006;52:779–781. doi: 10.1373/clinchem.2005.062596

24. López-Goñi I, García-Yoldi D, Marín CM, de Miguel MJ, Barquero-Calvo E, Guzmán-Verri C, et al. New Bruce-ladder multiplex PCR assay for the biovar typing of *Brucella suis* and the discrimination of *Brucella suis* and *Brucella canis*. Vet Microbiol. 2011;154:152–155. doi: 10.1016/j.vetmic.2011.06.035

25. Scott JC, Koylass MS, Stubberfield MR, Whatmore AM. Multiplex assay based on single-nucleotide polymorphisms for rapid identification of *Brucella* isolates at the species level. Appl Environ Microbiol. 2007;73:7331–7337. doi: 10.1128/AEM.00976-07

26. Reglamento de Vigilancia de la Salud. Decreto 37306-S-2012. Gaceta (Costa Rica), No 203, (27-8-2012). 2012; Available from: http://www.pgrweb.go.cr/scij/Busqueda/Normativa/Normas/nrm_texto_completo.aspx?param1=NRTC&param2=18&nValor1=1&nValor2=73471&nValor3=90176&strTipM=TC&lResultado=180&nValor4=1&strSelect=sel.

27. González-Barrientos R, Morales J-A, Hernández-Mora G, Barquero-Calvo E, Guzmán-Verri C, Chaves-Olarte E, et al. Pathology of striped dolphins (*Stenella coeruleoalba*) infected with *Brucella ceti*. J Comp Pathol. 2010;142:347–352. doi: 10.1016/j.jcpa.2009.10.017

28. Cannon RM, Roe RT. Livestock disease surveys: a field manual for veterinarians. Australian Government, Australia.1982.

29. Thrusfield M, Ortega C, de Blas I, Noordhuizen JP, Frankena K. WIN EPISCOPE 2.0: improved epidemiological software for veterinary medicine. Vet Rec. 2001;148:567–572.

30. Guzmán-Verri C, González-Barrientos R, Hernández-Mora G, Morales J- A, Baquero-Calvo E, Chaves-Olarte E, et al. *Brucella ceti* and brucellosis in cetaceans. Front Cell Infect Microbiol. 2012;2:3 doi: 10.3389/fcimb.2012.00003

31. Oliveira JB, Morales JA, González-Barrientos RC, Hernández-Gamboa J, Hernández-Mora G. Parasites of cetaceans stranded on the Pacific coast of Costa Rica. Vet Parasitol. 2011;182:319–328. doi: 10.1016/j.vetpar.2011.05.014

32. Suárez-Esquivel M, Baker KS, Ruíz-Villalobos N, Hernández-Mora G, Barquero-Calvo E, González-Barrientos R, et al. *Brucella* genetic variability in wildlife marine mammals populations relates to host preference and ocean distribution. Genome Biol Evol. 2017; (in press).
33. CATIE-Centro Agronómico Tropical para la Investigación y Enseñanza. La situación caprina en Costa Rica: Situación de la producción caprina en centroamérica y República Dominicana. 1987. Turrialba, Costa Rica. 1987;114:120.
34. FAOSTAT-Organización de las Naciones Unidas para la Alimentación y la Agricultura. Dirección de Estadística.2016: Available from: http://faostat3.fao.org/download/Q/QA/S.
35. Ducrotoy MJ, Conde-Álvarez R, Blasco JM, Moriyón I. A review of the basis of the immunological diagnosis of ruminant brucellosis. Vet Immunol Immunopathol. 2017;171:81–102.
36. Moreno E. Brucellosis in Central America. Vet Microbiol. 2002;90:31–8.
37. Rosales RR, Wingching JR. Sistemas de producción bufalina en Costa Rica. Cuantificación de la población y caracterización de los sistemas. Rev Agron Costarricense (Costa Rica) 2007;31:65–69.
38. Fosgate GT, Diptee MD, Ramnanan A, Adesiyun AA. Brucellosis in domestic water buffalo (*Bubalus bubalis*) of Trinidad and Tobago with comparative epidemiology to cattle. Trop Anim Health Prod. 2011;43:1479–1486. doi: 10.1007/s11250-011-9846-9
39. Borriello G, Capparelli R, Bianco M, Fenizia D, Alfano F, Capuano F, et al. Genetic resistance to *Brucella abortus* in the water buffalo (*Bubalus bubalis*). Infect Immun. 2006;74:2115–2120. doi: 10.1128/IAI.74.4.2115-2120.2006
40. Denny HR. A review of brucellosis in the horse. Equine Vet J. 1973;5:121–125.
41. Isidoro-Ayza M, Ruiz-Villalobos N, Pérez L, Guzmán-Verri C, Muñoz PM, Alegre F, et al. *Brucella ceti* infection in dolphins from the Western Mediterranean sea. BMC Vet Res. 2014;10:206 doi: 10.1186/s12917-014-0206-7
42. Whatmore AM, Dawson CE, Groussaud P, Koylass MS, King AC, Shankster SJ, et al. Marine mammal Brucella genotype associated with zoonotic infection. Emerg Infect Dis. 2008;14:517–518. doi: 10.3201/eid1403.070829

43. Moreno E. Retrospective and prospective perspectives on zoonotic brucellosis. Frontiers Microbiol. 2014;1–18. doi: 10.3389/fmicb.2014.00213
44. Chacón-Díaz C, Altamirano-Silva P, González-Espinoza G, Medina M- C, Alfaro-Alarcón A, Bouza-Mora L, et al. *Brucella canis* is an intracellular pathogen inducing a lower proinflammatory response than smooth zoonotic counterparts. Infect Immun. 2015; 83:4861–4870. doi: 10.1128/IAI.00995-15

CHAPTER 3

Bacterial Species Associated with Interdigital Phlegmon Outbreaks in Finnish Dairy Herds

Miia Kontturi[1], Reijo Junni[1], Heli Simojoki[1], Erja Malinen[2], Eija Seuna[2], Kirstine Klitgaard[3], Minna Kujala-Wirth[1], Timo Soveri[1] and Sinikka Pelkonen[2]

[1] Department of Production Animal Medicine, Faculty of Veterinary Medicine, University of Helsinki, Paroninkuja 20, 04920 Saarentaus, Finland

[2] Veterinary Bacteriology and Pathology unit, Finnish Food Authority, Mustialankatu 3, 00790 Helsinki, Finland

[3] National Veterinary Institute, Technical University of Denmark, Kemitorvet, Building 202, 2800 Kgs. Lyngby, Denmark

ABSTRACT

Background

Severe outbreaks of bovine interdigital phlegmon (IP) have occurred recently in several free stall dairy herds in Finland. We studied the aetiology of IP in

---

Citation: Kontturi, M., Junni, R., Simojoki, H. et al. Bacterial species associated with interdigital phlegmon outbreaks in Finnish dairy herds. BMC Vet Res 15, 44 (2019). https://doi.org/10.1186/s12917-019-1788-x.

such herds, and the association of bacterial species with the various stages of IP and herds of various morbidity of IP. Nineteen free stall dairy herds with IP outbreaks and three control herds were visited and bacteriological samples collected from cows suffering from IP ($n = 106$), other hoof diseases ($n = 58$), and control cows ($n = 64$). The herds were divided into high morbidity (morbidity ≥50%) and moderate morbidity groups (9–33%) based on morbidity during the first two months of the outbreak.

Results

F. necrophorum subspecies *necrophorum* was clearly associated with IP in general, and *T. pyogenes* was associated with the healing stage of IP. Six other major hoof pathogens were detected; *Dichelobacter nodosus, Porphyromonas levii, Prevotella melaninogenica, Treponema* spp. and *Trueperella pyogenes.* Most of the samples of acute IP (66.7%) harboured both *F. necrophorum* and *D. nodosus.* We found differences between moderate morbidity and high morbidity herds. *D. nodosus* was more common in IP lesion in high than in moderate morbidity herds.

Conclusions

Our result confirms that *F. necrophorum* subspecies *necrophorum* is the main pathogen in IP, but also *T. pyogenes* is associated with the healing stage of IP. Our results suggest that *D. nodosus* may play a role in the severity of the outbreak of IP, but further research is needed to establish other bacteriological factors behind these severe outbreaks.

BACKGROUND

During recent years, severe outbreaks of interdigital phlegmon (IP) have occurred in dairy herds in Finland, with sudden onset and of divergent morbidity. No preceding clear wound has been detected in the interdigital cleft of the IP cows. These new types of outbreaks have caused serious economic losses to affected dairy farms [1].

IP occurs usually as a sporadic infection of cattle. The herd incidence per lactation is generally 2–5% [2, 3], but studies of earlier outbreaks of IP report incidences of 17–25% during outbreaks [4, 5]. Common signs of IP are lameness; occasionally with an acute onset, a swelling of the interdigital area and the bulbs of the heels, and a fetid odour. A fissure with swollen protruding edges may appear along the interdigital cleft. In severe cases, systemic signs occur, including fever, recumbency, anorexia or decrease in milk production [6, 7]. IP reduces milk yield [8] and can necessitate early culling of the affected cow [8, 9].

Traditionally, *Fusobacterium necrophorum* is considered the major infective agent of IP [10,11,12] and is detected frequently in IP lesions. *F. necrophorum* is a common animal pathogen, producing several toxins that can injure tissues; leukotoxin, coded for the *lktA* gene, is considered a major virulence factor in cattle [13]. *lktA* is unique to *F. necrophorum* [14], and its detection has been used in cattle research as a reliable tool for the detection of *F. necrophorum* [15]. *F. necrophorum* is classified into two subspecies, *necrophorum* and *funduliforme.* Subspecies *necrophorum* is more frequently encountered in animal infections and in pure culture, whereas *funduliforme* is found in mixed infections and is considered less pathogenic [16].

In addition to *F. necrophorum,* several other bacteria such as, *Bacteroides melaninogenicus* [11, 12], *Dichelobacter nodosus* [7], *Porphyromonas levii* [17, 18], *Spirochetes* [5, 7], and *Trueperella pyogenes* [11] have been suggested to play a role in the pathogenesis of IP. Nevertheless, most of that research was done long ago and, for example, the taxonomical changes since then make interpretation of the results challenging. Also, a recent review describes that the role of various bacterial species in the pathogenesis of IP is still unresolved [6].

Recently the main research focus worldwide has been on digital dermatitis (DD) and treponemes and only a few studies have addressed IP and its bacteriology. However, because of numerous new type outbreaks of IP in dairy herds in Finland, we investigated the bacteriology of IP, including those bacteria earlier suggested to be involved in IP. The aim of the study was to investigate the bacteriology of IP in this new type of outbreaks; at various stages of IP, both acute and during the healing process, and compare the findings with healthy control cows. Moreover, we investigated whether these bacteriological findings differed between herds of various morbidities.

METHODS

Herds

During 2012–2015 we carried out a research project on infectious hoof diseases in Finland. As a part of the project, we made several farm visits to privately owned dairy herds affected by outbreaks of IP. Of the farms visited, 19 fulfilled the criteria for an outbreak of IP; at least three observed cases of IP within 1 week, and no previous history of IP in the herd for 10 years. The outbreak herds were later divided into two categories based on the incidence of IP within 2 months of the outbreak. Furthermore, we collected samples from control cows of three non-outbreak herds (IP free herds). All herds studied were housed in free stalls. The average herd size was 75 lactating cows (range 31–140, median 62) and the average milk yield was 9234 kg (8000–10,914 kg, median 9219 kg).

Cows

The primary selection criteria for inclusion of a cow in the study were lameness, prolonged lying-time, or a 'trouble report' from an automatic milking system. In the outbreak herds, we collected samples mainly from cows that had IP, but also from lesions apparently infected with bacteria. Such lesions included DD, interdigital dermatitis (ID), white line abscesses and sole ulcers. The IP lesions were classified as acute IP or healing IP. The diagnosis of acute IP was made if a symmetric swelling and possible ulceration appeared in the interdigital cleft. Healing IP was identified as proliferation tissue or apparent scar formation in the affected region. DD diagnosis was made according to Döpfer et al. [19]. We also sampled 1–5 control cows per IP outbreak herd. These were non-lame cows with no sign of IP, DD, ID, sole ulcer, or white line disease, and are hereafter referred to as control cows (IP herd). In control herds, we sampled 4–8 cows in each herd using the same criteria as for control cows (IP herd).

We sampled a single hoof from all control cows, but from 11 cows with IP or DD we took samples from two separate feet. Five outbreak herds were visited 2 or 3 times. During these visits, 10 IP cows were sampled repeatedly 2 or 3 times 11–34 days after the first sampling. These samples were additional and not included in the total number of hoof samples (total n=228). These resampled IP cows had clinical signs of IP at all sampling times. Table 1 presents the number of study herds of various morbidities, numbers of sampled cows and hoof samples in various disease groups.

Table 1: Dairy herds, cows and bacteriological samples of a study of interdigital phlegmon outbreaks in Finland

	Herd (n)	Cow (n)	Control (IP free herd)	Control (IP herd)	Acute IP	Healing stage IP	Other hoof disease
Number of herds	19	217					
High morbidity herd	7	65		13 (28.9%)	27 (45%)	11 (27.5%)	14 (26.4%)
Moderate morbidity herd	12	133		32 (71.1%)	33 (55%)	29 (72.50%)	39 (73.6%)
Non-outbreak herd	3	19	19 (100%)				
Number of cows		217	19/217 (8.8%)	45/217 (20.7%)	60/217 (27.7%)	40/217 (18.4%)	53/217 (24.4%)
Cows with antibiotic treatment							
None		151 (69.6%)	19 (100%)	45 (100%)	31 (51.7%)	7 (17.5%)	49 (92.5%)
Current		37 (17.1%)	0 (0%)	0 (0%)	21 (35%)	15 (37.5%)	1 (1.9%)
Previous		29 (13.4%)	0 (0%)	0 (0%)	8 (13.3%)	18 (45%)	3 (5.7%)
		Hoof sample (n)	Control (IP free herd)	Control (IP herd)	Acute IP	Healing stage IP	Other hoof disease
Number of hoof samples		228[a]	19 (8.3%)	45 (19.7%)	65[a] **(28.5%)**	41[a] **(18.0%)**	58[a] **(25.4%)**
Front feet		25 (11.0%)	0 (0%)	5 (11.1%)	12 (18.5%)	5 (12.2%)	3 (5.2%)
Hind feet		203 (89.0%)	19 (100%)	40 (88.9%)	53 (81.5%)	36 (87.8%)	55 (94.8%)
Hoof sample with antimicrobial treatment							
None		159 (69.7%)	19 (100%)	45 (100%)	36 (55.4%)	7 (17.1%)	52 (89.7%)
Current		38 (16.7%)	0 (0%)	0 (0%)	21 (32.3%)	16 (39.0%)	1 (1.7%)
Previous		31 (13.6%)	0 (0%)	0 (0%)	8 (12.3%)	18 (43.9%)	5 (8.6%)
Number of PCR tests		Hoof sample (n)	Control (IP free herd)	Control (IP herd)	Acute IP	Healing stage IP	Other hoof disease
Fusobacterium necrophorum		205	19 (100%)	43 (95.6%)	52 (80.0%)	37 (90.2%)	54 (93.1%)
Dichelobacter nodosus		205	19 (100%)	43 (95.6%)	52 (80.0%)	37 (90.2%)	54 (93.1%)
Porphyromonas levii		142	19 (100%)	41 (91.1%)	49 (75.4%)	33 (80.5%)	Not analyzed
Prevotella melaninogenica		142	19 (100%)	41 (91.1%)	49 (75.4%)	33 (80.5%)	Not analyzed
Treponema **group 2 & 3**		168	19 (100%)	42 (93.3%)	39 (60.0%)	36 (87.8%)	32 (59.3%)

Trueperella pyogenes		205	19 (100%)	43 (95.6%)	52 (80.0%)	37 (90.2%)	54 (93.1%)

[a]Two feet were sampled from 11 cows (5 acute IP, 1 healing stage IP, 5 other hoof diseases)

Numbers of sampled cows, numbers of hoof samples and a possible antimicrobial treatment in various disease groups; control cows in a herd with no outbreak of interdigital phlegmon, IP (IP free herd), control cows in a herd with an outbreak of IP (IP herd), acute interdigital phlegmon (Acute IP), IP at healing stage (Healing IP) and hoof diseases other than IP (Other). The group "Other" included hoof samples from digital dermatitis, interdigital dermatitis, white line abscess and sole ulcer. With antibiotic treatment, "None" signifies no current or previous antibiotic treatment during last month, "Current" signifies current antibiotic treatment or treatment within 6 days before sampling and "Previous" means previous treatment with antibiotics within 7–30 days prior to the sampling. "Number of PCR tests" **column means the number of samples that were successfully amplified in PCR**

Of the sampled cows (n=217) selected for the study, 58.5% were Ayrshire and 41.5% Holstein. Moreover 4.6% were heifers, 41.5% first parity cows, 22.1% second parity, 29.5% third or more parity cows and 36.9% were on early lactation (1–120 days in milk, DIM), 53.0% late lactation (121–305 DIM) and 7.8% were either dry cows or heifers. Information on parity and lactation stage was absent for 5 cows (2.3%).

Sampling Methods and Materials

Two veterinarians (MK, RJ) experienced in hoof diseases of cattle, evaluated the general condition and hoof health of the cows prior to sampling and recorded clinical diagnosis and antimicrobial treatment history. Every hoof was photographed at sampling, and diagnoses were standardized between the two veterinarians by evaluating some of the photographs together.

The sampling took place in a trimming chute; we lifted the foot up and spread the claws with an extensor. Then we washed the distal foot carefully with a hose, spouted with saline solution, and dried it with gauze. We collected the bacterial samples from the inflamed region using sterile swabs (FLOQSwabs), used them immediately for culturing, and took cytobrush samples from the same region for PCR analysis. We placed the cytobrushes (Medscand Medical Cytobrush Plus, CooperSurgical Inc., Germany) in sampling tubes (Micro tube 2 mL, Sarstedt, Germany) and froze them to –

20 °C in 24 h. We sampled the control cows similarly from the interdigital cleft. All bacterial samples in this study are hereafter referred as hoof samples. If needed, the farm veterinarian treated IP and DD cows with severe signs after sampling.

Bacteriological Culture

During the farm visits, we set up a field laboratory at the farm. It included culture media, disposable plastic loops (10 µL, Mekalasi Oy, Helsinki, Finland) and equipment to maintain anaerobic conditions. Fastidious Anaerobe Agar, FAA (LabM, Lancashire, UK) and Fusobacterium Neomycin Vancomycin, NV agar [20] were used for primary culture. NV media were provided by Kokkola laboratory ((Maintpartner OY, Kokkola, Finland) or the Finnish Food Safety Authority (Evira, Helsinki, Finland). Agar plates were prereduced in Genbox containers (Biomerieux, France). The agar plates were sealed to maintain anaerobic conditions (BD GasPak EZ, Becton, Dickinson and Company, USA and GENpag anaer, Biomerieux, France) within 2 hours of sampling and incubated anaerobically for 2 days at 37 °C.

Isolation and Identification of *Fusobacterium Necrophorum*

From cultures we picked greyish, umbonate colonies of various shapes and sizes typical of spp. *necrophorum*, and smaller, yellowish, and waxy colonies typical of spp. *funduliforme*. Both colony types expressed strong beta-haemolysis on FAA and NV agars. The colonies were identified using conventional bacteriological methods to species and subspecies level [20], and verified using PCR assays for *lkt*A and *haemagglutinin* (Table 2). Isolates were stored below −70 °C for further characterisation.

Table 2: PCR oligos and reaction conditions of a study of interdigital phlegmon outbreaks in Finnish dairy herds

PCR assay	Oligos	Annealing tempera-ture (°C)	PCR prod-uct (bp)	Reference
Dichelobacter nodosus	16S(F2): CGGGGTTATGTAGCTTGC	60	783	[43]
	16S(R2): TCGGTACCGAGTATTTC-TACCCAACACCT			

F. necrophorum leucotoxin	LT3 F: GGAGTAAGAGCAAC-TATGGGAGCAGCTAC	60	360	[44]
	LT3 R: CCCAATCCACCTTTTACAG-CAGCTCG			
F. necrophorum hemagglutinin	HAEM F: CATTGGGTTGGATAAC-GACTCCTAC	55	286	[45]
	HAEM R: CAATTCTTTGTCTAAGATG-GAAGCGG			
Trueper-ella pyogenes **pyolysin**	PLO F: TCATCAACAATCCCACGAAGAG	60	150	[46]
	PLO R: TTGCCTCCAGTTGACGCTTT			
Universal 16S	27f YM: 5'-AGAGTTTGATYMTGGCT-CAG-3′	53	~1500	[47]
	1492 r: 5'-TACCTTGTTACGACTT-3′			
Porphyromonas levii				
PORP01F01	GACCAAATCGTCGTACTTGACAAA	66.2	75	
PORP01R01	GCCTCGGCTGGCAGTAAG	66.4		
PORP-01P01FAM	ACTCTCATGGTTGCCTACTTCTA-CAATCTTTCC	71.3		
Prevotella melaninogenica				
PREV01F01	CCCGGCTGTTTAGAATACTTTGTCA	67.8	152	
PREV01R02	CTTTGCATGGGTGGTGTTGAT	67.2		
PREV01P-01FAM	AATTAATCGTCGTCCGATATCACCA-CATACAGAG	73.4		
Treponemas				
Group 1 (*T. medium*/*T. vin-centii* **–like)**	TmF 5′-GAATGCTCATCTGATGACGGTA-ATCGACG-3′	68	472–500	[21]
	TmR 5′-CCGGCCTTATCTAAGACCTTC-TACTAG-3′			
Group 2 (*T. phagedenis*-like)	TbF 5′-GAAATACTCAAGCTTAACTT-GAGAATTGC-3′	64	400	[21]
	TbR 5′-CTACGCTACCATATCTCTATA-ATATTGC-3′			
Group 3 (*T. denticola* /*T. putidum*-like)	TpF 5′-GGAGATGAGGGAATGC-GTCTTCGATG-3′	67	475	[21]
	TpR 5′-CAAGAGTCGTATTGCTACGCT-GATATATC-3′			
Universal 16S	16S F 5′-AGAGTTTGATCCTGG-3′	57	1526	[48]
	16S R 5′-TACCTTGTTACGACTT-3′			

DNA Extraction from the Cytobrush Samples

Total DNA was extracted from cytobrush samples with Qiagen Blood and Tissue Column kit (Qiagen Gmbh, Germany) following the manufacturer's instructions. The samples were eluted in 100 µL EB and stored at −20 °C. An aliquot of 2 µL was used as a template for PCR amplification. Bovine DNA in the preparations did not block the detection of bacterial target DNA by PCR. The PCR assays are listed in Table 2.

PCR for *Fusobacterium necrophorum*, *Dichelobacter nodosus* and *Trueperella pyogenes*

The PCR analyses were performed at the Finnish Food Safety Authority Evira. Of the 228 samples analyzed, 205 samples were successfully amplified. The PCR assays, oligos and conditions for PCR are shown in Table 2. PCR reactions consisted of 0.5 µM of each oligo, 200 µM dNTP (Thermo Fisher Scientific), 1.0 U Dynazyme polymerase, 1.5 mM $MgCl_2$ and 2 µL template in Dynazyme F-511 buffer (Thermo Fisher Scientific). PTC Thermal cycler (Thermo Fisher Scientific™) was used for ampilification. For *lktA* gene, the thermal profile consisted of 95 °C for 2 min followed by 35 cycles of 95 °C for 30 s, 60 °C for 30 s, 72 °C for 40 s, with a final extension at 72 °C for 5 min. For *haemagglutinin* gene, the thermal profile was 95 °C for 2 min, followed by 35 cycles of 95 °C for 30 s, 55 °C for 15 s and 72 °C for 30 s, with a final extension at 72 °C for 5 min. The PCR products were separated and visualized using electrophoresis and SybrSafe in 1.5% agarose gel.

PCR for *Porphyromonas levii and Prevotella Melaninogenica*

The PCR analyses were performed at ThermoFisher Scientific Vantaa, Finland. Control (IP herd and IP free herd) and IP samples (n = 142) were analysed. All PCR reactions contained 0.5 µM of primers and 0.25 µM of probes in 20 µL of final PCR volume. QuantStudio 5 Real-Time PCR System (Thermo Fisher Scientific™) was used for thermal cycling. The thermal profile consisted of 95 °C for 10 min followed by 40 cycles of 95 °C for 5 s, and 60 °C for 1 min. In-house programs were applied to design qPCR oligo sequences for *P. levii* and *P. melaninogenica* used in this study (Table 2). Inclusivity and exclusivity were confirmed in silico using all RefSeq (NCBI Reference Sequence Database) bacterial genomes as reference sequence data.

Commercial genomic DNA (gDNA) stocks from *P. melaninogenica* DSM26980 and *P. levii* DSM23370 were measured using a Qubit Fluorometer

(Qubit® 2.0 Fluorometer, Thermo Fisher Scientific™) and gDNA copy numbers were calculated using DNA Copy Number and Dilution Calculator (Thermo Fisher Scientific™). Both oligo sets were multiplexed with an internal amplification control (IAC) oligos and template DNA (eliminates false-negative results due to inhibition of the reaction) and compared to the singleplex reactions using a genomic DNA dilution series in triplicate. Amplification efficiency for both oligo sets was calculated from the multiplex reactions. No-template controls (NTC) were run with each multiplex to screen potential oligo cross-reactions. Sensitivity of the oligo sets was tested using a doubling dilution series of genomic DNA in 8 replicates. Specificity of both oligo sets was tested using the non-target panel of several bacteria. DNA samples were analysed with the two oligo sets using 2 μL of DNA. Positive controls and NTC's were included into each run.

PCR for *Treponema*

The PCR analyses were performed at Denmark Technical University. Altogether 168 cytobrush samples had enough DNA for the analysis. An initial PCR step using a universal bacterial oligo pair encompassing most of the 16S rRNA gene [21] was followed by nested PCR analysis using oligos specific for the three DD *Treponema* phylogroups as described by Evans et al. [21] (Table 2). In all PCR assays, a 25 μL reaction mixture contained 1.25 U AmpliTaq DNA polymerase (Applied Biosystems, CA, USA), 1.5 mM (universal oligos) or 3 mM (group specific oligos) $MgCl_2$ Solution (Applied Biosystems, USA), 100 μM of each dNTP (Amersham Biosciences, NJ, USA), 0.2 μM of each specific oligo, and 1 μL of the template in PCR Buffer II (Applied Biosystems, USA). Thermal cycling was performed in a T3 thermocycler (Biometra, Göttingen, Germany) as described by Evans et al. [21]. In each assay, water served as a negative control, and genomic DNA from each of the three *Treponema* groups as positive control. PCR products were separated on a 2% E-gel (Invitrogen, Carlsbad, 92,008 CA, USA), and visualized by UV fluorescence.

Bacterial Controls

The following type strains were used as controls in the PCR assays: *D. nodosus* ATCC 25549, *F. necrophorum* ssp. *necrophorum* ATCC 25286, *F. varium* ATCC 8501, *F. necrophorum* ssp. *funduliforme* DSM 19678, *T. pyogenes* ATCC 19411D, *P. levii* (DSM23370) and *P. melaninogenica* (DSM26980), *T. vincentii* (ATCC 35580), *T. phagedenis* (ATCC 27087) and

T. denticola (ATCC 3320). Our own *Arcanobacterium haemolyticum* isolate served as a negative control for *T. pyogenes* pyolysin.

Statistical Analysis

The bacteriological results and data recorded during the herd visits were entered Excel spreadsheets and the statistical analyses were carried out using Stata IC version 15.0 (Stata Corporation, Texas, USA). A *p*-value of <0.05 was considered statistically significant. The repeated samples were excluded from statistical analyses.

Two groups of cows served as controls in our study; control in IP free herd ($n=19$) and in IP herd ($n=45$), and were tested for statistical difference using chi square. All hoof samples were divided into four disease categories; 1) control, 2) acute IP, 3) healing IP, and 4) other hoof diseases. Antimicrobial treatments were divided into three categories; 1) no current or previous antimicrobial treatment during last month, 2) current antimicrobial treatment or treatment within 6 days before sampling and 3) previous treatment with antimicrobials within 7–30 days prior to the sampling. The outbreak herds ($n=19$) were divided into two categories 1) herds of high morbidity; ≥50% of the cows having IP and 2) herds of moderate morbidity; 9–33% of the cows with IP during the first 2 months of the outbreak. No herds had morbidity between these figures.

The effect of antimicrobial treatment to each bacterium was tested separately with a logistic regression model. The dependent variable was each bacterium separately and independent variables were disease categories 1–4 and antimicrobial treatment categories 1–3. Herd was included as a random factor in these models.

The possible association of culture results of fusobacteria and IP were tested using chi-squared test. The possible association of bacteria in IP samples and high or moderate morbidity outbreak of IP were tested using Fisher exact test; only cows without antimicrobial treatment were included in the analysis.

We studied the association of disease categories and various bacteria ($n=6$) in a multinomial logistic regression model. The herd had no effect on the results and was not included to the final model. The outcome of the model was disease categories (control, acute IP, healing IP) and variables were *F. necrophorum, D. nodosus, T. pyogenes, Treponema, P. levii* and

P. melaninogenica (all dichotomous, no presence/presence). The group of other hoof diseases was excluded from this analysis.

RESULTS

Association of *Fusobacterium necrophorum* Isolates in different Disease Categories

F. necrophorum ssp. *necrophorum* was detected by culture in 48/65 (73.8%) of the samples from acute IP and in 26/41 (63.4%) from healing IP and was clearly associated with IP ($p<0.01$) when both IP groups ($n=106$) were combined and compared with controls ($n=64$). All the *F. necrophorum* isolates, including both subspecies *necrophorum* and *funduliforme*, possessed the *lktA* gene. Figure 1 shows the prevalence of cultured fusobacteria in various disease categories; control cows (IP free herd, $n=19$), control cows (IP herd, $n=45$), acute IP ($n=65$), healing IP ($n=41$), other hoof diseases ($n=58$).

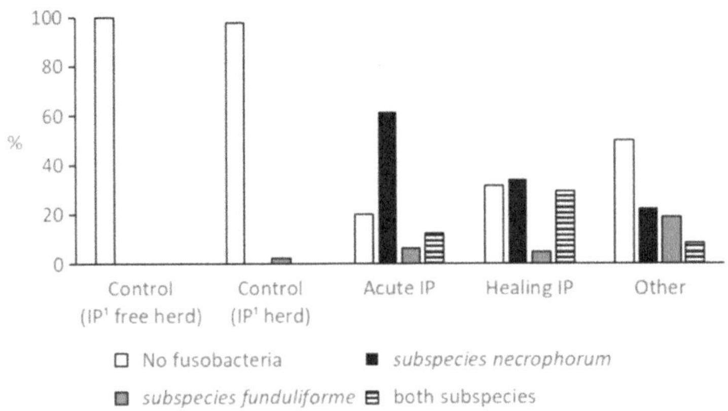

¹IP = interdigital phlegmon

Figure 1: Detection of *Fusobacterium necrophorum* ssp. *necrophorum* and ssp. *funduliforme* by culture in hoof samples from various disease categories. Samples ($n=228$) were collected from control cows (IP free herd, $n=19$), control cows (IP herd, $n=45$), acute interdigital phlegmon (Acute IP, $n=65$), during the healing process of IP (Healing IP, $n=41$) and from other hoof diseases than IP, including digital dermatitis, interdigital dermatitis, white line abscess and sole ulcer (Other, $n=58$).

The group of other hoof diseases (n=58) included samples from cases of DD, ID, sole ulcer and white line abscesses. In 20 DD samples, *F. necrophorum* ssp. *necrophorum* was detected in 7 (35.0%) samples. In other hoof diseases, including ID, white line abscesses and sole ulcers, ssp. *necrophorum* was detected in 11/38 (28.9%) of the samples.

Isolation of *Fusobacterium necrophorum* from Repeated Samples

The resampled hooves were culture negative for *F. necrophorum* ssp. *necrophorum* in a first sampling, but both were positive subsequently. One sample was positive at both samplings and seven samples were negative at the second sampling. One cow was sampled three times and after being positive for *F. necrophorum* ssp. *necrophorum* at the first sampling, it was negative at the second and positive at the third sampling. All cows except one were treated with antimicrobials between sampling times.

PCR Results

We obtained PCR results for *D. nodosus, F. necrophorum* and *T. pyogenes* from 205 hoof cytobrush samples, *P. levii* and *P. melaninogenica* from 142 and *Treponema* from 168 samples. Figure 1 shows the number of successful PCR tests in each disease category. Of 168 *Treponema* samples, 93 (55.4%) were positive for universal *Treponema* primer. None of the samples was positive for *Treponema* group 1 (*T. medium/ T. vincentii*-like). However, 28/168 (16.7%) were positive for *Treponema* group 2 (*T. phagedenis*-like) and 16/168 (9.5%) for *Treponema* group 3 (*T. putidum/ T. denticola*-like). *Treponema* group 3 was always detected simultaneously with *Treponema* group 2. Of these 16 samples were 4 acute IP, 3 healing IP, 8 DD and 1 other hoof disease.

PCR Results for Control Cows

D. nodosus was detected from 9/19 (47.4%) control cows (IP free herd) and 21/43 (48.8%) control cows (IP herd), *F. necrophorum* was detected in 0/19 (0%) and 4/43 (9.3%) of samples, *P. levii* in 1/19 (5.6%) and 3/41 (7.3%), *P. melaninogenica* in 0/19 (0%) and 2/41 (4.9%), *Treponema group* 2 and 3 in 4/19 (21.1%) and 6/42 (14.3%), and *T. pyogenes* in 1/19 (5.3%) and 0/43 (0%). No statistical differences were evident between the control groups regarding the bacteria detected and therefore data for each control group were combined for statistical analyses.

PCR Results for Samples of IP and other Hoof Diseases

Figure 2 presents the results of PCR analysis for various disease categories; control cows (n=62), acute IP (n=52), healing IP (n=37), and other hoof diseases (n=54). *P. levii* and *P. melaninogenica* were not analysed among the group of other hoof diseases. Several bacterial species were detected by PCR in numerous hoof samples (Table 3). The control cows were either PCR negative (26/59; 44.1%), or harboured *D. nodosus* alone (16/59; 27.1%) or in combination with *Treponema* group 2 and 3 (9/59; 15.3%). In most acute IP samples (24/36; 66.7%), *F. necrophorum* and *D. nodosus* were detected. They occurred with *P. levii* (4/36; 11.1%) or *Treponema* group 2 and 3 (4/36; 11.1%), and *F. necrophorum* alone was combined with *T. pyogenes* (4/36; 11.1%). For the healing stage of IP, the most frequently detected combinations were *F. necrophorum* and *T. pyogenes* (6/33, 18.2%), *F. necrophorum* alone (3/33; 9.1%) and *F. necrophorum*, *T. pyogenes* and *P. levii* (3/33, 9.1%).

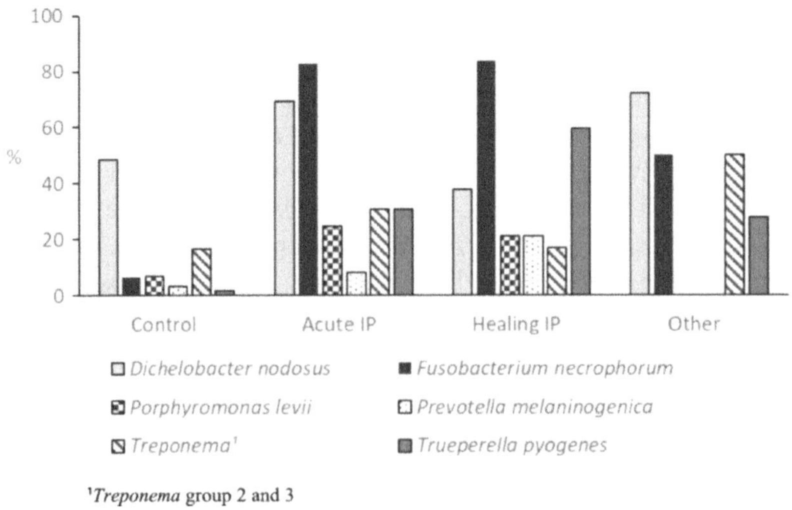

Figure 2: Detection of bacteria by PCR in hoof samples from various disease categories. The disease categories included; control cows (n=62), acute interdigital phlegmon (Acute IP, n=52), IP in a healing stage (Healing IP, n=37) and other hoof diseases than IP (Other, n=54). The group other hoof diseases included hoof samples from digital dermatitis, interdigital dermatitis, white line abscess and sole ulcer. Total number of hoof samples is 205, except with *P. levii* and *P. melaninogenica* (142) and *Treponema* group 2 and 3 (168).

Table 3: Combinations of bacterial species detected by PCR in various disease categories in interdigital phlegmon outbreaks in Finnish dairy herds

Bacterial combination	**Control**	**Acute IP**	**Healing IP**
n	59	36	33
No detected bacteria	26	2	
P. melaninogenica			1
P. levii	3		
Treponema [a]			1
T. pyogenes	1	1	
D. nodosus	16		2
D. nodosus, P. melaninogenica	1		
D. nodosus, Treponema	9	2	
D. nodosus, Treponema, P. levii		1	
D. nodosus, Treponema, T. pyogenes			1
F. necrophorum		1	3
F. necrophorum, P. melaninogenica			1
F. necrophorum, P. levii	1		2
F. necrophorum, Treponema		1	1

F. necrophorum, T. pyogenes		4	6
F. necrophorum, T. pyogenes, P. melaninogenica			2
F. necrophorum, T. pyogenes, P. levii			3
F. necrophorum, T. pyogenes, P. levii, P. melaninogenica			1
F. necrophorum, T. pyogenes, Treponema			1
F. necrophorum, D. nodosus	1	7	2
F. necrophorum, D. nodosus, P. melaninogenica			1
F. necrophorum, D. nodosus, P. levii		4	
F. necrophorum, D. nodosus, P. levii, P. melaninogenica		1	
F. necrophorum, D. nodosus, Treponema		4	
F. necrophorum, D. nodosus, P. melaninogenica, Treponema	1	1	
F. necrophorum, D. nodosus, T. pyogenes		2	2
F. necrophorum, D. nodosus, T. pyogenes, P. melaninogenica			1
F. necrophorum, D. nodosus, T. pyogenes, P. levii		1	1
F. necrophorum, D. nodosus, T. pyogenes, P. levii, P. melaninogenica		1	
F. necrophorum, D. nodosus, T. pyogenes, Treponema		2	1
F. necrophorum, D. nodosus, T. pyogenes, P. levii, Treponema		1	

[a]*Treponema* **includes** *Treponema* **group 2 and 3**

Various disease categories are control cows, interdigital phlegmon (IP) in an acute stage (Acute IP), and IP during a healing process (Healing IP). Total number of control and IP hoof samples is 128

Association of Disease Categories and Bacterial Species

We investigated the association of control samples, acute IP, and healing IP with the bacterial species detected by PCR (Table 4). *F. necrophorum* was associated distinctively with both stages of IP ($p<0.01$). *T. pyogenes* was found more often with the healing IP ($p=0.01$), but only a trend existed in the group of acute IP samples. Antimicrobial treatment affected detection of *D. nodosus* (current treatment OR=0.2, p=0.01, previous treatment OR=0.1, p<0.01) and *Treponema* group 2 and 3 (current treatment OR=0.1, p<0.01, previous treatment OR=0.1, $p=0.03$), but not detection of other bacteria.

Table 4: The multinomial logistic regression model for the association of various disease categories and presence of bacteria in outbreaks of interdigital phlegmon in Finnish dairy herds

Disease categories	*n*	**RRR**[a]	*p*-**value**	**95% CI**[b]
Control cows	59	Base outcome		
Acute IP	36			
Dichelobacter nodosus		2.1	0.36	0.44–9.88
Fusobacterium necrophorum		74.9	<0.01	14.31–391.71
Porphyromonas levii		1.7	0.62	0.22–12.43
Prevotella melaninogenica		0.7	0.80	0.04–12.67
Treponema[c]		3.8	0.11	0.75–19.33
Trueperella pyogenes		10.8	0.06	0.91–127.48
Constant[d]		0.04	<0.01	0.01–0.15
Healing IP	33			
Dichelobacter nodosus		0.4	0.26	0.08–1.95
Fusobacterium necrophorum		58.4	<0.01	10.29–332.00
Porphyromonas levii		1.1	0.96	0.13–8.73
Prevotella melaninogenica		3.0	0.44	0.19–47.02
Treponema[c]		2.2	0.40	0.35–13.76
Trueperella pyogenes		22.4	0.01	2.01–249.04
Constant[d]		0.08	<0.01	0.02–0.25

[a]RRR = relative risk ratio

[b]95% CI = 95% confidence interval

[c]*Treponema* **group 2 and 3**

[d]Constant is a baseline relative risk for each outcome

The disease categories were control cows, acute interdigital phlegmon (Acute IP) and IP in a healing stage (Healing IP). The herd had no effect on the results. In this model, the number of the hoof samples is 128

Bacterial Findings in High and Moderate Morbidity Herds

Of 19 outbreak herds, in 7 herds the morbidity was high (morbidity ≥50% during first 2 months of the outbreak) and 12 herds moderate (morbidity 9–33%). No herds had morbidity of 34–49%. We found no differences in detected bacteria in control samples of herds of various morbidity. We focused on acute IP samples and compared their bacteriology between these 7 high morbidity herds and 12 moderate morbidity herds. Bacterial species detected by PCR in hoof samples from acute IP in high and moderate morbidity herds are presented in Fig. 3 and combinations of bacterial species detected in Table 5.

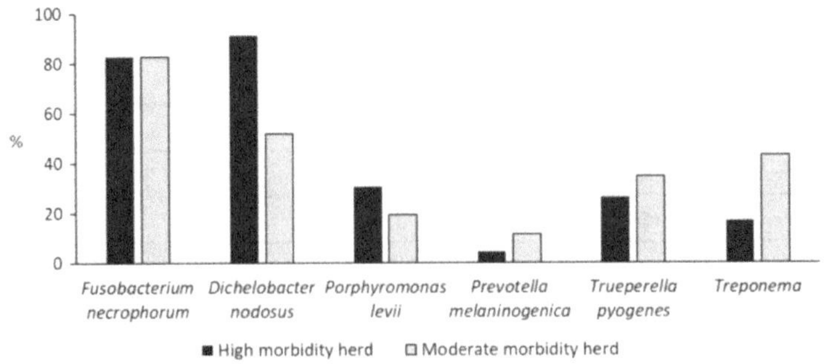

Figure 3: PCR results for hoof samples from acute interdigital phlegmon (IP) in herds with various morbidity. We visited high morbidity (morbidity ≥50% during the first two months of the outbreak) and moderate morbidity (morbidity 9–33%) herds. Number of hoof samples is 52, except with *P. levii* and *P. melaninogenica* ($n=49$) and *Treponema* ($n=39$). *Treponema* includes *Treponema* group 2 and 3.

Table 5: Combinations of bacterial species detected by PCR in hoof samples from acute interdigital phlegmon ($n=36$) in high morbidity (≥50%) and moderate morbidity (9–33%) Finnish dairy herds

Bacterial combination	**High**	**Moderate**
n	17	19
No detected bacteria		2
T. pyogenes		1
D. nodosus and Treponema [a]	1	1
D. nodosus, Treponema, P. levii	1	
F. necrophorum		1
F. necrophorum, Treponema		1
F. necrophorum, T. pyogenes	1	3
F. necrophorum, D. nodosus	7	
F. necrophorum, D. nodosus, P. levii	3	1
F. necrophorum, D. nodosus, P. levii, P. melaninogenica		1
F. necrophorum, D. nodosus, Treponema		4
F. necrophorum, D. nodosus, P. melaninogenica, Treponema		1
F. necrophorum, D. nodosus, T. pyogenes	2	
F. necrophorum, D. nodosus, T. pyogenes, P. levii		1
F. necrophorum, D. nodosus, T. pyogenes, P. levii, P. melaninogenica	1	
F. necrophorum, D. nodosus, T. pyogenes, Treponema	1	1
F. necrophorum, D. nodosus, T. pyogenes, P. levii, Treponema		1

[a]*Treponema* group 2 and 3

First, we analysed the association of culture results of fusobacteria in various morbidity herds. The presence of 2 *F. necrophorum* subspecies in acute IP samples from high ($n=31$) and from moderate morbidity herds ($n=34$) did not differ ($p=0.24$); of these samples, no fusobacteria were detected in 9 (29.0%) samples from high and in 4 (11.8%) samples from moderate morbidity herds. Subspecies *necrophorum* was detected in 18 samples (58.1%) from high and in 22 (64.7%) samples from moderate morbidity, ssp. *funduliforme* in 2 samples from both morbidity groups (6.5 and 5.9% respectively), and both subspecies in 2 samples (6.5%) from high and in 6 (17.7%) from moderate morbidity herds. Subsequently we compared the association of other bacteria in various morbidity herds; presence of *D. nodosus, F. necrophorum, P. levii, P. melaninogenica, Treponema* group 2 and 3, and *T. pyogenes* detected by PCR in acute IP samples from high and moderate morbidity herds is presented in Fig. 3.

The most common combination, *F. necrophorum* and *D. nodosus*, was found in 14/17 (82.4%) samples of acute IP from high and 10/19 (52.6%) samples from moderate morbidity herds (Table 5). *D. nodosus* was more often detected in IP in high than moderate morbidity herds ($p=0.05$, $n=35$).

DISCUSSION

F. necrophorum was found in this study as the main pathogen in IP. This is in line with previous studies [10,11,12]. Based on our results, it was ssp. *necrophorum* that was clearly associated with IP. We also detected *F. necrophorum* in DD and in other hoof diseases, but less frequently than in IP. Similarly, fusobacteria are detected in DD lesions in other studies [22, 23].

F. necrophorum is a normal inhabitant in the rumen of cattle [24]. Occasionally, it can be detected in the faeces, and thus it contaminates the environment [25]. In a study of DD microbiome, small number of fusobacteria were detected on healthy hooves [26]. Similarly, in our study *F. necrophorum* ssp. *necrophorum* was not detected on the skin of healthy hooves, even when a severe IP outbreak was evident in the herd. This indicates that *F. necrophorum* does not colonize the intact skin in large numbers. A moist environment or possible trauma has been mentioned as predisposing factors for IP in previous studies [7, 27]. Interestingly, in most of our acute IP study cows no hoof trauma was visible. In most of the study herds the free stall was also reasonably new and well-managed. As a result, we can speculate that *F. necrophorum* may have to interact with other bacteria to invade to the subcutaneous tissue in the interdigital cleft.

Unexpectedly in repeated sampling, fusobacteria were cultivated from IP lesions even though cows had been treated with antimicrobials and IP was at the healing stage. The clinical signs appeared to diminish after beginning of antimicrobial treatment, but *F. necrophorum* remained in the affected region. However, our bacteriological methods were not quantitative and therefore, we do not know the number of detected bacteria and whether the amount had diminished or not. In a small pilot study of outbreaks of IP in two herds, susceptibility of 27 *F. necrophorum* isolates to penicillin, tetracycline, cefuroxime and cefotaxime was determined by E-test. All isolates were found susceptible to tested antimicrobials [28]. Also other study reports that antimicrobial resistance is not characteristic of *F. necrophorum* in IP [29].

We detected *D. nodosus* from healthy hooves, IP and other hoof diseases. In most of the acute IP samples (66.7%), both *F. necrophorum* and *D. nodosus* were detected. A significant association was established with the presence of *D. nodosus* in IP lesions and high morbidity outbreak in the herd. This could indicate that the presence of *D. nodosus* affects the severity of IP. *D. nodosus* is associated with ID [30] and DD [30,31,32,33] and detected in healthy hooves [30]. It is hypothesised that *D. nodosus* could break down the epidermal barrier, creating a suitable environment for secondary invaders [32]. A recent study also suggests *D. nodosus* as a potentially important pathogen in DD [23]. Our qualitative investigation does not take account of the numbers of bacteria, which might differ in IP lesions compared with healthy hooves.

P. levii and *T. pyogenes* are detected with *F. necrophorum* in various cattle infections and evidence of interactions and possible synergism between these species is reported [34,35,36,37]. IP is induced using field strains of *F. necrophorum* and *B. melaninogenicus* [11]; the latter is reclassified as several *Porphyromonas* and *Prevotella* species [20]. Moreover, in other studies these bacteria are detected in IP samples [17, 18]. In addition to IP, *P. levii* is detected in DD lesions [22] and in an outbreak of necrotic vulvovaginitis [38]. Also, *T. pyogenes* is reported to occur in IP lesions [7, 11, 18]. In our study *T. pyogenes* was associated with a healing stage of IP and only a trend existed with acute IP, indicating that this pathogen has a secondary role in IP. Nevertheless, we were unable to establish an association between high morbidity and *P. levii* or *T. pyogenes*.

There are very few studies of the occurrence of treponemes in IP, but many concerning DD. Earlier studies revealed occurrence of *Spirochetes* in IP lesions [5, 7] but it remains uncertain whether the organisms were treponemes or not. Treponemes are regarded as the most important pathogens in DD [19, 22, 26, 39], and have been detected also in other hoof lesions, including toe necrosis, sole ulcer and white line disease [40, 41]. In our study, we detected *Treponema* group 2 and 3 in all disease categories, but more frequently in IP and in other hoof diseases; mainly DD. Interestingly all observed DD lesions were detected in herds of moderate morbidity (data not shown). To date ID and DD are not represented a major problem of cattle in Finland [42].

Of 217 cows sampled, 66 (30.4%) were currently being or had previously been treated with antimicrobials. It would have been unethical to leave the affected cows untreated until the sampling visit took place. Nevertheless, the possible effect of an antimicrobial treatment was taken into account in the analysis.

CONCLUSION

In the current study, we investigated several bacteria in new type of outbreaks of IP and possible bacterial dissimilarities in herds with various morbidity. We could detect all studied bacteria in IP lesions either alone or in various combinations but observed bacteriological differences in herds with various morbidity. The most substantial finding was the presence of *F. necrophorum* in IP lesions, and *T. pyogenes* at the healing stage of IP. Our results also suggest that *D. nodosus* may play a role in the severity of the outbreak of IP. It is also quite apparent that a correct diagnosis of IP cannot be made based on a single bacteriologic sample without a clinical inspection.

Virulence factors of *F. necrophorum* isolates and transmission of hoof pathogens among and within farms may represent an important subject that merits further research.

ACKNOWLEDGEMENTS

The authors thank the study farms and veterinarians, hoof trimmers and others who helped during the farm visits, Taina Lehto and Merja Pöytäkangas at the University of Helsinki for DNA isolation and Tiina Karla, Milja Tikkanen and Laura Vaahtoranta at Thermo Fisher Scientific Vantaa for PCR analysis of *Porphyromonas* and *Prevotella* species.

REFERENCES

1. Häggman J, Junni R, Simojoki H, Juga J, Soveri T. The costs of interdigital phlegmon in four loose-housed Finnish dairy herds. Acta Vet Scand. 2015;57:90.
2. DeFrain JM, Socha MT, Tomlinson DJ. Analysis of foot health records from 17 confinement dairies. J Dairy Sci. 2013;96:7329–39.
3. Oberbauer AM, Berry SL, Belanger JM, McGoldrick RM, Pinos-Rodriquez JM, Famula TR. Determining the heritable component of dairy cattle foot lesions. J Dairy Sci. 2013;96:605–13.
4. David GP. Severe foul-in-the-foot in dairy cattle. Vet Rec. 1993;132:567–8.
5. Doherty M, Bassett H, Markey B, Healy A, Sammin D. Severe foot lameness in cattle associated with invasive spirochaetes. Ir Vet J. 1998;1:195–8.
6. Van Metre DC. Pathogenesis and treatment of bovine foot rot. Vet Clin N Am-Food A 2017; 33:183–194.
7. Gupta RB, Fincher MG. Bruner DW. A study of the etiology of foot-rot in cattle. Cornell Vet. 1964;54:66–77.
8. Hernandez J, Shearer JK, Webb DW. Effect of lameness on milk yield in dairy cows. J Am Vet Med Assoc. 2002;220:640–4.
9. Booth CJ, Warnick LD, Gröhn YT, Maizon DO, Guard CL, Janssen D. Effect of lameness on culling in dairy cows. J Dairy Sci. 2004;87:4115–22.
10. Flint JC, Jensen R. Pathology of necrobacillosis of the bovine foot. Am J Vet Res. 1951;12:5–13.
11. Berg JN, Loan RW. *Fusobacterium necrophorum* and *Bacteroides melaninogenicus* as etiologic agents of foot rot in cattle. Am J Vet Res. 1975;36:1115–22.
12. Clark BL, Stewart DJ, Emery DL. The role of *Fusobacterium necrophorum* and *Bacteroides melaninogenicus* in the aetiology of interdigital necrobacillosis in cattle. Aust Vet J. 1985;62:47–9.
13. Tan ZL, Nagaraja TG, Chengappa MM. *Fusobacterium necrophorum* infections: virulence factors, pathogenic mechanism and control measures. Vet Res Commun. 1996;20:113–40.
14. Oelke A, Nagaraja TG, Wilkerson MJ, Stewart GC. The leukotoxin operon of *Fusobacterium necrophorum* is not present in other species

of *Fusobacterium*. Anaerobe. 2005;11:123–9.

15. Zhou H, Bennett G, Hickford JGH. Variation in *Fusobacterium necrophorum* strains present on the hooves of footrot infected sheep, goats and cattle. Vet Microbiol. 2009;135:363–7.
16. Lechtenberg K, Nagaraja T, Leipold H, Chengappa MM. Bacteriologic and histologic studies of hepatic abscesses in cattle. Am J Vet Res. 1988;49:58–62.
17. Morck DW, Olson ME, Louie TJ, Koppe A, Quinn B. Comparison of ceftiofur sodium and oxytetracycline for treatment of acute interdigital phlegmon (foot rot) in feedlot cattle. J Am Vet Med Assoc. 1998;212:54.
18. Sweeney M, Watts J, Portis E, Lucas M, Nutsch R, Meeuwse D, Bade D, Oliver V, Morc D, Shinabarger D, Poppe S, Peterson M, Sweeney D, Knechtel M, Zurenko G. Identification of *Porphyromonas levii* isolated from clinical cases of bovine interdigital necrobacillosis by 16S rRNA sequencing. Vet Ther. 2009;10:4.
19. Döpfer D, Koopmans A, Meijer FA, Szakáll I, Schukken YH, Klee W, Bosma RB, Cornelisse JL, van Asten AJAM, ter Huurne AAHM. Histological and bacteriological evaluation of digital dermatitis in cattle, with special reference to spirochaetes and *Campylobacter faecalis*. Vet Rec. 1997;140:620–3.
20. Jousimies-Somer H, Summanen P. Recent taxonomic changes and terminology update of clinically significant anaerobic gram-negative Bacteria (excluding spirochetes). Clin Infect Dis. 2002;35(Suppl 1):S21.
21. Evans N, Brown J, Demirkan I, Singh P, Getty B, Timofte D, Daan Vink W, Murray R, Blowey R, Birtles R, Hart C, Carter S. Association of Unique, isolated Treponemes with bovine digital dermatitis lesions. J Clin Microbiol. 2009;47:689–96.
22. Nielsen MW, Strube ML, Isbrand A, Al-Medrasi WDHM, Boye M, Jensen TK, Klitgaard K. Potential bacterial core species associated with digital dermatitis in cattle herds identified by molecular profiling of interdigital skin samples. Vet Microbiol. 2016;186:139–49.
23. Moreira T, Filho E, Carvalho A, Strube M, Nielsen M, Klitgaard K, Jensen T. Pathology and bacteria related to digital dermatitis in dairy cattle in all year round grazing system in Brazil. PLoS One. 2018;3:13.
24. Robinson TJ, Jasper DE, Guilbert HR. The isolation of *Spherophorus necrophorus* from the rumen together with some feed lot data on

abscess and telangiectasis. J Anim Sci. 1951;10:733–41.

25. Smith GR, Thornton EA. Effect of disturbance of the gastrointestinal microflora on the faecal excretion of *Fusobacterium necrophorum* biovar a. Epidemiol Infect. 1993;110:333–7.
26. Johnson DW, Dommert AR, Kiger DG. Clinical investigations of infectious foot rot of cattle. J Am Vet Med Assoc. 1969;155:1886–91.
27. Junni R, Kontturi M, Seuna E, Kujala M, Simojoki H, Malinen E, Pekkanen K, Pelkonen S & Soveri T. Outbreaks of interdigital phlegmon in Finland. Proceedings of the 17th international symposium and 9th international conference on lameness in ruminants, Bristol, UK. 2013;59.
28. Bergsten C, Jansson Mörk M, Pringle M, Persson Y. Can interdigital phlegmon (foot rot) be treated without antibiotics? Proceedings of the 19th international symposium and 11th international conference on lameness in ruminants, Munich, Germany 2017;315.
29. Knappe-Poindecker M, Gilhuus M, Jensen TK, Klitgaard K, Larssen RB, Fjeldaas T. Interdigital dermatitis, heel horn erosion, and digital dermatitis in 14 Norwegian dairy herds. J Dairy Sci. 2013;96:7617–29.
30. Capion N, Boye M, Ekstrøm CT, Jensen TK. Infection dynamics of digital dermatitis in first-lactation Holstein cows in an infected herd. J Dairy Sci. 2012;95:6457–64.
31. Rasmussen M, Capion N, Klitgaard K, Rogdo T, Fjeldaas T, Boye M, Jensen TK. Bovine digital dermatitis: possible pathogenic consortium consisting of *Dichelobacter nodosus* and multiple *Treponema* species. Vet Microbiol. 2012;160:151–61.
32. Sullivan LE, Evans NJ, Blowey RW, Grove-White DH, Clegg SR, Duncan JS, Carter SD. A molecular epidemiology of treponemes in beef cattle digital dermatitis lesions and comparative analyses with sheep contagious ovine digital dermatitis and dairy cattle digital dermatitis lesions. Vet Microbiol. 2015;178:77–87.
33. Ruder CA, Sasser RG, Williams RJ, Ely JK, Bull RC, Butler JE. Uterine infections in the postpartum cow: II. Possible synergistic effect of *Fusobacterium necrophorum* and *Corynebacterium pyogenes*. Theriogenology. 1981;15:573–80.
34. Nagaraja TG, Beharka AB, Chengappa MM, Carroll LH, Raun AP, Laudert SB, Parrott JC. Bacterial flora of liver abscesses in feedlot cattle fed tylosin or no tylosin. J Anim Sci. 1999;77:973.

35. Kan IY, Blum S, Elad D. Synergism between *Porphyromonas levii* and *Arcanobacterium pyogenes* in a murine abscess model. Isr J Vet Med. 2009;64:62–5.
36. Karstrup CC, Pedersen HG, Jensen TK, Agerholm JS. Bacterial invasion of the uterus and oviducts in bovine pyometra. Theriogenology. 2017;93:93–8.
37. Elad D, Friedgut O, Alpert N, Stram Y, Lahav D, Tiomkin D, Avramson M, Grinberg K, Bernstein M. Bovine necrotic Vulvovaginitis associated with *Porphyromonas levii*. Emerg Infect Dis. 2004;10:505–7.
38. Klitgaard K, Boye M, Capion N, Jensen T. Evidence of multiple *Treponema* phylotypes involved in bovine digital dermatitis as shown by 16S rRNA gene analysis and fluorescence in situ hybridization. J Clin Microbiol. 2008;46:3012–20.
39. Zinicola M, Lima F, Lima S, Machado V, Gomez M, Döpfer D, Guard C, Bicalho R. Altered microbiomes in bovine digital dermatitis lesions, and the gut as a pathogen reservoir. PLoS One. 2015;10:3.
40. Evans NJ, Blowey RW, Timofte D, Isherwood DR, Brown JM, Murray R, Paton RJ, Carter SD. Association between bovine digital dermatitis treponemes and a range of 'non-healing' bovine hoof disorders. Vet Rec. 2011;168:214.
41. Sykora S, Kofler J, Glonegger-Reichert J, Dietrich J, Auersperg G, Brandt S. *Treponema* DNA in bovine 'non-healing' versus common sole ulcers and white line disease. Vet J. 2015;205:417–20.
42. Kontturi M, Kujala M, Junni R, Malinen E, Seuna E, Pelkonen S, Soveri T, Simojoki H. Survey of interdigital phlegmon outbreaks and their risk factors in free stall dairy herds in Finland. Acta Vet Scand. 2017;59:1.
43. Moore LJ, Wassink GJ, Green LE, Grogono-Thomas R. The detection and characterization of *Dichelobacter nodosus* from cases of ovine footrot in England and Wales. Vet Microbiol. 2005;108:57–67.
44. Ludlam H, Milner N, Brazier J, Davies I, Perry K, Marriott R, Donachie L, Curran M. *lktA*-encoded leukotoxin is not a universal virulence factor in invasive *Fusobacterium necrophorum* infections in animals and man. J Med Microbiol. 2009;58:529–30.
45. Aliyu S, Marriott R, Curran MD, Parmar S, Bentley N, Brown N, Brazier J, Ludlam H. Real-time PCR investigation into the importance of *Fusobacterium necrophorum* as a cause of acute pharyngitis in general practice. J Med Microbiol. 2004;53:1029–35.

46. Silva E, Gaivão M, Leitão S, Jost BH, Carneiro C, Vilela CL, Lopes da Costa L, Mateus M. Genomic characterization of *Arcanobacterium pyogenes* isolates recovered from the uterus of dairy cows with normal puerperium or clinical metritis. Vet Microbiol. 2008;132:111–8.
47. Frank JA, Reich CI, Sharma S, Weisbaum JS, Wilson BA, Olsen GJ. Critical evaluation of two primers commonly used for amplification of bacterial 16S rRNA genes. Appl Environ Microbiol. 2008;74:2461–70.
48. Rurangirwa F, Dilbeck P, Crawford T, McGuire T, McElwain T. Analysis of the 16S rRNA gene of micro-organism WSU 86-1044 from an aborted bovine foetus reveals that it is a member of the order *Chlamydiales*: proposal of *Waddliaceae* fam. Nov., *Waddlia chondrophila* gen. Nov., sp. nov. Int J Syst Bacteriol. 1999;49:577–81.

SECTION 2: BOVINES

CHAPTER 4

Direct Detection and Differentiation of Pathogenic Leptospira Species Using a Multi-Gene Targeted Real Time PCR Approach

Ana Sofia Ferreira[1,2], Pedro Costa[1,3], Teresa Rocha[1] , Ana Amaro[1] , Maria Luı´sa Vieira[3] , Ahmed Ahmed[5] , Gertrude Thompson[2,4], Rudy A. Hartskeerl[5] , João Inácio[1,6]

[1] Instituto Nacional de Investigação Agrária e Veterinária, I.P. (INIAV, I.P.), Unidade Estratégica de Investigação e Serviços em Produção e Saúde Animal, Lisboa, Portugal, School of Pharmacy

[2] Instituto de Cie^ncias Biome´dicas de Abel Salazar, Universidade do Porto, Porto, Portugal

[3] Unidade de Microbiologia Me´dica, Instituto de Higiene e Medicina Tropical, Universidade Nova de Lisboa, Lisboa, Portugal

[4] Research Center in Biodiversity and Genetic Resources (CIBIO-ICETA), Universidade do Porto, Porto, Portugal

[5] WHO/ FAO/OIE and National Collaborating Centre for Reference and Research on Leptospirosis, KIT Biomedical Research, Amsterdam, The Netherlands

[6] School of Pharmacy and Biomolecular Sciences, University of Brighton, Brighton, United Kingdom

Citation: Ferreira AS, Costa P, Rocha T, Amaro A, Vieira ML, Ahmed A, et al. (2014) Direct Detection and Differentiation of Pathogenic Leptospira Species Using a Multi-Gene Targeted Real Time PCR Approach. PLoS ONE 9(11): e112312. https://doi.org/10.1371/journal.pone.0112312.

ABSTRACT

Leptospirosis is a growing public and veterinary health concern caused by pathogenic species of *Leptospira*. Rapid and reliable laboratory tests for the direct detection of leptospiral infections in animals are in high demand not only to improve diagnosis but also for understanding the epidemiology of the disease. In this work we describe a novel and simple *TaqMan*-based multi-gene targeted real-time PCR approach able to detect and differentiate *Leptospira interrogans*, *L. kirschneri*, *L. borgpeteresenii* and *L. noguchii*, which constitute the veterinary most relevant pathogenic species of *Leptospira*. The method uses sets of species-specific probes, and respective flanking primers, designed from *ompL1* and *secY* gene sequences. To monitor the presence of inhibitors, a duplex amplification assay targeting both the mammal *β-actin* and the leptospiral *lipL32* genes was implemented. The analytical sensitivity of all primer and probe sets was estimated to be <10 genome equivalents (GE) in the reaction mixture. Application of the amplification reactions on genomic DNA from a variety of pathogenic and non-pathogenic *Leptospira* strains and other non-related bacteria revealed a 100% analytical specificity. Additionally, pathogenic leptospires were successfully detected in five out of 29 tissue samples from animals (*Mus* spp., *Rattus* spp., *Dolichotis patagonum* and *Sus domesticus*). Two samples were infected with *L. borgpetersenii*, two with *L. interrogans* and one with *L. kirschneri*. The possibility to detect and identify these pathogenic agents to the species level in domestic and wildlife animals reinforces the diagnostic information and will enhance our understanding of the epidemiology of leptopirosis.

INTRODUCTION

Leptospirosis is a growing and underestimated public health and veterinary concern, caused by pathogenic spirochetes belonging to the family *Leptospiracea*, genus *Leptospira* [1], [2]. The disease is an important cause of abortion, stillbirths, infertility, poor milk production and death amongst livestock, harboring a significant economic impact [3]–[5]. Its transmission requires circulation of the agents among domestic and wild animal reservoirs, with rodents recognized as the most important sources that establish persistent renal carriage and urinary shedding of *Leptospira*. Humans are incidental hosts acquiring a systemic infection upon direct or indirect exposure to the urine, blood or tissue of an infected animal. Farmers, veterinarians, sewer workers, pet keepers, rodent catchers and those persons participating in aquatic leisure activities are more prone to acquire the disease.

Conventional classification of *Leptospira* is based on serological criteria, using the serovar as the basic taxon. To date over 250 pathogenic serovars separated into 25 serogroups are known [6]. The serological classification system is complemented by a genotypic one, in which 21 genetic species are currently recognized, including pathogenic, intermediate and non-pathogenic (or saprophytic) species [7]–[10]. Genetic species boundaries hardly correlate with the serological classification [8].

Serological approaches are used commonly for diagnosis of leptospirosis in animals. The reference method is the Microscopic Agglutination Test (MAT), which has the advantage of being specific for serogroups [3] but has several drawbacks of being laborious and requiring a panel of viable *Leptospira* cultures. Isolation of leptospires, from suspect clinical specimens, constitutes the definitive diagnosis but is also technically demanding, time consuming and subject to contamination and high rates of failure [4]. Isolates are traditionally classified to the serovar level by the Cross Agglutinin Absorption Test (CAAT) [8] which is cumbersome for routine use and is only performed in a few reference laboratories worldwide.

Rapid and reliable laboratory tests for the direct detection of leptospiral infections in animals are in high demand, particularly to support suitable control measures. Serology does not corroborate well with the presence of pathogenic viable leptospires in the kidneys or urine and detection of the agents is necessary to identify healthy animal carriers. Molecular-based assays have been previously described for detecting leptospires in clinical samples. Most approaches are PCR-based and target specific genes or polymorphisms in the genome of pathogenic leptospires. Several real time PCR assays have been described predominantly for use with human samples such as whole-blood, serum or urine [11]–[17] but only few have been plentifully validated [18],[19]. A few assays were evaluated or used for detecting *Leptospira* in kidney tissue, blood, urine and other clinical specimens from animals such as sheep [20], dogs [21], [22], pigs [5], deer [23], flying foxes [24] and rodents [25], [26]. Most assays rely on SYBR green detection chemistry and only differentiate between pathogenic and non-pathogenic leptospires, lacking the ability to distinguish between different species. Nevertheless, speciation of infecting *Leptospira* from clinical material may be important for determining the clinical significance, the probable source of infection, to distinguish sporadic cases from possible outbreaks and to better access the epidemiology of the disease.

In the present work we have developed a novel and simple *TaqMan*-based multi-gene targeted real-time PCR approach yielding high sensitivity and specificity for the direct detection and differentiation of the most relevant pathogenic *Leptospira* species in animal samples, suitable for introduction into the routine diagnostics of veterinary laboratories.

MATERIALS AND METHODS

Bacterial Strains

Eighty five reference strains and clinical and environmental isolates of *Leptospira* spp. belonging to pathogenic, intermediate and non-pathogenic phylogenetic clades were used in this study (Table 1). Strains were obtained from the collection maintained by the *Instituto Nacional de Investigação Agrária e Veterinária* (INIAV), Portugal, which is the Portuguese reference laboratory for animal diseases, from the Leptospirosis Laboratory at the *Instituto de Higiene e Medicina Tropical* (IHMT/UNL), Portugal, and from the WHO/FAO/OIE and National Leptospirosis Reference Centre in Amsterdam, The Netherlands. Strains were grown in liquid Ellinghausen-McCullough-Johnson-Harris (EMJH) medium for up to 7 days.

Table 1: ***Leptospira*** **strains used in the present study and results of the real time PCR assays using the species-specif**ic probes and flanking primers

Species	Serogroup	Serovar	Strain	Source[1]	Set 1[2]	Set 2[3]	Set 3[4]	Set 4[5]	Set 5[6]
L. interrogans	Australis	Muenchen	München C 90	KIT	+	+	-	-	-
	Australis	Australis	Ballico	KIT	+	+	-	-	-
	Australis	Bratislava	Jez Bratislava	INIAV	+	+	-	-	-
	Autumnalis	Autumnalis	Akiyami A	INIAV	+	+	-	-	-
	Bataviae	Bataviae	Van Tienem	INIAV	+	+	-	-	-
	Canicola	Canicola	Hond Utrecht IV	INIAV	+	+	-	-	-
	Djasiman	Djasiman	Djasiman	KIT	+	+	-	-	-
	Hebdomadis	Hebdomadis	Hebdomadis	KIT	+	+	-	-	-
	Hebdomadis	Kremastos	Kremastos	KIT	+	+	-	-	-
	Icterohaemorrhagiae	Birkini	Birkin	KIT	+	+	-	-	-
	Icterohaemorrhagiae	Copenhageni	M20	INIAV	+	+	-	-	-

	Icterohaemorrhagiae	Icterohaemor-rhagiae	RGA	INIAV	+	+	-	-	-
	Icterohaemorrhagiae	Lai	Lai	KIT	+	+	-	-	-
	Pomona	Pomona	Pomona	INIAV	+	+	-	-	-
	Pyrogenes	Pyrogenes	Salinem	INIAV	+	+	-	-	-
	Sejroe	Hardjo type Prajitno	Hardjopra-jitno	IHMT	+	+	-	-	-
L. borg-petersenii	Ballum	Ballum	Mus 127	INIAV	+	-	+	-	-
	Ballum	Castellonis	Castellon 3	KIT	+	-	+	-	-
	Hebdomadis	Jules	Jules	KIT	+	-	+	-	-
	Hebdomadis	Worsfoldi	Worsfold	KIT	+	-	+	-	-
	Javanica	Ceylonica	Piyasena	KIT	+	-	+	-	-
	Javanica	Poi	Poi	INIAV	+	-	+	-	-
	Javanica	Zhenkang	L 82	KIT	+	-	+	-	-
	Mini	Mini	Sari	IHMT	+	-	+	-	-
	Pyrogenes	Kwale	Julu	KIT	+	-	+	-	-
	Sejroe	Hardjo type bovis	Sponselee	KIT	+	-	+	-	-
	Sejroe	Hardjo type bovis	L550	KIT	+	-	+	-	-
	Sejroe	Hardjo type bovis	JB197	KIT	+	-	+	-	-
	Sejroe	Nyanza	Kibos	KIT	+	-	+	-	-
	Sejroe	Sejroe	M84	KIT	+	-	+	-	-
	Tarassovi	Kisuba	Kisuba	KIT	+	-	+	-	-
	Tarassovi	Tarassovi	Mitis John-son	INIAV	+	-	+	-	-
L. kirsch-neri	Australis	Ramisi	Musa	KIT	+	-	-	+	-
	Autumnalis	Bulgarica	Nicolaevo	KIT	+	-	-	+	-
	Autumnalis	Butembo	Butembo	KIT	+	-	-	+	-
	Cynopteri	Cynopteri	3522C	IHMT	+	-	-	+	-
	Grippotyphosa	Grippoty-phosa type Moskva	Moskva V	IHMT	+	-	-	+	-
	Grippotyphosa	Ratnapura	Wum-alasena	KIT	+	-	-	+	-
	Grippotyphosa	Vanderho-edeni	Kipod 179	KIT	+	-	-	+	-
	Icterohaemorrhagiae	Bogvere	LT 60-69	KIT	+	-	-	+	-
	Pomona	Mozdok	5621	KIT	+	-	-	+	-

	Pomona	Mozdok	Portugal 1990	INIAV	+	-	-	+	-
	Pomona	Tsaratsovo	B 81/7	KIT	+	-	-	+	-
L. noguchii	Australis	Nicaragua	1011	KIT	+	-	-	-	+
	Autumnalis	Fortbragg	Fort Bragg	KIT	+	-	-	-	+
	Bataviae	Argentini-ensis	Peludo	KIT	+	-	-	-	+
	Djasiman	Huallaga	M 7	KIT	+	-	-	-	+
	Louisiana	Louisiana	LSU 1945	KIT	+	-	-	-	+
	Panama	Panama	CZ 214	INIAV	+	-	-	-	+
	Pomona	Proechimys	1161 U	KIT	+	-	-	-	+
	Pyrogenes	Myocastoris	LSU 1551	KIT	+	-	-	-	+
	Shermani	Carimagua	9160	KIT	+	-	-	-	+
L. santa-rosai	Ballum	Peru	MW 10	KIT	+	-	-	-	-
	Bataviae	Balboa	735 U	KIT	+	-	-	-	-
	Bataviae	Kobbe	CZ 320	KIT	+	-	-	-	-
	Grippotyphosa	Canalzonae	CZ 188	KIT	+	-	-	-	-
	Hebdomadis	Borincana	HS 622	KIT	+	-	-	-	-
	Hebdomadis	Maru	CZ 285	KIT	+	-	-	-	-
	Javanica	Fluminense	Aa 3	KIT	+	-	-	-	-
	Mini	Beye	1537 U	KIT	+	-	-	-	-
	Sarmin	Rio	Rr 5	KIT	+	-	-	-	-
	Sejroe	Guaricura	Bov.G.	KIT	+	-	-	-	-
	Shermani	Babudieri	CI 40	KIT	+	-	-	-	-
	Shermani	Shermani	1342 K	KIT	+	-	-	-	-
	Tarassovi	Atchafalaya	LSU 1013	KIT	+	-	-	-	-
L. weilii	Celledoni	Celledoni	Celledoni	INIAV	+	-	-	-	-
	Celledoni	Mengding	M 6906	KIT	+	-	-	-	-
	Javanica	Coxi	Cox	KIT	+	-	-	-	-
	Javanica	Mengma	S 590	KIT	+	-	-	-	-
	Javanica	Mengrun	A 102	KIT	+	-	-	-	-
	Mini	Hekou	H 27	KIT	+	-	-	-	-
	Pyrogenes	Menglian	S 621	KIT	+	-	-	-	-
	Sarmin	Sarmin	Sarmin	KIT	+	-	-	-	-
	Tarassovi	Topaz	94-79970/3	KIT	+	-	-	-	-
	Tarassovi	Vughia	LT 89-68	KIT	+	-	-	-	-
L. alexan-deri	Hebdomadis	Manzhuang	A 23	KIT	nd	-	-	-	-
	Javanica	Mengla	A 85	KIT	nd	-	-	-	-
	Manhao	Manhao 3	L 60	KIT	nd	-	-	-	-

	Mini	Yunnan	A 10	KIT	nd	-	-	-	-
L. meyeri	Ranarum	Ranarum	ICF	KIT	nd	-	-	-	-
	Semaranga	Semaranga	Veldrat Semaranga 173	KIT	nd	-	-	-	-
L. inadai	Manhao	Lincang	L 14	KIT	nd	-	-	-	-
L.fainei	Hurstbridge	Hurstbridge	BUT 6T	KIT	nd	-	-	-	-
L.biflexa	Andaman	Andamana	CH 11	KIT	-	-	-	-	-
	Semaranga	Patoc	Patoc I	KIT	-	-	-	-	-

[1]INIAV - Instituto Nacional de Investigação Agrária e Veterinária, Lisbon, Portugal. IHMT - Instituto de Higiene e Medicina Tropical, Lisbon, Portugal. KIT - Royal Tropical Institute, Amsterdam, The Netherlands;

[2]Set 1 targets the *lipL32* gene of pathogenic *Leptospira* spp.;

[3]Set 2 targets the *secY* gene of *L. interrogans*;

[4]Set 3 targets the *ompL1* gene of *L. borgpetersenii*;

[5]Set 4 targets the *secY* gene of *L. kirschneri*;

[6]Set 5 targets the *secY* gene of *L. noguchii*; nd - not done; Amplification (+) or no amplification (−).

Culturing *Leptospira* from tissue samples was performed as described by the OIE Manual of Diagnostic Tests and Vaccines for Terrestrial Animals [27]. Other bacterial strains were provided by INIAV for assessing the analytical specificity of the amplification reactions, representing the species: *Acinetobacter baumannii* (LNIV 1628/12), *Bacillus licheniformis* (VLA 1831), *Klebsiella pneumoniae* (VLA 1643), *Salmonella* Dublin (VLA 1272), *Streptococcus agalactiae* (VLA 33), *Proteus mirabilis* (LNIV 2269/II), *Yersinia enterocolitica* (VLA 1884), *Staphylococcus aureus* (VLA 1032), *Pseudomonas aeruginosa* (VLA 67), *Arcanobacterium pyogenes* (VLA 1321) and *Listeria monocytogenes* (VLA 1774).

Spiked Tissue Samples

A sample of kidney tissue from a bovine was used for testing as spiked sample. The kidney was acquired from a local official slaughterhouse (Raporal, Portugal), obtained from a bovine intended for normal human consumption, with no signs of leptospirosis. The bovine was not killed specifically for the purpose of this study. Approximately 200 mg portions of kidney tissue were excised with a sterile scalpel and homogenized with 5 ml of PBS buffer in a sterile plastic bag (Whirl-Pak bags) using a stomacher lab-blender. Kidney samples were individually spiked with the following strains, in order to determine the analytical detection sensitivity: *Leptospira interrogans* (serovar Autumnalis, strain Akiyami), *L. kirschneri* (serovar Mozdok, strain Portugal 1990) [28], *L. noguchii* (serovar Panama, strain CZ 214K) and *L. borgpetersenii* (serovar Tarassovi, strain Mitis Johnson). All the strains were grown at 29°C and the concentrations of leptospires were determined using a Petroff-Hausser counting chamber and adjusted to 10^8 cells/ml with PBS buffer. For each strain, tenfold serial dilutions from 10^7 to 10^0 cells/ml were prepared in PBS buffer and 0.1 ml aliquots were used to spike 0.9 ml of tissue homogenates. Tissue homogenate spiked with 0.1 ml PBS buffer was used as negative control. DNA extraction was performed as described in the paragraph "Genomic DNA extraction" below.

Tissue Samples

INIAV IP is the Portuguese Reference Laboratory for animal diseases and provides diagnostic services to national veterinary authorities and private clients. Twenty seven dead wild rodents (25 *Mus* spp. and 2 *Rattus* spp.) were sent to the INIAV laboratory during the year 2011 for analysis and further used in this study (Table 2). The rodents were captured in the Lisbon Zoo under routine operations for rodent population control, by the local veterinary authorities. No animals were sacrificed for the only purposes of research. Additionally, a Patagonian mara (*Dolichotis patagonum*), also from the zoo, and a swine (*Sus domesticus*) stillbirth fetus, from a private client, both suspect of dying with leptospirosis, were submitted for analysis to our reference laboratory and later included in this study (Table 2). On arrival to the laboratory, animals were given a reference number and sent to the pathology where kidney, liver and/or lung tissue samples were collected. Specimens were then analysed using culture-based methods according to the OIE standard procedures for leptospirosis [27]. Briefly, specimens were aseptically collected at necropsy, immediately emulsified in sterile buffered saline solution in a 10% tissue suspension, two to three drops were inoculated in a first tube of medium and two more tubes were similarly inoculated with increasing 10-fold dilutions of the tissue suspension. For the tissue culture, a semisolid *Leptospira* EMJH medium was used by adding 0.1% agar to commercial EMJH (Difco), to which rabbit serum (0.4%) and 5-Fluorouracil (100 μg/ml) were further added [27].

Table 2: Results of the bacteriological culture and of the real time amplification assays for the tissue samples analyzed in the present study

Sample	Origin	Set Actin[1]	Set 1[2]	Set 2[3]	Set 3[4]	Set 4[5]	Set 5[6]	Bacteriological analysis[7]
12-17433-Z1	*Mus* sp.	+	+	-	+	-	-	*L. borgpetersenii*
12-18078-Z6	*Mus* sp.	+	+	-	+	-	-	*L. borgpetersenii*
12-18458-Z13	*Mus* sp.	+	-	-	-	-	-	Negative
12-18458-Z14	*Mus* sp.	+	-	-	-	-	-	Negative
12-19472-Z15	*Mus* sp.	+	-	-	-	-	-	Negative
12-20553-Z16	*Mus* sp.	+	-	-	-	-	-	Negative
12-22955-Z17	*Mus* sp.	+	-	-	-	-	-	Negative
12-22955-Z18	*Mus* sp.	+	-	-	-	-	-	Negative
12-22955-Z19	*Mus* sp.	+	-	-	-	-	-	Negative
12-22955-Z20	*Mus* sp.	+	-	-	-	-	-	Negative
12-22955-Z22	*Mus* sp.	+	-	-	-	-	-	Negative
12-22955-Z23	*Mus* sp.	+	-	-	-	-	-	Negative
12-22955-Z24	*Mus* sp.	+	-	-	-	-	-	Negative
12-22955-Z25	*Mus* sp.	+	-	-	-	-	-	Negative
12-22955-Z26	*Mus* sp.	+	-	-	-	-	-	Negative
12-22955-Z27	*Mus* sp.	+	-	-	-	-	-	Negative
12-22955-Z28	*Mus* sp.	+	-	-	-	-	-	Negative
12-22955-Z29	*Mus* sp.	+	-	-	-	-	-	Negative
12-22955-Z30	*Mus* sp.	+	-	-	-	-	-	Negative
12-22955-Z31	*Mus* sp.	+	-	-	-	-	-	Negative
12-22955-Z32	*Mus* sp.	+	-	-	-	-	-	Negative
12-22955-Z33	*Mus* sp.	+	-	-	-	-	-	Negative
12-22955-Z34	*Mus* sp.	+	-	-	-	-	-	Negative
12-22955-Z36	*Mus* sp.	+	-	-	-	-	-	Negative
12-22955-Z37	*Rattus* sp.	+	+	+	-	-	-	*L. interrogans*
12-22955-Z38	*Mus* sp.	+	-	-	-	-	-	Negative
12-22955-Z39	*Rattus* sp.	+	-	-	-	-	-	Negative
11-36840	*Dolichotis patagonum*	+	+	+	-	-	-	*L. interrogans*
12-494	*Sus domesticus* (fetus)	+	+	-	-	+	-	*L. kirschneri*

[1]Set Actin targets the *β-actin* gene of mammals,

[2]Set 1 targets the *lipL32* gene of pathogenic *Leptospira*;

[3]Set 2 targets the *secY* gene of *L. interrogans*;

[4]Set 3 targets the *ompL1* gene of *L. borgpetersenii*;

[5]Set 4 targets the *secY* gene of *L. kirschneri*;

[6]Set 5 targets the *secY* gene of *L. noguchii*;

[7]The analysis of the partial sequences of the *secY* gene of each isolate allowed to identify the *Leptospira* species; Amplification (+) or no amplification (−). DNA was extracted directly from tissues homogenates as described below.

Genomic DNA Extraction

Genomic DNA was extracted from both bacterial liquid cultures and tissue homogenates using the QIAamp DNA extraction kit according to the manufacturer's instructions (Qiagen, Hilden, Germany), with a final elution volume of 200 μl. The DNA concentration from the pure cultures was estimated spectrophotometrically using a Nanodrop 1000 spectrophotometer (Nanodrop Technologies, Wilmington, DE) and standardized to a concentration of 10^4 genome equivalents (GE)/μl for use in the reactions. The number of GE was estimated using an average genome size of 4.6 Mb [29]. Genomic DNA suspensions were stored at −20°C until further use.

Design of *TaqMan* Probes and Flanking Primers

DNA sequences of representative strains and species of *Leptospira* were retrieved from NCBI-GenBank and aligned using the ClustalW algorithm implemented in the program MegAlign (vers. 5.03) (DNAStar, USA). Primers and dual labeled hydrolysis probes (*TaqMan* probes) were designed to target selected species-specific genetic polymorphisms of the following pathogenic *Leptospira* spp.: *L. interrogans*, *L. borgpetersenii*, *L. kirschneri* and *L. noguchii* (Table 3). Probes and primers specificities were assessed *in silico* using the BLAST tools from NCBI-GenBank. All probes and primers were synthesized by MWG Biotech (Germany).

Table 3: Primers and probes used in this study targeting selected genes of pathogenic species of *Leptospira*

Set	Primer/ Probe	Sequence (5′- 3′)	Annealing temperature	Complementary target species
Set	F_Actin	GGC TCY ATY *CTG GCC TC*	60°C	*β-actin* gene of mammals
Actin [1]	R_Actin	GCA YTT GCG GTG SAC RAT G		
	P_Actin	Cy5.5 (Quasar 705) -TAC TCC TGC TTG CTG ATC CAC ATC-BHQ2		

Set 1 [2]	45F	*AAG CAT TAC CGC TTG TGG TG*	60°C	*lipL32* gene of pathogenic *Leptospira* spp.
	286R	*GAA CTC CCA TTT CAG CGA TT*		
	taq-189P	FAM-AAA GCC AGG ACA AGC GCC G-BHQ1		
Set 2	PFLint2	CTT GAG CCT GCG CGT TAY C	63°C	*secY* gene of *L. interrogans*
	PRLint2	*CCG ATA ATT CCA GCG AAG ATC*		
	TaqLint2	TET-CTC ATT TGG TTA GGA GAA CAG ATC A-BHQ1		
Set 3	F_bpn	*GAT TCG GGT TAC AAT TAG ACC*	65°C	*ompL1* gene of *L. borgpetersenii*
	R_bpn1	*TTG ATC TAA CCG GAC CAT AGT*		
	TqM_bpn	Cy5.5 (Quasar 705) -TAC TAA GGA TGG TTT GGA CGC TGC-BHQ2		
Set 4	F_nery	*CTG GCT TAA TCA ATG CTT CTG*	60°C	*secY* gene of *L. kirschneri*
	R_nery	*CTC TTT CGG TGA TCT GTT CC*		
	TqM_nery	Texas Red-CAG TTC CAG TTG TAA TAG ATA AGA TTC-BHQ2		
Set 5	FLnog2	*TCA GGG TGT AAG AAA GGT TC*	63°C	*secY* gene of *L. noguchii*
	RLnog2	*CAA AAT TAA AGA AGA AGC AAA GAT*		
	TaqLnog	FAM-CGA TTG GCT TTT TGC TTG AAC CATC-BHQ1		

[1]Retrieved from Costa et al. [31];

[2]Retrieved from Stoddard et al. [16].

Real-time PCR Assays

We have implemented the following assay format for testing DNA templates extracted from biological samples: (i) a first duplex amplification step aiming the detection of pathogenic *Leptospira* spp. (by targeting the leptospiral *lipL32* gene; Table 3) and including an internal control to monitor the presence of potential amplification inhibitors (by targeting the mammal *β-actin* gene; Table 3); (ii) if pathogenic leptospires are detected in the first reaction, these may be further discriminated by testing each of the *L. interrogans*, *L. borgpetersenii*, *L. kirschneri* and *L. noguchii* targeted probes/primers (Table 3). The CFX96 real-time PCR detection system (Bio-Rad, USA) was used for all assays. The amplification reactions were optimized individually for all the probes and associated primers using the SsoFast Probes Supermix (Bio-Rad, USA), according to the manufacturer's instructions. Each reaction was conducted in a total volume of 20 µl consisting of 1× SsoFast Probes Supermix, 400 nM of each primer, 150 nM of *TaqMan* probe, DNase free water (GIBCO) and 5 µl of DNA template solution (extracted from pure cultures or tissues samples). Non-template negative controls (with PCR grade water) were included in each run to rule out the possibility of cross-contamination. The assay thermal conditions were as follows: 95°C for 2 min, followed by 45 cycles of 5 s at 95°C and 15 s at the optimized annealing temperature for each probe (Table 3). The thermal cycling conditions for the duplex amplification targeting *β-actin* and *lipL32* were 95°C for 2 min, followed by 45 cycles of 5 s at 95°C and 35 s at 60°C. Reproducibility of the assays was assessed by repeating the assays at least twice. Data analyses were performed by the detection system of the real-time PCR equipment, according to the manufacturer›s instructions.

Analytical Specificity and Sensitivity

In order to determine if each set of probe and associated primers was specific for the respective *Leptospira* target species, the amplification assays were tested on DNA templates extracted from different strains belonging to pathogenic, intermediate and non-pathogenic *Leptospira* species (Table 1), and from other non-related bacteria previously mentioned in "bacterial strains" section. The analytical sensitivity of the amplification assays (limits of detection – LODs) were determined using 10-fold serial dilutions of genomic DNA extracted from pure cultures of *L. interrogans* (serovar Autumnalis, strain Akiyami), *L. kirschneri* (serovar Mozdok, strain Portugal 1990), *L. noguchii* (serovar Panama, strain CZ 214K) and *L. borgpetersenii* (serovar Tarassovi, strain Mitis Johnson). LODs on tissue samples were assessed using DNA extracted from the serially diluted spiked macerates. Each template was tested in triplicate.

Sequencing

Leptospira isolates obtained from tissue samples were identified by comparative sequence analysis of a 245 bp region of the *secY* gene, as described by Victoria *et al.* [30]. Briefly, the region of interest was amplified using primers SecYII (*5'-GAA TTT CTC TTT TGA TCT TCG-3'*) and SecYIV (*5'-GAG TTA GAG CTC AAA TCT AAG-3'*), which amplify *secY* sequences from all pathogenic strains of *Leptospira*. PCR amplifications were performed on a C1000 thermocycler (Bio-Rad) using the following program: an initial step of denaturation for 5 min at 95°C, followed by 34 cycles consisting of annealing, 45 sec at 54°C, extension, 2 min at 72°C, and denaturation, 30 sec at 94°C. Nucleotide sequences were determined, using the same primers, by commercially available sequencing services. Nucleotide sequence analysis and comparison with other relevant reference sequences were performed using the BLAST suite at NCBI-GenBank and aligned using Clustal X or MEGA software (version 5.0).

RESULTS

Design of Probes and Primers

Species-specific sets of primers and probes targeting *L. interrogans*, *L. borgpetersenii*, *L. kirschneri* and *L. noguchii* are listed in Table 3. These sets of probes and primers contained sufficient polymorphisms to warrant '*in silico*' species specific amplification.

Analytical Specificity and Sensitivity

Execution of the PCRs on DNA extracted from various bacteria, revealed a highly specific amplification from any of the pathogenic strains belonging to the respective target *Leptospira* spp., i.e. *L. interrogans*, *L. kirschneri*, *L borgpetersenii* and *L. noguchii*. None of the other strains yielded a positive amplification reaction (Table 1; Fig. 1A). The analytical sensitivity (LOD) of the amplification assays were found to be between 1 and 10 genome copies in the PCR mixture for each probe and primer set.

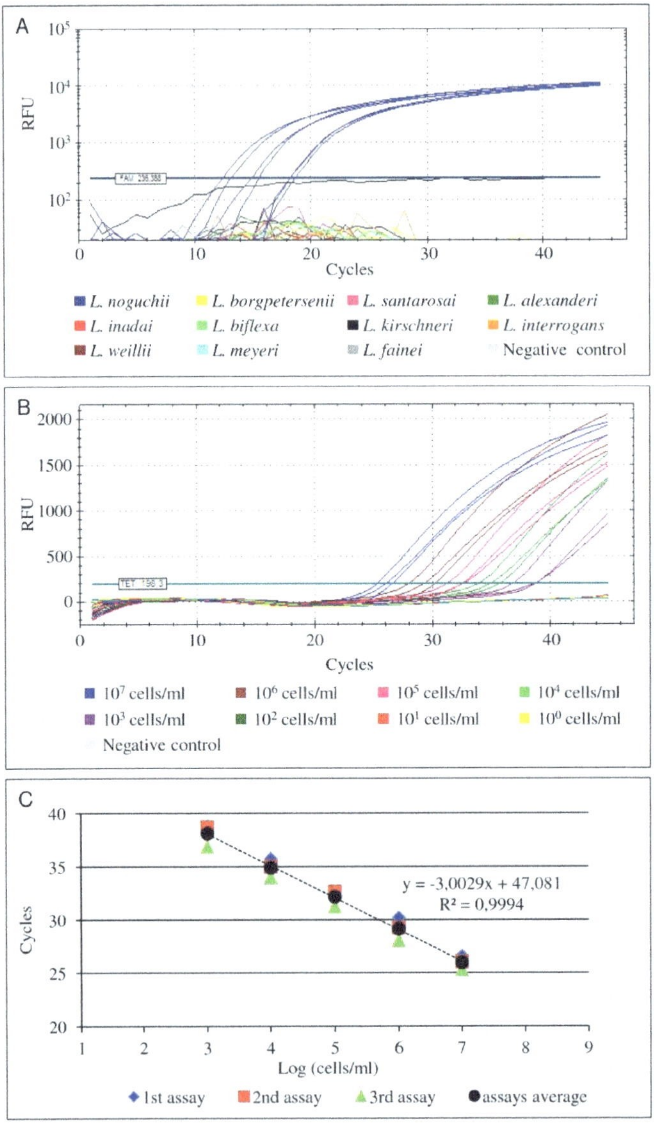

Figure 1: Illustration of the real-time PCR amplification curves obtained during the optimization of the assays.

(A) Specificity tests of the *L. noguchii* targeted amplification assay using the TaqLnog probe combined with the flanking primers FLnog2 and RLnog2. Blue amplification curves represent *L. noguchii* strains. All other non-target strains yielded no amplification results. (B) Estimation of the limit of detection of the amplification assay targeting *L. interrogans* (serovar

Autumnalis, strain Akiyami) using DNA extracted directly from spiked bovine kidney samples as template as a typical example of all *Leptospira* probe and primer sets. The amplification curves obtained from different ten-fold serial dilutions of the target *Leptospira* are represented by different colours. Unspiked tissue homogenate (grey line) was used as negative control. (C) Standard curve obtained from the analysis of the amplification curves mentioned in the previous panel B. RFU - Relative Fluorescence Units.

Spiked Tissue Samples

The LOD of the PCRs on spiked tissue samples was similar for all probe/primers sets targeting the respective target species, and estimated to be 10^3 leptospires/ml of tissue homogenate (≈ per 20 mg of tissue) (Fig. 1B). Furthermore, the same LOD was estimated for the *lipL32*-targeted probe/primers when used in duplex amplification reactions with the mammal *β-actin* probe (not shown).

Clinical Tissue Samples

DNA extracted from 27 kidney samples of wild rodents were analysed with the *lipL32* and mammal *β-actin* targeted duplex assay (Table 2; Fig. 2A). Leptospiral DNA was detected in three samples, as demonstrated by a positive amplification of the *lipL32* gene region (Table 2; Fig. 2A). Furthermore, the partial *β-actin* gene was amplified from all samples, showing that the PCR reactions were not significantly inhibited by potential contaminants. When tested with each of the *L. interrogans*, *L. borgpetersenii*, *L. kirschneri* and *L. noguchii* targeted probes/primers, only these three samples showed amplification (Table 2; Fig. 2B). Two of these DNA samples were identified as *L. borgpetersenii* and one sample as *L. interrogans*. Testing a pooled sample of kidney and liver tissues from a Patagonian mara, and a lung sample from an aborted swine fetus with the duplex PCR revealed a positive amplification for both samples (Table 2). Subsequent testing with the species-specific sets of probes and primers showed that the Patagonian mara was infected with *L. interrogans* and the swine fetus with *L. kirschneri*.

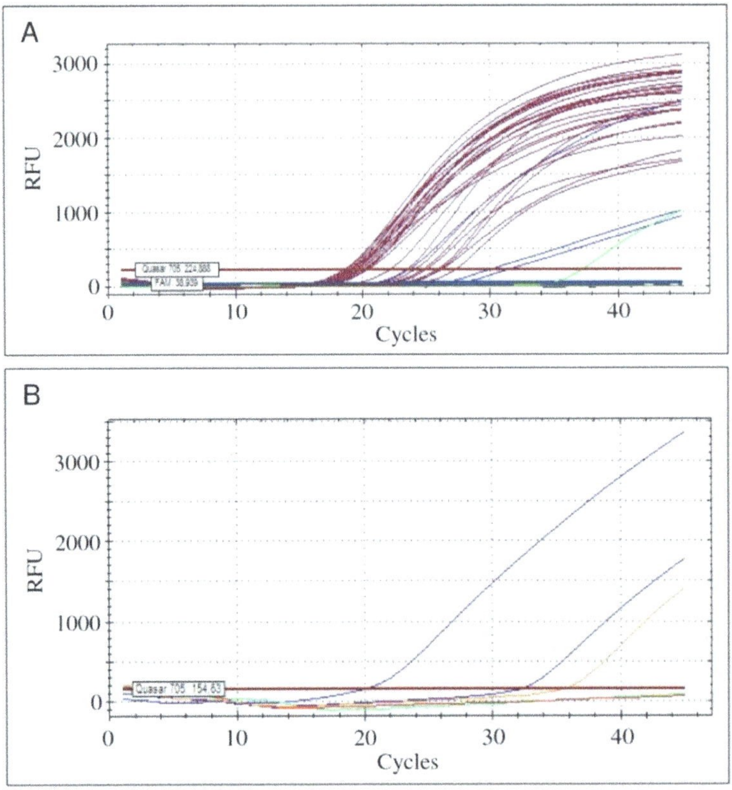

Figure 2: Illustration of the real-time PCR amplification curves obtained during the testing of naturally-infected tissue samples

(A) Results of the *β-actin* and *lipL32* targeted duplex amplification assay when testing representative samples from the wild rodents. The partial *β-actin* gene was amplified from all tissue samples (dark pink lines). Leptospiral DNA was detected in three samples by a positive amplification of the *lipL32* gene (blue lines). A spiked positive control with *L. interrogans* (serovar Autumnalis, strain Akiyami) is shown (green line). (B) From the previous leptospiral positive amplification results, two samples were assessed as infected with *L. borgpetersenii* using the respective targeted amplification assay with probe TqM_bpn and flanking primers F_bpn and R_bpn1 (blue lines). The positive and negative controls are illustrated by the orange and red lines, respectively.

Leptospira isolates were only cultured from the samples that also yielded PCR-positive results, thus confirming the presence of viable leptospires

(Table 2).

Molecular speciation through analysis of the partial sequences of the *secY* gene was in concordance with the results obtained by the species-specific PCRs. Two isolates were identified as *L. borgpetersenii* (from wild rodents; GenBank accession numbers KM066006 and KM066007), one as *L. kirschneri* (from the swine fetus; accession number KM066009) and two as *L. interrogans* (from a wild rodent and the Patagonian mara; accession numbers KM066008 and KM066010, respectively).

DISCUSSION

In this work we present a two step real-time PCR strategy to infer the presence of pathogenic leptospires in clinical and veterinary samples. In the first step, we assess if an animal tissue sample is infected with a pathogenic leptospire by targeting its *lipL32* gene. The *lipL32* gene encodes an outer membrane lipoprotein that is confined to pathogenic *Leptospira* species [16]. The second step identifies the four most common and veterinary relevant pathogenic *Leptospira* species, *L. interrogans*, *L. borgpetersenii*, *L. kirschneri* and *L. noguchii* using dedicated sets of probes and primers.

Probes and flanking primers were developed by *in silico* analysis and further tested for their practical utility on DNA extracted from cultured bacteria, spiked tissues and clinical specimens. The amplification assays have proved to be specific to the respective targeted species, with no cross-reactions when non-pathogenic leptospires or other pathogens were tested. The amplification of the *β-actin* gene was included in the initial *lipL32*-based PCR to assess the presence of amplification inhibitors in tissue samples [31]. However, the abundant presence of *β-actin* gene copies in DNA samples extracted from tissues may ensure some amplification even when low levels of potential inhibitors are present (but amplification curves are usually weaker and anomalous). The analytical sensitivity deduced for the amplification assays, i.e. 1 to 10 GE on DNA extracted from cultured leptospires and 10^3 leptospires/ml tissue homogenate, were similar to the ones of other previous studies concerning the molecular detection of leptospires [15]–[17], [19], [22].

The panel of species-specific probes and flanking primers may be extended with the design of novel oligonucleotides, e.g. for use in regions where the occurrence of additional species of pathogenic leptospires is common. As far as we know, this is the first report describing a strategy capable of clearly identify four most frequently found pathogenic *Leptospira* species based on the use of *TaqMan* probes.

From 27 kidney samples of wild rodents, and samples from a Patagonian mara and a porcine fetus suspected of leptospirosis, three rodent samples and the samples from the Patagonian mara and fetus all yielded a positive PCR test for the presence of pathogenic leptospires. In concordance, these samples were also positive by culture. Culture provides proof of infection and thus is an ideal reference standard. Consequently, these results are consistent with a 100% clinical sensitivity and specificity of the PCR. Subsequent prospective analysis of a larger sample set would allow substantiating this conclusion.

Phylogenetic identification of the cultures also allowed supporting the findings obtained with the species-specific PCRs. Indeed, speciation by phylogeny was in all cases in concordance with the results obtained via the PCR method.

Initially, we anticipated that more samples would be positive by the real time PCR assay than by culture [5], [32]–[34]. Recently, Fornazari *et al.* [20] reported that quantitative PCR presented the highest sensitivity among several techniques to detect leptospires in tissues samples, the bacteriological culture being the least sensitive. Apparently, our procedure of culturing, using macerated fresh tissue has been highly effective. Alternatively, it cannot be excluded that the bacterial load of the tissues might have been very high. Nevertheless, the low rate of positive animals (11%) is not too discrepant from the prevalence values found in other studies where leptospiral DNA was detected in rodents tissues by PCR-based assays, which ranged from 13% to 20% [25], [35], [36]. Furthermore, as far as we know, the region of Lisbon, where the rodents were captured, is not usually regarded as having major leptospirosis problems [2], which may also reflect a lower prevalence of the agent in reservoirs such as wild rodents. We anticipate that our assays may be useful in studies inferring the prevalence of pathogenic leptospires in wild rodents and other animals, with the advantage of differentiating the infecting *Leptospira* species.

The amplification assays described were able to detect pathogenic leptospires in samples of animal tissues, such as kidney or lung. Although the analysis of this kind of samples is not essential for an early diagnosis of leptospirosis, it has a great value in situations such as epidemiological and post-mortem investigations. The last situation is very well illustrated in this work with the detection of pathogenic leptospires in tissues of a Patagonian mara and a swine fetus. Both animals were suspect of having leptospirosis, which was confirmed by this study. The porcine fetus was infected with a strain belonging to *L. kirschneri*. Pigs may be infected by several *Leptospira* species (and serovars) that may cause infertility, fetal death and abortion. *Leptospira kirschneri* has been reported but seems to be less frequently found in pigs in Portugal than other species [37]. The Patagonian mara, a relatively large rodent that lived in the local zoo, was found to be infected with *L. interrogans*. To our knowledge, this is the first report describing the molecular detection or the isolation of a pathogenic leptospire from that rodent, which proved to have died of leptospirosis. Zoos are often infested with rats that are notorious reservoirs of *L. interrogans*. We hypothesise that this Patagonian mara has been infected by rats as the primary infection reservoir, which would support the potential hazard of rodents in zoos for both (exotics) animals and public.

The amplification assay described in this work is able to indentify the four most relevant pathogenic species of *Leptospira* infecting farm and wild animals. While the approach can be extended to other *Leptospira* species, it is important to continually evaluate the specificity of previously designed probes and primers and, if necessary, modify and improve the sequences, in order to ensure an effective and specific detection and identification of the circulating *Leptospira* species.

CONCLUSIONS

The molecular assays presented in this work allow the detection and identification of four relevant pathogenic species of *Leptospira*, directly from animal tissues. The assays proved to be specific and sensitive, and much faster than the bacteriological culture, reducing the time for confirmatory leptospirosis diagnosis. The assays are amenable to future automation possibilities and will reinforce the diagnostic information and enhance our knowledge about the epidemiology of leptopirosis.

ACKNOWLEDGMENTS

Madalena Monteiro (INIAV, I.P.) is acknowledged for excellent technical assistance in performing pathological analysis. Lurdes Clemente and Ana Botelho are acknowledged for providing non-*Leptospira* bacterial strains. Narciso Lapão, Nuno Gaspar and António Crespo are acknowledged for their help in providing samples and respective information.

REFERENCES

1. Dupouey J, Faucher B, Edouard S, Richet H, Kodjo A, et al. (2014) Human leptospirosis: An emerging risk in Europe?. Comp. Immunol. Microbiol. Infects Dis 37: 77–83.
2. Vieira ML, Gama-Simões MJ, Collares-Pereira M (2006) Human leptospirosis in Portugal: A retrospective study of eighteen years. Int J Infect Dis 10: 378–386.
3. Adler B, Moctezuma AP (2010) Leptospira and leptospirosis. Vet. Microbiol 140: 287–296.
4. Bomfim MR, Barbosa-Stancioli EF, Koury MC (2008) Detection of pathogenic leptospires in urine from naturally infected cattle by nested PCR. Vet J 178: 251–256.
5. Fearnley C, Wakeley PR, Gallego-Beltran J, Dalley C, Williamson S, et al. (2008) The development of a real-time PCR to detect pathogenic *Leptospira* species in kidney tissue. Res Vet Sci 85: 8–16.
6. Picardeau M (2013) Diagnosis and epidemiology of leptospirosis. Med Mal Infect 43: 1–9.
7. Brenner DJ, Kaufmann AF, Sulzer KR, Steigerwalt AG, Rogers FC, et al. (1999) Further determination of DNA relatedness between serogroups and serovars in the family Leptospiraceae with a proposal for *Leptospira alexanderi* sp. nov. and four new *Leptospira* genomospecies. Int J Syst Bacteriol 49: 839–858.
8. Cerqueira GM, Picardeau M (2009) A century of *Leptospira* strain typing. Infect Genet Evol 9: 760–768.
9. Morey RE, Galloway RL, Bragg SL, Steigerwalt AG, Mayer LW, et al. (2006) Species-specific identification of Leptospiraceae by 16S rRNA gene sequencing. J Clin Microbiol 44: 3510–3516.
10. Saito M, Villanueva SY, Kawamura Y, Iida K, Tomida J, et al. (2013) *Leptospira idonii* sp. nov., isolated from environmental water. Int J Syst Evol Microbiol 63: 2457–2462.
11. Agampodi SB, Matthias MA, Moreno AC, Vinetz JM (2012) Utility of quantitative polymerase chain reaction in leptospirosis diagnosis: association of level of leptospiremia and clinical manifestations in Sri Lanka. Clin Infect Dis 54: 1249–1255.
12. Bourhy P, Bremont S, Zinini F, Giry C, Picardeau M (2011) Comparison of real-time PCR assays for the detection of pathogenic *Leptospira* spp. in blood and identification of variations in target sequences. J Clin

Microbiol 49: 2154–2160.

13. Levett PN, Morey RE, Galloway RL, Turner DE, Steigerwalt AG, et al. (2005) Detection of pathogenic leptospires by real-time quantitative PCR. J Med Microbiol 54: 45–49.
14. Merien F, Portnoi D, Bourhy P, Charavay F, Berlioz-Arthaud A, et al. (2005) A rapid and quantitative method for the detection of *Leptospira* species in human leptospirosis. FEMS Microbiol Lett 249: 139–147.
15. Smythe LD, Smith IL, Smith GA, Dohnt MF, Symonds ML, et al. (2002) A quantitative PCR (TaqMan) assay for pathogenic *Leptospira* spp. BMC Infect Dis 2: 13.
16. Stoddard RA, Gee JE, Wilkins PP, McCaustland K, Hoffmaster AR (2009) Detection of pathogenic *Leptospira* spp. through TaqMan polymerase chain reaction targeting the *lipL32* gene. Diagn Microbiol Infect Dis 64: 247–255.
17. Villumsen S, Pedersen R, Borre MB, Ahrens P, Jensen JS, et al. (2012) Novel TaqMan PCR for detection of *Leptospira* species in urine and blood: Pit-falls of in silico validation. J Microbiol Methods 91: 184–90.
18. Ahmed A, Engelberts MF, Boer KR, Ahmed N, Hartskeerl RA (2009) Development and validation of a real-time PCR for detection of pathogenic *Leptospira* species in clinical materials. PLoS One 4: e7093.
19. Slack A, Symonds M, Dohnt M, Harris C, Brookes D, et al. (2007) Evaluation of a modified Taqman assay detecting pathogenic *Leptospira* spp. against culture and *Leptospira*-specific IgM enzyme-linked immunosorbent assay in a clinical environment. Diagn Microbiol Infect Dis 57: 361–366.
20. Fornazari F, da Silva RC, Richini-Pereira VB, Beserra HE, Luvizotto MC, et al. (2012) Comparison of conventional PCR, quantitative PCR, bacteriological culture and the Warthin Starry technique to detect *Leptospira* spp. in kidney and liver samples from naturally infected sheep from Brazil. J Microbiol Methods 90: 321–326.
21. Ahmed A, Klaasen HLBM, van der Veen M, van der Linden H, Goris MGA, et al. (2012) Evaluation of real-time PCR and culturing for the detection of leptospires in canine samples. Adv. Microbiol 2: 162–170. [Google Scholar]
22. Rojas P, Monahan AM, Schuller S, Miller IS, Markey BK, et al. (2010) Detection and quantification of leptospires in urine of dogs: a maintenance host for the zoonotic disease leptospirosis. Eur J Clin

Microbiol Infect Dis 29: 1305–1309.

23. Subharat S, Wilson PR, Heuer C, Collins-Emerson JM (2011) Evaluation of a SYTO9 real-time polymerase chain reaction assay to detect and identify pathogenic *Leptospira* species in kidney tissue and urine of New Zealand farmed deer. J Vet Diagn Invest 23: 743–752.
24. Cox TE, Smythe LD, Leung LKP (2005) Flying foxes as carriers of pathogenic *Leptospira* species. J. Wildlife Dis 41: 753–757.
25. Levieuge A, Aboubaker MH, Terrier O, Drancourt M, Davoust B (2010) Real-time PCR detection of *Leptospira* sp. in rodents from Toulon harbour (France). Revue Méd. Vét 161: 264–266. [Google Scholar]
26. Palaniappan RU, Chang YF, Chang CF, Pan MJ, Yang CW, et al. (2005) Evaluation of lig-based conventional and real time PCR for the detection of pathogenic leptospires. Mol Cell Probes 19: 111–117.
27. OIE 2008. Manual of Diagnostic Tests and Vaccines for Terrestrial Animals (Chapter 2.1.9), World Organization for Animal Health.
28. Rocha T (1990) Isolation of *Leptospira interrogans* serovar *mozdok* from aborted swine fetuses in Portugal. Vet Record 126: 602.
29. Nascimento AL, Verjovski-Almeida S, Van Sluys MA, Monteiro-Vitorello CB, Camargo LE, et al. (2004) Genome features of *Leptospira interrogans* serovar Copenhageni. Braz J Med Biol Res 37: 459–477.
30. Victoria B, Ahmed A, Zuerner RL, Ahmed N, Bulach DM, et al. (2008) Conservation of the S10-spc-alpha locus within otherwise highly plastic genomes provides phylogenetic insight into the genus *Leptospira* . PLoS One 3: e2752.
31. Costa P, Ferreira AS, Amaro A, Albuquerque T, Botelho A, et al. (2013) Enhanced detection of tuberculous mycobacteria in animal tissues using a semi-nested probe-based real-time PCR. PLoS ONE 8: e81337.
32. Boqvist S, Montgomery JM, Hurst M, Thu HT, Engvall EO, et al. (2003) *Leptospira* in slaughtered fattening pigs in southern Vietnam: presence of the bacteria in the kidneys and association with morphological findings. Vet Microbiol 93: 361–368.
33. Grooms DL, Bolin CA (2005) Diagnosis of fetal loss caused by bovine viral diarrhea virus and *Leptospira* spp. Vet Clin North Am Food Anim Pract 21: 463–472.
34. Lilenbaum W, Varges R, Ristow P, Cortez A, Souza SO, et al. (2009) Identification of *Leptospira* spp. carriers among seroreactive goats and

sheep by polymerase chain reaction. Res Vet Sci 87: 16–19.

35. Foronda P, Martin-Alonso A, Del Castillo-Figueruelo B, Feliu C, Gil H, et al. (2011) Pathogenic *Leptospira* spp. in wild rodents, Canary Islands, Spain. Emerg Infect Dis 17: 1781–1782.
36. Latifah I, Rahmat MS, Hayarti KB, Paramasvaran S, Azizah MR, et al. (2012) Prevalence of leptospiral DNA among wild rodents from a selected area in Beguk Dam Labis, Segamat, Johor, Malaysia. Malays J Pathol 34: 157–159.
37. Rocha T (1998) A review of leptospirosis in farm animals in Portugal. Rev Sci Tech 17: 699–712.

CHAPTER 5

Prevalence of Bacterial Genotypes and Outcome of Bovine Clinical Mastitis due to Streptococcus Dysgalactiae and Streptococcus Uberis

Åsa Lundberg[1,2], Ann Nyman[1] , Helle Ericsson Unnerstad[1] and Karin Persson Waller[1,2]

[1]Department of Animal Health and Antimicrobial Strategies, National Veterinary Institute, SE-751 89 Uppsala, Sweden
[2] Department of Clinical Sciences, Swedish University of Agricultural Sciences, SE-750 07 Uppsala, Sweden

ABSTRACT

Background

Streptococcus dysgalactiae and *Streptococcus uberis* are common causes of clinical mastitis (CM) in dairy cows. In the present study genotype variation

Citation: Lundberg, Å., Nyman, A., Unnerstad, H.E. et al. Prevalence of bacterial genotypes and outcome of bovine clinical mastitis due to Streptococcus dysgalactiae and Streptococcus uberis. Acta Vet Scand 56, 80 (2014). https://doi.org/10.1186/s13028-014-0080-0

of *S. dysgalactiae* and *S. uberis* was investigated, as well as the influence of bacterial species, or genotype within species, on the outcome of veterinary-treated CM (VTCM). Isolates of *S. dysgalactiae* (n = 132) and *S. uberis* (n = 97) were genotyped using pulsed-field gel electrophoresis. Identical banding patterns were called pulsotypes. Outcome measurements used were cow composite SCC, milk yield, additional registered VTCMs and culling rate during a four-month follow-up period.

Results

In total, 71 *S. dysgalactiae* pulsotypes were identified. Nineteen of the pulsotypes were isolated from more than one herd; the remaining pulsotypes were only found once each in the material. All *S. uberis* isolates were of different pulsotypes. During the follow-up period, the SCC of *S. dysgalactiae*-cows was significantly lower than the SCC of *S. uberis*-cows ($P < 0.05$). Median SCC of *S. dysgalactiae*-cows was 71 500 cells/ml and of *S. uberis*-cows 108 000 cells/ml. No other differences in outcome parameters could be identified between species or genotypes.

Conclusions

Identical *S. dysgalactiae* genotypes were isolated from more than one herd, suggesting some spread of this pathogen between Swedish dairy herds. The genetic variation among *S. uberis* isolates was substantial, and we found no evidence of spread of this pathogen between herds. The milk SCC was lower during the follow-up period if *S. dysgalactiae* rather than *S. uberis* was isolated from the case, indicating differences in treatment response between bacterial species.

Keywords: *Streptococcus dysgalactiae* subsp. *dysgalactiae*, *Streptococcus uberis*, Dairy cow, Veterinary-treated clinical mastitis, Genotypes, Somatic cell count, Long-term mastitis outcome, Pulsed-field gel electrophoresis, PFGE

BACKGROUND

Clinical mastitis (CM) is a common disease among dairy cows. A number of pathogens can cause CM, but the relative importance of different pathogens differs between regions and countries in the world. In Sweden, *Streptococcus dysgalactiae* subsp. *dysgalactiae* and *Streptococcus uberis* are the third and fourth most common bacteria found in CM accounting for 15.6% and 11.1%, respectively, of the cases [1].

Since the introduction of molecular bacteriology in the field of mastitis, new possibilities for studying the epidemiology of udder pathogens have emerged. *S. uberis* has been extensively studied at the molecular level, and pulsed-field gel electrophoresis (PFGE) has been used for genotypic characterization of *S. uberis* isolates from Europe [2], Australia [3-6], New Zealand [7], and South America [8]. A number of studies have confirmed that this pathogen predominantly shows a heterogeneous genotype pattern consistent with environmentally spread bacteria [2,3,5-7], although there is evidence that *S. uberis* also can be spread between cows as a contagious pathogen within a herd [4,9-11]. There are a few studies reporting that the same genotype can be identified in more than one herd [4,10] indicating possible contagious spread with cows or equipment between herds. Molecular typing has also revealed differences between *S. uberis* strains in clinical manifestation [4,12] and duration of infection [13].

Although often considered as an environmental pathogen [14,15], studies have also indicated that *S. dysgalactiae* may be considered a contagious pathogen [16,17]. Compared with *S. uberis*, only a few studies on genotyping of *S. dysgalactiae* isolates have been performed to our knowledge [5,6,18]. Within a herd, more than one strain is usually present, but a few strains are often found in multiple cows suggesting spread between cows [5,6]. In addition, genetically related isolates have been found on multiple farms, suggesting either contagious spread between herds [6] or a common environmental source [5].

Prudent use of antimicrobials to treat CM is an important part of mastitis control programs. Common aims of treatment are clinical and bacteriological cure, and a return to normal milk somatic cell count (SCC). Bacteriological cure rate have in field studies been reported to be 65–90% for *S. dysgalactiae* and 45–90% for *S. uberis* when using treatment with benzyl penicillin or related compounds [19-21]. Detailed information on clinical cure rate is scarce, but has been reported to be 73% for *S. dysgalactiae* and 77% for *S. uberis* in one of the studies [19]. In a study on treatment of heifers for CM caused by *S. dysgalactiae* around calving, 15.4% of the quarters were nonfunctional and another 36.3% had an increased SCC in the milk and/or an intramammary infection 30 days after treatment [22]. The studies cited above used a follow-up period of two to four weeks after the end of treatment. To our knowledge, studies on differences in treatment outcome between *S. dysgalactiae* and *S. uberis* over a longer time period, or between genotypes within those species, have not been performed.

The understanding of infection epidemiology is essential to prevent and control mastitis. The distribution of *S. dysgalactiae* and *S. uberis* and genotypes within species may vary between regions and countries [23,24], but genotyping studies of those udder pathogens have, to our knowledge, not been performed in the Nordic countries, and not on a national level. The main aim of this study was, therefore, to explore the genotype variation of *S. dysgalactiae* and *S. uberis* associated with bovine CM in Sweden using streptococcal isolates collected in a national survey on CM. The study also aimed to investigate if bacterial species, or genotype within species, influences the outcome of veterinary-treated CM (VTCM), as measured by cow composite SCC, milk production, additional registered VTCMs and culling rate during a four month follow-up period. Differences in cow factors, geography and seasonality between CM cases due to *S. dysgalactiae* and *S. uberis* were also studied.

METHODS

Bacterial Isolates

Isolates of *S. dysgalactiae* (n = 164) and *S. uberis* (n = 117) were collected in a national survey on the prevalence of udder pathogens in bovine CM [1]. Case selection and diagnostic procedures have been described elsewhere [1]. In short, milk samples from cases of CM were collected by field veterinarians in 51 veterinary practices distributed all over Sweden during a one-year period. The samples were cultured at 37°C for 16–24 hours on blood agar plates and evaluated in accordance with the routines of the veterinary practices. Cultured plates were sent to the National Veterinary Institute, Uppsala, Sweden where the growth was verified by routine laboratory tests. The time from day of sampling until arrival at the laboratory was 3.0 days (range 1–7 days), on average. Streptococci were identified by colony morphology, CAMP-reaction, and 12 biochemical reactions (hippurate, aesculine, salicine, sorbitol, mannitol, raffinose, lactose, saccharose, inuline, trehalose, starch and glycerine). For suspected *S. dysgalactiae* isolates not identified by the 12 biochemical reactions, Lancefield grouping (Streptex, Murex Biotech Limited, Dartford, UK) was used. To differentiate enterococci from *S. uberis*, growth of red colonies on SlaBa plates (Slanetz & Bartley Medium, Oxoid Ltd., Basingstoke, England) was evaluated. Isolates were stored frozen in trypticase soy broth containing 15% glycerol.

Preparation and Digestion of Bacterial DNA

Isolates were thawed and cultured overnight at 37°C on 5% bovine blood agar supplemented with 0.05% esculine. One micro liter of *S. dysgalactiae* colony material was suspended in 250 µl lysis buffer (1 M NaCl, 10 mM Tris [pH 8.0], 200 mM EDTA, 0.5% Sarcosyl, 0.2% natrium deoxycholate; National Veterinary Institute, Uppsala, Sweden), mixed with 250 µl molten Agarose Prep (Amersham Biosciences, Uppsala, Sweden) and poured into plug molds (BioRad, CA, USA). Once solidified the plugs were placed in a buffer containing 5 ml lysis buffer and 200 µl lysozyme (from a stock solution of 20 mg/ml; Roche Diagnostics Scandinavia AB, Bromma, Sweden) and gently shaken overnight at 37°C. The buffer was then replaced by a proteolysis buffer containing 2.5 ml lysis buffer and 100 µl Proteinas K (from a stock solution of 50 U/ml; Roche Diagnostics Scandinavia AB, Bromma, Sweden) and the plugs were incubated for approximately 24 h at 56°C with gentle shaking. Plugs were washed twice with 10 ml Super-Q-water (National Veterinary Institute, Uppsala, Sweden) for 20 min in 56°C, and four times with 10 ml TrisEDTA (TE) buffer (10 mM Tris [pH 8.0], 1 mM EDTA; National Veterinary Institute, Uppsala, Sweden) for at least 20 min in 56°C. For long term storage, plugs were kept at 4°C in 1 ml TE buffer. Before digestion, plugs were sliced in halves, and one half was equilibrated in 200 µl CutSmart buffer (NewEngland Biolabs, MA, USA) for 30 min on gentle shaking at 25°C. The rest of the plug was returned to long term storage in TE buffer. After equilibration, each plug was digested at 25°C overnight in a digestion solution of 100 µl CutSmart buffer and 10 U of the restriction enzyme *SmaI* (NewEngland Biolabs, MA, USA).

Preparation and digestion of *S. uberis* DNA was performed using the same protocol as for *S. dysgalactiae* DNA, with the exception that digestion was performed in a digestion solution containing 5 U of *SmaI*.

Pulsed-field Gel Electrophoresis

DNA fragments were separated using a clamped homogenous electric field device (CHEF-DR II, BioRad) with pulse times of 5–15 sec over 10 h, and 15–60 sec over 13 h at 6 V, and run through a 1.2% gel of Agarose NA (GE Healthcare, Uppsala, Sweden) in 0.5 × Tris-borat EDTA-buffer (0.9 M Tris-borat, 20 mM EDTA; National Veterinary Institute, Uppsala, Sweden). Then, gels were stained with GelRed (Biotium, CA, USA).

For each bacterial species, one isolate with an easily interpreted pattern was selected among our collected isolates, as an internal reference and was run every 5th to 6th lane on all gels. In addition, a Lambda Ladder PFG Marker (New England Biolabs, MA, USA) was used in the first and last lane on each gel.

The protocol was repeated for untypeable isolates.

Dendrogram Analysis

Macrorestriction patterns of *S. dysgalactiae* and *S. uberis* were analyzed separately. The dendrogram analyses were performed using BioNumerics software (BioNumerics Version 7.1; Applied Maths, Inc 2014; Austin, TX, USA). Similarity was computed using the Dice coefficient and an unweighted pair group method with arithmetic mean (UPGMA), with optimization set to 1.5% and the tolerance value set to 1.25%. Isolates were considered to be of the same cluster if the similarity level was above 80%, and of the same pulsotype when banding patterns were identical. *S. dysgalactiae* clusters were identified by capital letters, and pulsotypes within cluster with a number suffix (e.g. E1, E2, F1, F2). *S. uberis* clusters were identified by Roman numerals. Only clusters and pulsotypes represented by more than one isolate per genotype received an identity. The remaining genotypes were referred to as "singles".

Data Editing and Statistical Analyses

Prevalence of Genotypes

In order to study the presence and distribution of different genotypes on a national level and to ensure epidemiological independence among isolates, the first isolate per species from a monoinfected cow (i.e. cows where only one bacterial species was found, and where only one udder quarter was infected) collected from each herd was included. However, if a farm had no samples collected from a monoinfected cow, the first isolate per species collected from that herd was included. Given those criteria, 135 isolates of *S. dysgalactiae* and 103 isolates of *S. uberis* were included in this part of the study. Descriptive statistics were used to present prevalence of genotypes.

Differences in cow factors, season and geography

Differences in breed of cow, parity, days in milk (DIM), season and geographic region of Sweden between *S. dysgalactiae* and *S. uberis* cases were presented using descriptive statistics.

Breed of cow was divided into three categories: Swedish Holstein (SH), Swedish Red and White (SR) and mixed breed/other breed (where the most common observation was a mix between SH and SR).

Parity was categorized into first, second, third, and fourth and higher lactations. Stage of lactation was categorized into first month after calving or later. Seasons were categorized into pasture season (May to August), early housing season (September to December) and late housing season (January to April). The highest subdivision of geographic regions of Sweden (eastern Sweden, southern Sweden including the islands, and northern Sweden) according to the Nomenclature of Territorial Units for Statistics (NUTS1) was used when studying geographic occurrence of strains and genotypes [25]. Associations between bacterial species and categorical variables were investigated using Fisher's exact test, as were associations between bacterial genotypes and categorical variables.

Outcome measurements and cow records

To investigate the impact of bacterial species or genotype on outcome of VTCM, only isolates from monoinfected cows were included (*S. dysgalactiae*: n = 98; *S. uberis*: n = 73). Cow composite SCC and milk yield at test milking, recurring or new cases of VTCM (VTCMadd), and culling due to mastitis during the follow-up period were used as outcome measurements. The follow-up period was zero to 120 days after VTCM for the parameters SCC, milk yield, and culling due to mastitis, and fourteen to 120 days for VTCMadd. Day zero was defined as the day the cultured plate arrived at the National Veterinary Institute. Somatic cell count and milk yield were measured at monthly milk recordings, and this data was obtained from the Swedish Official Milk Recording Scheme (SOMRS; Swedish Dairy Association, Stockholm, Sweden).

From SOMRS, cow records including breed, parity, date of calvings, and date and cause for culling were also obtained. The criterion for culling due to mastitis was fulfilled if either the primary or secondary officially recorded reason for culling was mastitis or increased SCC. Disease recordings were collected from the Swedish Animal Disease Recording System (SADRS) through SOMRS. A record in the SADRS of the original VTCM was present for 92% of the cows. VTCMadd was defined as an additional record of VTCM in the SADRS during the follow-up period.

Breed, parity and calving date was also recorded in a questionnaire by the veterinarian at milk sample submission. All cows where SOMRS records and questionnaire records were consistent were eligible for inclusion in the study. In case of minor discrepancies between official cow records and questionnaire records, individual assessment of the cow's records was made before a decision was made to include or exclude the cow. In general, the official records were considered more likely to be correct. In cases of major discrepancies, the cow was included using questionnaire records only in the prevalence and risk factor part of the study, and was excluded completely from the outcome part of the study.

Eight cows were dried off or calved during the follow-up period. For these cows, only data from monthly milk recordings before dry-off was included.

Differences in outcome were calculated between the two different streptococci, between different pulsotypes and clusters within species and between groups of pulsotypes and clusters within species based on prevalence. Definitions of groups used has previously been described in Lundberg *et al.* [26].

Differences in SCC and milk yield between bacterial species, between genotypes or between groups of genotypes were first tested in univariable mixed-effect linear regression models with repeated measurements of monthly SCC or milk yield within cow during the follow-up period as outcome variable. Mixed-effect models were used to take into account that repeated measurements of SCC or milk yield for a certain cow might be more similar than measurements of SCC or milk yield for different cows, hence, the random effect in the mixed models was repeated measurement of SCC or milk yield and an independent covariance structure (equal variances for random effects, all covariances are zero) was used. To obtain normally distributed residuals, SCC was transformed using the Box-Cox power transformation $(SCC^{-0.1263839}-1)/-0.1263839$. Bacterial species or genotype (pulsotypes, clusters, or groups of pulsotypes or clusters) were used as explanatory variables to the outcome. The cow parameters breed, parity and DIM at monthly milk recordings were included in the models as independent variables because of their possible impact on SCC and milk yield. When differences in milk yield were tested, SCC was included among independent variables and vice versa. To reduce the full models, a manual stepwise backward model selection procedure was used and only variables with a p-value of ≤0.05 were included in the final models. Two-way interactions between the significant main effects were tested. Possible confounders were considered in all models. A variable was considered as a confounder if the point estimates of the coefficients in a model changed >20% with the potential confounder present. The model fit of the multivariable analyses was tested by visual examination of diagnostic plots according to Dohoo *et al*. [27].

In addition, the proportions of cows with a SCC below 200 000 cells/ml at all monthly milk recordings during the follow-up period was calculated for *S. dysgalactiae* and *S. uberis* as well as for different genotypes or genotype groups within species. The follow-up period was shortened to 14 to 120 days to allow for SCC to normalize after therapy, and differences in proportions were tested with the Fisher's exact test. Only cows with data recorded for at least two monthly milk recordings were included in these calculations.

Proportions of cows with a VTCMadd registered and proportion of cows that were culled due to mastitis during the follow-up period were presented using descriptive statistics.

All statistical analyses were performed using Stata 13 (StataCorp, 2014; Stata Statistical Software: Release 13.1; College Station, TX, USA: StataCorp LP).

RESULTS

Macrorestriction Patterns and Prevalence of Genotypes

One hundred thirty two of the *S. dysgalactiae* isolates and 97 of the *S. uberis* isolates could be genotyped using our protocol.

Pulsed-field gel electrophoresis of *S. dysgalactiae* yielded 7 to 14 fragments in the considered size range of 45 to 500 kb. (Additional file 1) shows a dendrogram of the results. Analysis of the *S. dysgalactiae* isolates revealed 71 pulsotypes. Nineteen of the pulsotypes could be found in two to 13 herds each (Figure 1). The remaining 52 pulsotypes were only found once each in the material. Sixty-eight of the pulsotypes could be compiled into nine clusters. Three pulsotypes could not be clustered with other pulsotypes. Six of the clusters were considered rare, each represented by two to six isolates, while three of the clusters (E, F, and G) were considered common, each represented by 30 to 40 isolates. Those three clusters accounted for 82% of the isolates (Table 1).

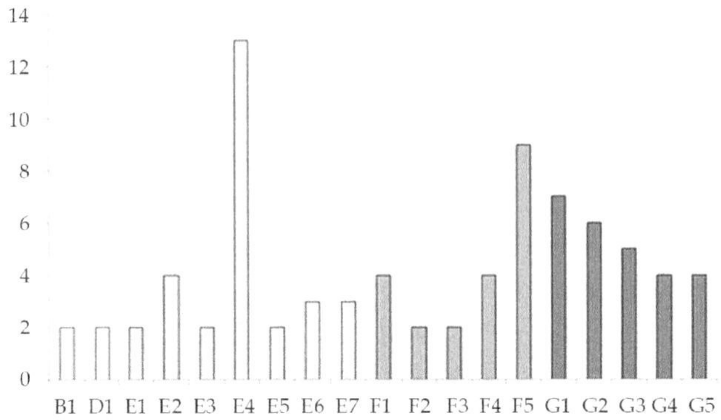

Figure 1: Numbers of *Streptococcus dysgalactiae* isolates of different pulsotypes. Numbers of epidemiologically independent *Streptococcus dysgalactiae* isolates of different pulsotypes from cases of veterinary-treated clinical mastitis in Sweden (pulsotypes only isolated once (n = 52) are not included). Pulsotypes belonging to the same cluster share the same color.

Table 1: Numbers (n) and percentages (%) of epidemiologically independent *Streptococcus dysgalactiae* isolates from cases of veterinary-treated clinical mastitis in Sweden divided into clusters after genotyping by pulsed-field gel electrophoresis and clustering at the level of 80% genetic similarity

Cluster	n (%)
A	3 (2.3)
B	6 (4.6)
C	2 (1.5)
D	5 (3.8)
E	40 (30.3)
F	30 (22.7)
G	38 (28.8)
H	3 (2.3)
I	2 (1.5)
Singles	3 (2.3)
Total	132 (100)

Pulsed field gel electrophoresis of *S. uberis* yielded 10 to 13 fragments in the 40–660 kb size range. (Additional file 2) shows a dendrogram of the results. All 97 *S. uberis* isolates were found to be of different pulsotypes. Forty-five isolates belonged to 21 clusters; the remaining isolates could not be assigned to a cluster (Additional file 2). Each cluster was represented by two or three isolates.

Differences in Cow Factors, Season and Geography between *S. dysgalactiae* and *S. uberis*

Descriptive statistics are given in Table 2. A larger proportion of *S. uberis*-isolates than of *S. dysgalactiae*-isolates came from cows of the SH breed compared to cows of the SR breed, but the difference was not statistically significant ($P = 0.079$). The distribution of parities and stage of lactation was similar among *S. dysgalactiae* and *S. uberis* isolates. About half of the isolates came from cases of VTCM occurring during the first month of lactation, with the rest spread out throughout the lactation.

Table 2: Numbers (n) and percentages (%) of *Streptococcus dysgalactiae* and *Streptococcus uberis* isolates from cases of veterinary-treated clinical mastitis in cows of different breeds, parities, and stages of lactation, occurring in different seasons and geographic regions of Sweden

Variable	Class	*S. dysgalactiae*, n (%)	*S. uberis*, n (%)
Breed[1]	SR	67 (49.6)	38 (36.9)
	SH	62 (45.9)	58 (56.3)
	Other	6 (4.4)	7 (6.8)
Parity	First	56 (41.5)	42 (40.8)
	Second	30 (22.2)	22 (21.4)
	Third	21 (15.6)	18 (17.5)
	Fourth or higher	28 (20.7)	21 (20.4)
Stage of lactation	<30 days postpartum	68 (50.4)	44 (42.3)
	≥30 days postpartum	67 (49.6)	59 (57.3)
Season	Early housing[2]	48 (35.6)	43 (41.8)
	Late housing[3]	47 (34.8)	23 (22.3)
	Pasture[4]	40 (29.6)	37 (35.9)
Region	Southern Sweden	91 (67.4)	59 (57.3)
	Eastern Sweden	26 (19.3)	21 (20.4)
	Northern Sweden	18 (13.3)	23 (22.3)

[1]SR=Swedish Red, SH=Swedish Holstein; [2]September–December; [3]January–April; [4]May–August.

There was a significant difference in occurrence of *S. dysgalactiae* and *S. uberis* over seasons ($P < 0.05$; Table 2). The occurrence of *S. uberis* was lowest during the late housing season; no fluctuations throughout the year could be seen for *S. dysgalactiae*.

The majority of both *S. dysgalactiae* and *S. uberis* isolates came from Southern Sweden. A larger proportion of *S. uberis*-isolates than of *S. dysgalactiae*-isolates came from Northern Sweden, and the opposite was found for Southern Sweden, but the difference was not significant ($P = 0.08$). The geographic distribution of the isolates within the country is given in Figure 2.

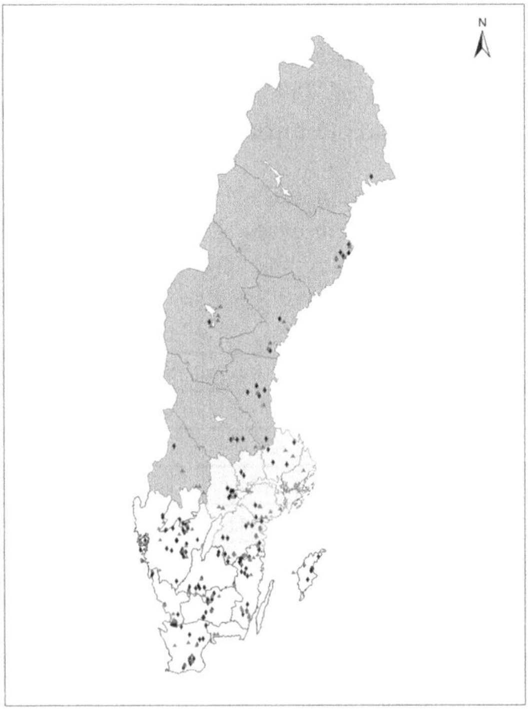

Figure 2: Geographic distribution of cases of clinical mastitis. Geographic distribution of cases of veterinary-treated clinical mastitis caused by *Streptococcus dysgalactiae* (black diamonds) and *Streptococcus uberis* (grey triangles).

Differences in Cow Factors, Season and Geography between Bacterial Genotypes within Species

For *S. dysgalactiae*, the number of isolates per pulsotype and most clusters were small, and statistical analyses of differences between groups were therefore not performed. Some pulsotypes could, however, only be found in one breed, lactation stage or parity, or during one season or in a specific region. The distribution of clusters, or groups of pulsotypes or clusters, did not indicate any trends related to cow factors, season or geography.

Due to few cows in each *S. uberis* cluster, statistical analyses of differences between clusters were not performed and no trends could be identified.

Differences in Outcome between *S. Dysgalactiae* and *S. Uberis*

For 163 monoinfected cows (95 *S. dysgalactiae*-cows and 68 *S. uberis*-cows) results from at least one monthly milk recording after the VTCM were available. In total, 541 monthly SCC and milk yield recordings were included (3.3 recordings per cow, on average).

During the follow-up period after the VTCM, the SCC of *S. dysgalactiae*-cows was significantly lower than the SCC of *S. uberis*-cows ($p < 0.05$; Figure 3). The median SCC during the follow-up period was 71 500 cells/ml (1st quartile: 35 500; 3rd quartile: 203 500) and 108 000 cells/ml (1st quartile: 47 000; 3rd quartile: 319 000 for *S. dysgalactiae*-cows and *S. uberis*-cows, respectively. The difference in SCC between species remained in the multivariable model when the influence of milk yield, parity, breed and DIM on SCC also was considered (Table 3).

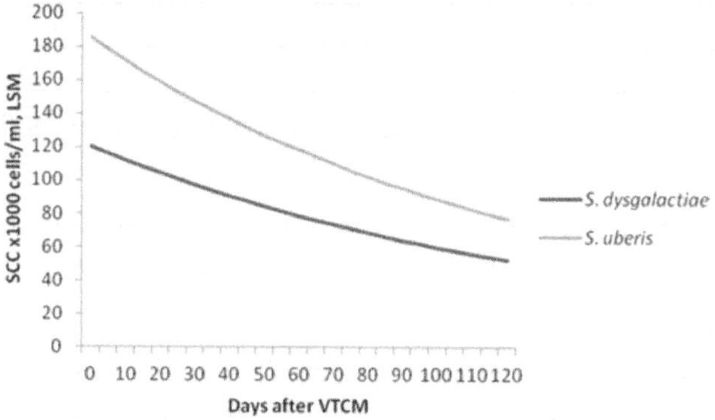

Figure 3: Long-term outcome after clinical mastitis as measured by SCC. Least square means (LSM) of cow somatic cell counts (SCC; ×1 000 cells/ml) at monthly milk recordings after veterinary-treated cases of clinical mastitis (VTCM) from the results of a multivariable mixed-effect linear regression model investigating associations between *Streptococcus dysgalactiae* and *Streptococcus uberis*.

Table 3: Final hierarchical multivariable linear-regression analysis of variables significantly ($P<0.05$) associated with somatic cell count (SCC; transformed using the BoxCox power transformation) at monthly milk recordings during a follow-up period of 120 days after a veterinary treated clinical mastitis

Variable [1]	β	**S.E. (** β **)**	P **-value**	**LSM** [2]	**95% CI (LSM)**
Intercept	3.996	0.17	-	-	-
Bacterial finding					
S. dysgalactiae	Ref.[3]	-	-	86.6	70.1, 107.5
S. uberis	0.260	0.093	0.005	138.6	106.3, 182.6
Other factors					
Parity					
First	Ref.	-	-	75.9	59.8, 97.1
Second	0.219	0.113	0.051	112.0	82.8, 153.4
Third	0.169	0.137	0.218	102.3	68.7, 155.6
Fourth or higher	0.433	0.136	0.001	260.5	109.1, 260.5
Milk yield	−0.002	0.0005	0.000	-	-
At Q1: 23.8 kg	-	-	-	132.8	108.1,164.2
At Q2: 28.7 kg	-	-	-	109.8	92.0, 131.4
At Q3: 33.4 kg	-	-	-	91.8	76.3, 111.0
Days in milk	−0.001	0.0004	0.001	-	-
At Q1: 67 days	-	-	-	129.4	105.0, 160.5
At Q2: 115 days	-	-	-	115.6	96.4, 139.3
At Q3: 188 days	-	-	-	97.8	81.5, 117.7

[1]Q1, Q2, and Q3 = first quartile, second quartile/median, and third quartile, respectively.

[2]Least square means (LSM) of Box-Cox transformed SCC back-transformed to original scale (×1 000/ml) presented by variable.

[3]Ref. = Reference category.

Eighty-four *S. dysgalactiae*-cows and 58 *S. uberis*-cows had more than one registered recording of SCC during the period of 14 to 120 days after VTCM. The proportion of *S. dysgalactiae*-cows with a SCC below 200 000 cells/ml at all recordings during this period was 58%. The corresponding percentage for *S. uberis*-cows was 43%. The difference between species was not significant ($P = 0.09$).

The average daily milk yield at monthly milk recordings during the follow-up period was 28.5 kg (range: 9.2–50.8 kg) and 29.3 kg (range: 9.8–

53.6 kg) for *S. dysgalactiae*-cows and *S. uberis*-cows, respectively, and did not differ significantly between species.

Eleven (12%) *S. dysgalactiae*-cows had a VTCMadd registered and six (6%) were culled during the follow-up period. The corresponding numbers for *S. uberis*-cows were six (9%) with VTCMadd and six (8%) culled.

Differences in Outcome between Bacterial Genotypes within Species

No univariable associations between milk yield or SCC and bacterial species or genotype or groups of genotypes could be found. However, there seemed to be some indication that the proportion of *S. dysgalactiae* cows having SCC constantly below 200 000 cells/ml during the follow-up period varied between pulsotypes (Table 4), although the numbers of isolates in each group were too small for statistical analyses.

Table 4: Numbers of cows with a somatic cell count (SCC) below 200 000 cells/ml at all monthly milk recordings, and number of cows with SCC above 200 000 cells/ml at least one monthly milk recording 14 to 120 days after veterinary-treated clinical mastitis, caused by different clusters or pulsotypes1 of *Streptococcus dysgalactiae*

Cluster/Pulsotype	<200 000 cells/ml	>200 000 cells/ml	Total
A	2	1	3
B	2	3	5
B1	0	2	2
D	1	1	2
E	14	10	24
E2	1	2	3
E4	5	2	7
E5	1	1	2
E6	1	2	3
E7	2	1	3
F	11	6	17
F1	1	2	3
F4	4	0	4
F5	5	1	6
G	13	12	25
G1	1	5	6

G2	3	1	4
G3	2	1	3
G4	1	2	3
G5	2	1	3
H	1	1	2

[1]Cows infected with unique clusters/pulsotypes are not included in the table.

The number of cows within each pulsotype or cluster of *S. dysgalactiae* and *S. uberis*, or within each group of pulsotypes or clusters, that had a VTCMadd registered or that were culled during the follow-up period was too small for statistical analysis and no trends could be identified.

DISCUSSION

Genotype Distribution

In line with most previous studies [2,3,5-7], we found that the genotype pattern of *S. uberis* is heterogeneous in Sweden and we found no evidence of contagious spread of *S. uberis* between herds as all isolates investigated were of different pulsotypes. In contrast, identical pulsotypes of *S. dysgalactiae* were found in multiple herds in the present study, although 39% of the isolates belonged to unique pulsotypes. The findings of identical strains of *S. dysgalactiae* in different herds is in line with previous studies [5,6,11]. A common environmental source could explain why the same genotype was found in multiple herds, but this hypothesis seems less likely for many of the genotypes since identical isolates were found in separate parts of the country. Trade of infected animals is another, and perhaps more likely, way of spread of the pathogen between herds and it has been shown that the trade of livestock is expansive within Sweden [28]. For local spread (kilometers) the fly *Hydrotaea irritans*, which can act as a vector for *S. dysgalactiae* [29], could also be involved. It is possible that some genotypes contain virulence factors that make them especially apt to spreading between cows. No such attributes of the strains were investigated in this study, but previous studies have shown strain differences in virulence factors, possibly connected to spread between cows, for both *S. dysgalactiae* and *S. uberis* [30].

Given the existing information and the original study design, it was not possible to study the genetic within-herd variation among the udder pathogens. Previous studies of both *S. dysgalactiae* and *S. uberis* have, however, shown that several strains can be found in a herd and that in some instances one or a few of these strains can be found in other herds as well [4,6,10]. It is therefore possible that we, when including only one sample per herd, might have underestimated the spread of genotypes between herds for both *S. dysgalactiae* and *S. uberis* in the present study.

The moderate and large genotypic variation found for *S. dysgalactiae* and *S. uberis*, respectively, in the present study differs markedly from a previous study on isolates of *Staphylococcus aureus* from the same national survey [26]. In that study, almost two thirds of the *S. aureus* isolates belonged to two pulsotypes, emphasizing the contagious nature of *S. aureus* as opposed to the two streptococcal species investigated in the present study.

Differences in Outcome of VTCM between Bacterial Species and Genotypes

Cows veterinary-treated for CM caused by *S. dysgalactiae* had a lower SCC during the follow-up period, than did cows treated for infections caused by *S. uberis*. Possible explanations for this finding could be a stronger inflammatory response to *S. uberis* at the initial infection [31] or difference in bacteriological cure rates between species [19].

In the present study, we did not have any information on SCC at the time of VTCM other than the California Mastitis Test (CMT) score performed by the investigating veterinarian. The CMT is, however, an imprecise test, and most cases of CM are categorized as either 4 or 5 on the Nordic scale (i.e. SCC>~800 000 cells/ml) making CMT unsuitable for use in the present study.

Unfortunately, no information about bacteriological cure after VTCM was available in our study material. National guidelines at the time of sampling stated that infections during lactation caused by *S. dysgalactiae* and *S. uberis* should be treated by parenteral benzyl penicillin for 4–5 days [32]. The isolates in the present study were tested for susceptibility to penicillin *in vitro* [33]. All of the *S. dysgalactiae* isolates and 94% of the *S. uberis* isolates tested were susceptible, therefore it seems unlikely that any possible difference in bacteriological cure was due to resistance to penicillin.

No statistically significant differences in SCC between genotypes, or groups of genotypes, within bacterial species could be found in this study. The number of genotypes in relation to the total number of isolates was high, which made comparisons difficult. The discriminatory power of PFGE is high, making it probable that isolates with identical PFGE patterns are in fact genetically related. Tenover *et al.* [34] proposed that genotypes with a one to three band difference (consistent with one genetic event) are closely related and that genotypes with a four to six band difference are possibly related. This was, however, proposed for hospital outbreaks and to our knowledge, similar guidelines for national surveillance material do not exist. It is therefore not certain that isolates with the 80% level of genetic similarity used for clustering in the present study are related. This could explain why differences in our outcome parameters could not be found between *S. dysgalactiae* clusters. To further study if differences in outcome of VTCM could be attributed to bacterial genotypes of *S. dysgalactiae* or *S. uberis*, a study including a larger number of natural infections could be used, resulting in more isolates per pulsotype or cluster and enabling statistical comparisons. In experimental infections, significant differences in pathogenicity between different strains of *S. dysgalactiae* [35], as well as between different strains of *S. uberis* [36], have been shown in the past, and another option may be to study long-term outcome after experimental infections of isolates of different genotypes. However, it is also possible that isolates of the same pulsotype or cluster express different virulence factors. Further studies in this area are therefore warranted.

We found no differences in milk yield between bacterial species, or between genotypes within species. As milk yield and SCC are correlated, i.e. the milk yield decreases with increasing SCC [37] and we found a difference in SCC between streptococcal species, a difference in milk yield between *S. dysgalactiae* and *S. uberis* during the follow-up period would have been expected. It is possible; however, that the difference in SCC between species was too small to influence the milk yield significantly. It is also possible that *S. uberis* infections are more common in high yielding cows, but since information on milk yield before VTCM was not available in the present study this could not be controlled for in the comparison between bacterial species.

Species and Genotype Differences in Cow Factors, Season and Geography

The results indicate that breed, season and geographical region, but not parity and stage of lactation, might differ between cases of VTCM due to *S. dysgalactiae* and *S. uberis* in Sweden. However, the results should be interpreted with caution as the study was not designed specifically for evaluation of the effect of cow factors, season and geography on the risk for infection with different streptococci. For example, information about other, possibly confounding, factors, like herd size, barn type and breed distribution in herds and over geographic regions, were not available in this study and thus could not be controlled for in the statistical analysis. The study material was, however, representative for VTCM on a national level as the selection of cases was based on the distribution of cows throughout the country.

The lower and higher proportion of *S. uberis* in the late housing season and in SH cows, respectively, compared to *S. dysgalactiae* is in line with the previous study where the occurrence of VTCM caused by *S. uberis* was compared to VTCM caused by all other pathogens [1]. The seasonal pattern is also similar to the results of a Canadian study, where the incidence rate of CM caused by *S. uberis* was low during the winter and peaked in August [38]. However, in Canada there seemed to be a decline in *S. dysgalactiae* incidence rate during the summer as well [38], a pattern not seen in our material. Moreover, regional differences have been described for *S. dysgalactiae* in Canada [24], but have to our knowledge not been described for *S. uberis*.

As already mentioned, parity did not differ between cases of *S. dysgalactiae* and *S. uberis* mastitis. Unpublished observations suggest, however, that mastitis cases due to the two streptococcal species investigated in the present study are more prevalent in first parity cows compared to clinical cases of *Staphylococcus aureus* mastitis studied in a previous paper [26]. The CM cases of the two studies originated from the same national survey.

The effect of DIM on the distribution of CM caused by *S. uberis* in this study seem to differ from that reported from New Zealand where *S. uberis* is predominantly isolated during the first month of lactation [39]. These differences between countries may be due to different production systems.

CONCLUSIONS

Identical *S. dysgalactiae* genotypes were found in cases of bovine CM from more than one herd, indicating some spread of infections between Swedish dairy herds, but a large proportion of the banding patterns were only found once. In contrast, we found no evidence of spread of *S. uberis* genotypes between herds. The genetic variation among *S. uberis* isolates was substantial. The results suggest that, in Sweden, *S. uberis* should be considered as an environmental pathogen, while *S. dysgalactiae* probably can act either as a contagious or an environmental pathogen, at least when considering spread between herds. The only outcome of veterinary-treated CM that differed significantly between bacterial species or genotypes was the milk SCC, which was lower during a four month follow-up period after veterinary treatment of CM if *S. dysgalactiae* rather than *S. uberis* was isolated from the case.

ACKNOWLEDGEMENTS

The authors would like to thank the Swedish Dairy Association/Växa Sverige for contributing data, and Maria Persson for performing most of the laboratory work. Thanks also to Anna Aspán at the National Veterinary Institute and Karin Östensson at the Swedish University of Agricultural Sciences. The project was funded by the Swedish Research Council for Environment, Agricultural Sciences and Spatial Planning (Formas), and the Swedish Farmers' Foundation for Agricultural Research.

AUTHORS' CONTRIBUTIONS

All authors participated in the design of the study. ÅL participated in the genotyping, carried out the statistical analyses and drafted the manuscript. AN participated in the statistical analyses. HEU participated in the bacteriological analyses. KPW coordinated the study and helped draft the manuscript. All authors read and approved the final manuscript.

REFERENCES

1. Ericsson Unnerstad H, Lindberg A, Persson Waller K, Ekman T, Artursson K, Nilsson-Ost M, Bengtsson B. Microbial aetiology of acute clinical mastitis and agent-specific risk factors. Vet Microbiol. 2009;137:90–97. doi: 10.1016/j.vetmic.2008.12.005.
2. Khan IU, Hassan AA, Abdulmawjood A, Lämmler C, Wolter W, Zschöck M. Identification and epidemiological characterization of *Streptococcus uberis* isolated from bovine mastitis using conventional and molecular methods. J Vet Sci. 2003;4:213–224.
3. Abureema S, Smooker P, Malmo J, Deighton M. Molecular epidemiology of recurrent clinical mastitis due to *Streptococcus uberis*: evidence of both an environmental source and recurring infection with the same strain. J Dairy Sci. 2014;97:285–290. doi: 10.3168/jds.2013-7074.
4. Phuektes P, Mansell PD, Dyson RS, Hooper ND, Dick JS, Browning GF. Molecular epidemiology of *Streptococcus uberis* isolates from dairy cows with mastitis. J Clin Microbiol. 2001;39:1460–1466. doi: 10.1128/JCM.39.4.1460-1466.2001.
5. Baseggio N, Mansell PD, Browning JW, Browning GF. Strain differentiation of isolates of streptococci from bovine mastitis by pulsed-field gel electrophoresis. Mol Cell Probes. 1997;11:349–354. doi: 10.1006/mcpr.1997.0126.
6. Wang SM, Deighton MA, Capstick JA, Gerraty N. Epidemiological typing of bovine streptococci by pulsed-field gel electrophoresis. Epidemiol Infect. 1999;123:317–324. doi: 10.1017/S0950268899002745.
7. McDougall S, Parkinson TJ, Leyland M, Anniss FM, Fenwick SG. Duration of infection and strain variation in *Streptococcus uberis* isolated from cows' milk. J Dairy Sci. 2004;87:2062–2072. doi: 10.3168/jds.S0022-0302(04)70024-7.
8. Lasagno MC, Reinoso EB, Dieser SA, Calvinho LF, Buzzola F, Vissio C, Bogni CI, Odierno LM. Phenotypic and genotypic characterization of *Streptococcus uberis* isolated from bovine subclinical mastitis in Argentinean dairy farms. Rev Argent Microbiol. 2011;43:212–217.
9. Zadoks RN, Allore HG, Barkema HW, Sampimon OC, Gröhn YT, Schukken YH. Analysis of an outbreak of *Streptococcus uberis* mastitis. J Dairy Sci. 2001;84:590–599. doi: 10.3168/jds.S0022-0302(01)74512-2.

10. Coffey TJ, Pullinger GD, Urwin R, Jolley KA, Wilson SM, Maiden MC, Leigh JA. First insights into the evolution of *Streptococcus uberis*: a multilocus sequence typing scheme that enables investigation of its population biology. Appl Environ Microbiol. 2006;72:1420–1428. doi: 10.1128/AEM.72.2.1420-1428.2006.
11. Rato MG, Bexiga R, Nunes SF, Cavaco LM, Vilela CL, Santos-Sanches I. Molecular epidemiology and population structure of bovine *Streptococcus uberis*. J Dairy Sci. 2008;91:4542–4551. doi: 10.3168/jds.2007-0907.
12. Pullinger GD, Coffey TJ, Maiden MC, Leigh JA. Multilocus-sequence typing analysis reveals similar populations of *Streptococcus uberis* are responsible for bovine intramammary infections of short and long duration. Vet Microbiol. 2007;119:194–204. doi: 10.1016/j.vetmic.2006.08.015.
13. Zadoks RN, Gillespie BE, Barkema HW, Sampimon OC, Oliver SP, Schukken YH. Clinical, epidemiological and molecular characteristics of *Streptococcus uberis* infections in dairy herds. Epidemiol Infect. 2003;130:335–349. doi: 10.1017/S0950268802008221.
14. Todhunter DA, Smith KL, Hogan JS. Environmental streptococcal intramammary infections of the bovine mammary gland. J Dairy Sci. 1995;78:2366–2374. doi: 10.3168/jds.S0022-0302(95)76864-3.
15. Gillespie BE, Jayarao BM, Pankey JW, Oliver SP. Subtyping of *Streptococcus dysgalactiae* and *Streptococcus uberis* isolated from bovine mammary secretions by DNA fingerprinting. Zentralblatt Für Veterinärmedizin Reihe B J Vet Med Ser B. 1998;45:585–593.
16. Barkema HW, Schukken YH, Lam TJ, Beiboer ML, Benedictus G, Brand A. Management practices associated with the incidence rate of clinical mastitis. J Dairy Sci. 1999;82:1643–1654. doi: 10.3168/jds.S0022-0302(99)75393-2.
17. Fox LK, Gay JM. Contagious mastitis. Vet Clin North Am Food Anim Pract. 1993;9:475–487.
18. Bert F, Branger C, Poutrel B, Lambert-Zechovsky N. Differentiation of human and animal strains of *Streptococcus dysgalactiae* by pulsed-field gel electrophoresis. FEMS Microbiol Lett. 1997;150:107–112. doi: 10.1016/S0378-1097(97)00105-5.
19. Kalmus P, Simojoki H, Orro T, Taponen S, Mustonen K, Holopainen J, Pyörälä S. Efficacy of 5-day parenteral versus intramammary

benzylpenicillin for treatment of clinical mastitis caused by gram-positive bacteria susceptible to penicillin in vitro. J Dairy Sci. 2014;97:2155–2164. doi: 10.3168/jds.2013-7338.

20. McDougall S, Agnew KE, Cursons R, Hou XX, Compton CRW. Parenteral treatment of clinical mastitis with tylosin base or penethamate hydriodide in dairy cattle. J Dairy Sci. 2007;90:779–789. doi: 10.3168/jds.S0022-0302(07)71562-X.
21. Taponen S, Dredge K, Henriksson B, Pyyhtiä A-M, Suojala L, Junni R, Heinonen K, Pyörälä S. Efficacy of intramammary treatment with procaine penicillin G vs. procaine penicillin G plus neomycin in bovine clinical mastitis caused by penicillin-susceptible, gram-positive bacteria–a double blind field study. J Vet Pharmacol Ther. 2003;26:193–198. doi: 10.1046/j.1365-2885.2003.00473.x.
22. Waage S, Skei HR, Rise J, Rogdo T, Sviland S, Odegaard SA. Outcome of clinical mastitis in dairy heifers assessed by reexamination of cases one month after treatment. J Dairy Sci. 2000;83:70–76. doi: 10.3168/jds.S0022-0302(00)74857-0.
23. Douglas VL, Fenwick SG, Pfeiffer DU, Williamson NB, Holmes CW. Genomic typing of *Streptococcus uberis* isolates from cases of mastitis, in New Zealand dairy cows, using pulsed-field gel electrophoresis. Vet Microbiol. 2000;75:27–41. doi: 10.1016/S0378-1135(00)00184-X.
24. Olde Riekerink RGM, Barkema HW, Kelton DF, Scholl DT. Incidence rate of clinical mastitis on Canadian dairy farms. J Dairy Sci. 2008;91:1366–1377. doi: 10.3168/jds.2007-0757.
25. European Union Regulation (EC) no 1059/2003 of the European parliament and of the council of 26 May 2003 on the establishment of a common classification of territorial units for statistics (NUTS) Off J EU. 2003;1059/2003:L154/151–L154/141.
26. Lundberg A, Aspán A, Nyman A, Unnerstad HE, Waller KP. Associations between bacterial genotype and outcome of bovine clinical *Staphylococcus aureus* mastitis. Acta Vet Scand. 2014;56:2. doi: 10.1186/1751-0147-56-2.
27. Dohoo I, Martin W, Stryhn H. Veterinary Epidemiologic Research. 2. VER Inc: Charlottetown, Prince Edward Islands, Canada; 2010.
28. Widgren S, Frössling J. Spatio-temporal evaluation of cattle trade in Sweden: description of a grid network visualization technique. Geospatial Health. 2010;5:119–130.

29. Chirico J, Jonsson P, Kjellberg S, Thomas G. Summer mastitis experimentally induced by *Hydrotaea irritans* exposed to bacteria. Med Vet Entomol. 1997;11:187–192. doi: 10.1111/j.1365-2915.1997.tb00312.x.
30. Frost AJ, Wanasinghe DD, Woolcock JB. Some factors affecting selective adherence of microorganisms in the bovine mammary gland. Infect Immun. 1977;15:245–253.
31. Schepers AJ, Lam TJ, Schukken YH, Wilmink JB, Hanekamp WJ. Estimation of variance components for somatic cell counts to determine thresholds for uninfected quarters. J Dairy Sci. 1997;80:1833–1840. doi: 10.3168/jds.S0022-0302(97)76118-6.
32. Ekman T, Franklin A, Hallén Sandgren C, Jonsson P. Proposed treatment of clinical mastitis in cows [in Swedish] Sven Vet Tidn. 1995;47:713–718.
33. Bengtsson B, Unnerstad HE, Ekman T, Artursson K, Nilsson-Ost M, Waller KP. Antimicrobial susceptibility of udder pathogens from cases of acute clinical mastitis in dairy cows. Vet Microbiol. 2009;136:142–149. doi: 10.1016/j.vetmic.2008.10.024.
34. Tenover FC, Arbeit RD, Goering RV, Mickelsen PA, Murray BE, Persing DH, Swaminathan B. Interpreting chromosomal DNA restriction patterns produced by pulsed-field gel electrophoresis: criteria for bacterial strain typing. J Clin Microbiol. 1995;33:2233–2239.
35. Higgs TM, Neave FK, Bramley AJ. Differences in intramammary pathogenicity of four strains of *Streptococcus dysgalactiae*. J Med Microbiol. 1980;13:393–399. doi: 10.1099/00222615-13-3-393.
36. Tassi R, McNeilly TN, Fitzpatrick JL, Fontaine MC, Reddick D, Ramage C, Lutton M, Schukken YH, Zadoks RN. Strain-specific pathogenicity of putative host-adapted and nonadapted strains of *Streptococcus uberis* in dairy cattle. J Dairy Sci. 2013;96:5129–5145. doi: 10.3168/jds.2013-6741.
37. Hand KJ, Godkin A, Kelton DF. Milk production and somatic cell counts: a cow-level analysis. J Dairy Sci. 2012;95:1358–1362. doi: 10.3168/jds.2011-4927.
38. Olde Riekerink RGM, Barkema HW, Stryhn H. The effect of season on somatic cell count and the incidence of clinical mastitis. J Dairy Sci. 2007;90:1704–1715. doi: 10.3168/jds.2006-567.

39. McDougall S. Intramammary treatment of clinical mastitis of dairy cows with a combination of lincomycin and neomycin, or penicillin and dihydrostreptomycin. N Z Vet J. 2003;51:111–116. doi: 10.1080/00480169.2003.36349.

CHAPTER 6

Estimation of Sensitivity and Specificity of Bacteriology, Histopathology and PCR for the Confirmatory Diagnosis of Bovine Tuberculosis Using Latent Class Analysis

Aure´lie Courcoul[1]., Jean-Louis Moyen[2]., Laure Bruge` re[2] , Sandy Faye[2] , Sylvie He´nault[3] , He´le`ne Gares[2] , Maria-Laura Boschiroli[3]

[1] Epidemiology Unit, Paris-Est University, Anses, Laboratory for Animal Health, Maisons-Alfort, France

[2] Regional Analysis and Research Laboratory of Dordogne (LDAR24), Coulounieix-Chamiers, France

[3] Bovine Tuberculosis Reference Laboratory, Paris-Est University, Anses, Laboratory for Animal Health, Bacterial Zoonoses Unit, MaisonsAlfort, France

ABSTRACT

Bacteriology and histopathology are the most commonly used tests used for official confirmatory diagnosis of bovine tuberculosis (bTB) in cattle in

---

Citation: Courcoul A, Moyen J-L, Brugère L, Faye S, Hénault S, Gares H, et al. (2014) Estimation of Sensitivity and Specificity of Bacteriology, Histopathology and PCR for the Confirmatory Diagnosis of Bovine Tuberculosis Using Latent Class Analysis. PLoS ONE 9(3): e90334. https://doi.org/10.1371/journal.pone.0090334.

most countries. PCR is also being used increasingly because it allows a fast diagnosis. This test could be applied as a supplement to or replacement for current bTB confirmatory diagnostic tests but its characteristics have first to be evaluated. The aim of this study was to estimate and compare sensitivities and specificities of bacteriology, histopathology and PCR under French field conditions, in the absence of a gold standard using latent class analysis. The studied population consisted of 5,211 animals from which samples were subjected to bacteriology and PCR (LSI VetMAX™ *Mycobacterium tuberculosis* Complex PCR Kit, Life Technologies) as their herd of origin was either suspected or confirmed infected with bTB or because bTB-like lesions were detected during slaughterhouse inspection. Samples from 697 of these animals (all with bTB-like lesions) were subjected to histopathology. Bayesian models were developed, allowing for dependence between bacteriology and PCR, while assuming independence from histopathology. The sensitivity of PCR was higher than that of bacteriology (on average 87.7% [82.5–92.3%] versus 78.1% [72.9–82.8%]) while specificity of both tests was very good (on average 97.0% for PCR [94.3–99.0%] and 99.1% for bacteriology [97.1–100.0%]). Histopathology was at least as sensitive as PCR (on average 93.6% [89.9–96.9%]) but less specific than the two other tests (on average 83.3% [78.7–87.6%]). These results suggest that PCR has the potential to replace bacteriology to confirm bTB in samples submitted from suspect cattle.

INTRODUCTION

Bovine tuberculosis (bTB), caused principally by *Mycobacterium bovis*, is an on-going issue in several parts of Europe. Although eradication and control programs have been in place for many decades, the prevalence of bTB is currently increasing both in non-bTB-free and in officially bTB-free EU countries [1]. In this context, detection and confirmation of bTB infection have to be fast and reliable to ensure efficient surveillance and control programs. Knowledge about accuracy of the tests used to detect and confirm infection seems, then, essential. Detailed information on the characteristics of tests used for bTB detection (such as skin tests or γ-interferon tests) is available [2]–[4]. However, the sensitivity and specificity of tests used for bTB confirmatory diagnosis are far less known. In the EU, the presence of *M. bovis* in a sample may be officially demonstrated by histopathology and confirmed by bacterial culture. Histopathology is less specific than bacterial culture: Varello et al. [5] demonstrated that the average specificity of histological methods was 92.3% when compared with bacteriology as reference test. This often requires, when a sample is histopathology positive, confirmation of the infection by bacteriology. In France, cattle herds are subjected to movement restrictions if they present positive screening tests results. In order to lift the restriction, either reactors have to be tested negative to a second skin test or, if they are culled, they have to be tested negative to confirmatory tests (in this case, the herd has sometimes also to be tested negative to a second skin test to lift the restriction). Bacteriology, which takes at least three weeks but more often up to three months, has thus important constraints on the field. Besides, its sensitivity is not perfect, around 85% in samples with visible lesions (European Union Reference Laboratory for Bovine Tuberculosis, personal communication). In this setting, an alternative to obtain a faster infection confirmation or invalidation could be PCR. When infection is confirmed, it would allow a responsive herd management (i.e. whole or partial herd culling, tracing of contact herds, investigation of surrounding wildlife, etc.). When infection is not confirmed, movement restrictions could quickly be lifted.

The LSI VetMAX™ *Mycobacterium tuberculosis* Complex PCR Kit, distributed by Life Technologies, enables the detection of *M. tuberculosis* complex (MTBC) organisms. It detects the IS*6110* specific insertion sequence found in the genome of all MTBC members. This PCR may be used as a supplement to or replacement for current bTB confirmatory diagnostic tests (i.e. bacterial culture and histopathology) but, to our knowledge, it has not yet been evaluated from samples obtained as part of routine bTB surveillance and control programs, which seems a pre-requisite for a sensible use.

The performance of a diagnostic test is classically evaluated against a perfect test defined as a gold standard. As previously stated, none of the tests used to confirm bTB infection has 100% sensitivity and 100% specificity and can thus be considered as a gold standard. When a reference test is unavailable, a Bayesian formulation of the latent class analysis can be an option to assess the sensitivity and specificity of diagnostic tests [6]. The disease status of the tested individuals is designated "latent" (existing but not presently evident or realized) and none of the tests are considered a reference test. In latent class analysis, three assumptions known as the Hui-Walter paradigm are generally made [7]: (1) two or more populations with differing prevalences are required; (2) the sensitivity and specificity of the tests are constant across all populations; and (3) the tests are conditionally independent. For tests based on the same biological basis, this latter assumption can be relaxed [2], [8], [9]. In our case, it seems reasonable to assume that bacteriology and PCR are dependant tests as they both detect the mycobacterium directly.

The aim of this study was therefore to estimate, using latent class analysis and accounting for conditional dependence between bacterial culture and PCR, the sensitivity and the specificity of the diagnostic tests (bacterial culture, histopathology and PCR) carried out in France on lymph nodes in order to confirm or invalidate bTB diagnosis on suspect animals (e.g. animals positive to skin tests, animals with macroscopic bTB-like lesions or animals from infected herds).

MATERIALS AND METHODS

Ethical Statement

bTB is a notifiable disease for which there are control and surveillance campaigns in France. Official methods for diagnosis of this disease are culture, PCR and histopathology. Therefore, all the samples included in this study are issued from animals analyzed within an official context. No purpose killing of animals was performed for this study. All samplings were in complete agreement with national and European regulations. No ethical approval was necessary.

Sample Collection and Population Stratification

The source population was the animals whose samples were submitted to the "Laboratoire d'Analyse et de Recherche de Dordogne" (LDAR 24) for confirmation of *M. bovis* infection between 2008 and 2012. These animals were analyzed because (i) they presented non negative skin tests and/or γ-interferon tests results and were slaughtered for diagnostic purposes, (ii) they were slaughtered as a result of the total or partial slaughter of their herd of origin after confirmation of the herd's infection, or (iii) they presented macroscopic bTB-like lesions at routine abattoir inspection. Samples from the totality of animals submitted to the laboratory for bTB confirmatory diagnosis between 2008 and 2012, except for 30 animals for which PCR and bacteriology results or herd of origin were not available, were retrospectively enrolled in the study. The study sample included 5,211 individuals from 1,325 French cattle herds in 51 departments (over 96 in metropolitan France).

Tracheobronchial, retropharyngeal and mediastinal lymph nodes were sampled on each animal. Bacteriology and PCR were performed at the Laboratoire d'Analyse et de Recherche de Dordogne (LDAR 24). An animal was considered bacteriology positive (PCR positive respectively) if at least one of the samples was bacteriology positive (PCR positive respectively). Among these 5,211 animals, samples from 697 individuals belonging to 358 herds were also subjected to histopathology upon request either of the official veterinary services or the LDAR 24. All the animals tested using the three tests (bacterial culture, histopathology and PCR) presented macroscopic bTB-like lesions. On the contrary, the 4,514 remaining animals had no bTB-like lesions.

The herd incidence of infection over the 2000–2007 period in the cantons from where the animals originated was used to retrospectively define three populations with differing expected animal prevalences (see Figure S1). In France, a canton is a small geographical area (on average, 141 km^2) which includes several villages. The first population (population A) comprised animals from cantons with an incidence superior to 0.25 reported cases per 100 herd-years over the 2000–2007 period. The second population (population B) comprised animals from cantons with an incidence superior to 0 but inferior or equal to 0.25 reported cases per 100 herd-years over the same period. The third population (population C) comprised animals from cantons where no bTB outbreak had been reported between 2000 and 2007.

Diagnostic Tests

Preparation of the samples for PCR and culture

PCR and culture were performed on the above mentioned individual lymph nodes, presenting or not lesions at the abattoir. Lymph nodes were analyzed individually. Samples were submitted to the LDAR 24 chilled (approximately 4°C) within 48 hours after their sampling at the abattoir. Prior to PCR or culturing, 2 to 5 g of sampled tissue were crushed with a 4% H_2SO_4 solution to decontaminate the tissue. After 10 min, the acid was neutralized by adding a 6% NaOH solution.

PCRs

A commercial kit (LSI VetMAX™ *Mycobacterium tuberculosis* Complex PCR Kit 2 wells) was used. The targeted sequence was IS*6110*, which is present in all species of the *M. tuberculosis* complex [10]. After mechanical lysis of tissue, DNA was extracted by using the QIAamp DNA mini kit (Qiagen) or by Magvet MV384 (Life Technologies) with a King Fisher KF96 automate, following the manufacturer's instructions. Then, 5 μl of the extracted DNA was mixed with 20 μl of reaction mix and the reaction was carried out at 50°C for 2 minutes (1 cycle), followed by one cycle of 10 min at 95°C and 40 cycles of 15 s at 95°C and 1 min at 60°C. Results were interpreted as negative, positive (CT≤38) or invalidated, following the manufacturer's recommendations and by comparison with negative and positive controls. Even if laboratory personnel were aware that sampled animals were suspect of bTB infection, they delivered PCR results as soon as they were available, without knowing bacteriology and histopathology test results.

Bacteriology

Bacterial culture was performed following the protocol established by the French NRL (NF U 47–104) for isolation of *Mycobacterium bovis*. After decontamination, the supernatant was seeded on two different media: Löwestein-Jensen and Colestos [11]. All seeded media were incubated at 37°C+/−3°C for 3 months and examined every 2 weeks. The isolated *Mycobacterium tuberculosis* complex colonies were confirmed by DNA amplification as described by Hénault et al. [12] and *M. bovis* was confirmed by Luminex spoligotyping as described by Zhang et al. [13].

Histopathology

When histopathology was requested by the official veterinary services, part of the samples was fixed in formalin by the LDAR 24 and sent at room temperature to one of the two accredited laboratory for histopathology (Regional Analysis Laboratory of Côte d'Armor [LDA22] and Pathologic Anatomy Laboratory of the National Veterinary School in Lyon [ENVL]). These laboratories performed hematoxylin-eosine histopathology and Ziehl Neelsen staining on each sample.

STATISTICAL MODEL

Two analyses were carried out using a Bayesian formulation of the latent class model. The first analysis aimed at assessing sensitivity and specificity of bacteriology and PCR from the tests results of the 5,211 animals enrolled in the study. The second one aimed at assessing sensitivity and specificity of bacteriology, histopathology and PCR from the test results of the 697 animals subjected to the three tests (these latter animals had bTB-like lesions).

First Analysis

Bacteriology and PCR were not considered independent as they both detect MTBC directly. Two tests are considered to be conditionally independent when knowledge of the outcome of the first one gives no information about the outcome of the second one, conditional on the true disease status [14]. Even if PCR detects DNA whereas bacteriology detects viable bacteria, their results are expected to be dependent, relying on an animal's true infection status. As explained by Gardner et al. [15], conditional dependence of test sensitivities occurs when the second (first) test has different sensitivities for infected animals that test-positive and for those that test-negative on the first (second) test. This was considered here to be the case for both tests sensitivities and specificities. Therefore dependence between bacteriology and PCR was accounted for. However, the correlation given infection status between these two tests was unknown. Considering a model for these two tests with three sampled populations, there were therefore nine parameters to be estimated (three prevalences, two sensitivities, two specificities and two covariances between tests). Although this model yields 9 degrees of freedom (three from each population), it lacks identifiability, i.e. the parameters cannot be estimated based on data alone [16]. For the Bayesian, identifiability is not obligatory if good prior information is available. However, in our case, we had no reliable information neither on the prevalence of infection at the animal level in populations A, B or C, nor on the characteristics of bacteriology or PCR under French field conditions. In order to ensure that the model was identifiable, the conditional covariances between bacteriology and PCR were fixed. Thus, there were only seven parameters to estimate. The model used for this first analysis was basically the same as the one used in Paul et al. [8]. We ran eight different models using covariances expressed as a proportion (i.e. 0.0, 0.1, 0.2, 0.3, 0.4, 0.5, 0.6 and 0.7) of the maximum conditional covariances for the test sensitivities and specificities. When this proportion was zero, bacteriology and PCR were assumed independent, but this was no longer the case as this proportion increased. These eight models were compared using the Deviance Information Criterion (DIC).

Second Analysis

As previously, bacteriology and PCR were considered dependent tests given infection status. However, histopathology was assumed conditionally independent of the two other tests as this method analyzes morphological changes in tissues. As in the first analysis, covariances were fixed and expressed as a proportion of the maximum conditional covariances for the test sensitivities and specificities. Eight values of proportion were tested from 0.0 to 0.7 and the models were compared using the DIC. Each model was identifiable, with nine parameters to estimate (three prevalences, three sensitivities and three specificities).

Bayesian Computation

For all the parameters to be estimated, we used uninformative priors in the shape of a uniform distribution on the interval between 0 and 1 (modeled using the beta (1, 1) distribution). The analyses were implemented in Openbugs version 3.2.2 [17]. For each model, the first 10 000 iterations were discarded as burn-in and the next 10 000 were used for posterior inference. Two chains were run from different initial values. To assess convergence of these chains, we visually inspected the time-series plots of selected variables as well as the Gelman-Rubin diagnostic plots.

RESULTS

Respectively 61.7% and 22.7% of 5,211 sampled animals came from the departments of Côte d'Or and Dordogne (see Figure S2). 48.7% of them belonged to from herds without any recent history of bTB (i.e. without any declaration of herd infection between 2000 and 2012). The 5,188 animals for which age, sex and breed data were available were mainly female (81.8%) and on median 3.9 years old. Respectively 20.9% of them were less than 18 months old and 26.6% more than 6 years old. The most represented cattle breeds were Charolais (60.2% of them), Limousin (17.2%) and Prim'Holstein (6.5%).

First Analysis

Among the 5,211 animals tested by bacteriology and PCR, 2,617 (50.2%), 715 (13.7%) and 1,879 (36.1%) came from population A, B and C respectively. The cross-tabulated distribution of the outcome of the two tests (bacteriology and PCR) is shown in Table 1.

Table 1: Number of animals in analysis 1 according to (i) their test results for bacteriology (bac) and PCR and (ii) their population of origin

Population	POS_{bac}/POS_{PCR}	POS_{bac}/NEG_{PCR}	NEG_{bac}/POS_{PCR}	NEG_{bac}/NEG_{PCR}	Total
A	230	29	68	2290	2617
B	29	3	12	671	715
C	52	1	9	1817	1879
Total	311	33	89	4778	5211

POS: positive/NEG: negative.

The test characteristics and infection prevalences in populations A, B and C estimated from this data are presented in Table 2. The model assuming independence between bacteriology and PCR (i.e. proportion of maximal conditional covariance equal to zero) had the smallest DIC, suggesting this model as the most fitted. However, the DIC of the eight models are very close, suggesting an equivalent fit. Bacteriology and PCR had very high estimated specificities that remained almost unchanged whatever the value of the proportion of conditional covariance between the two tests. On the contrary, the sensitivities estimates decreased as this proportion increased but sensitivity of PCR was always higher than that of bacteriology (except for models with proportions of conditional covariance ≥0.5 where posterior credible intervals of the two sensitivities overlapped; however, these models had a lower fit). Figure 1 shows the posterior distributions for sensitivities and specificities of the two tests in the model assuming independence between bacteriology and PCR (model with the smallest DIC).

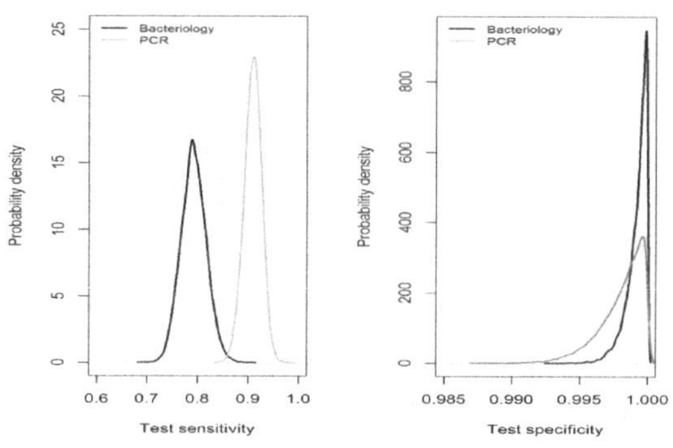

Figure 1: Posterior distributions for sensitivities and specificities of bacteriology and PCR.

These estimations are based on the model assuming independence between bacteriology and PCR (model with the smallest DIC).

Table 2: Mean estimates and 95% posterior credible intervals (CI) of the sensitivity (Se) and specificity (Sp) of bacteriology (bact) and PCR and population specific animal prevalences of the eight models assuming different proportions of maximum conditional covariances (analysis 1)

Conditional covariance[a]	Test characteristics (95% CI)				Prevalences (%)			DIC[b]
	Se_{bact}	Sp_{bact}	Se_{PCR}	Sp_{PCR}	A	B	C	
0	**79.2 (74.4–84.4)**	**99.9 (99.7–100)**	**90.9 (87.3–94.3)**	**99.8 (99.4–100.0)**	**12.5 (11.1–13.9)**	**6.1 (4.3–8.1)**	**3.3 (2.4–4.2)**	**64.94**
0.1	78.3 (73.4–83.4)	99.9 (99.6–100)	89.9 (85.9–93.7)	99.8 (99.3–100.0)	12.6 (11.2–14.1)	6.2 (4.4–8.2)	3.3 (2.5–4.2)	65.05
0.2	77.2 (72.3–82.3)	99.9 (99.6–100)	88.6 (84.1–92.8)	99.8 (99.3–100.0)	12.8 (11.3–14.3)	6.2 (4.4–8.3)	3.3 (2.5–4.3)	65.17
0.3	75.7 (70.8–81.1)	99.9 (99.6–100)	87.0 (81.9–91.9)	99.8 (99.3–100.0)	13.0 (11.5–14.6)	6.3 (4.5–8.5)	3.4 (2.5–4.4)	65.18
0.4	73.7 (68.4–79.1)	99.9 (99.6–100)	84.7 (78.6–90.5)	99.8 (99.3–100.0)	13.4 (11.7–15.1)	6.5 (4.6–8.7)	3.5 (2.5–4.5)	65.17
0.5	71.0 (65.2–77.1)	99.8 (99.4–100)	81.6 (74.4–88.5)	99.7 (99.2–100.0)	13.8 (12.1–15.7)	6.7 (4.6–9.0)	3.6 (2.6–4.7)	65.43
0.6	66.9 (59.8–73.7)	99.8 (99.3–100)	76.9 (67.7–85.4)	99.7 (99.2–100.0)	14.7 (12.5–16.9)	7.1 (4.8–9.6)	3.7 (2.6–5.0)	65.45
0.7[c]	59.8 (49.8–68.9)	99.7 (99.1–100)	68.9 (55.9–81.0)	99.6 (98.8–100.0)	16.3 (13.2–19.9)	7.8 (5.1–11.1)	4.0 (2.5–5.6)	65.66

[a]Proportion of maximum upper limit of conditional covariance.

[b]Deviance information criterion.

[c]For this model, a thin interval of 3 was defined in order to ensure that the MCMC chains are no longer autocorrelated.

Note: the results of the model with the lowest DIC are in bold.

Second Analysis

Among the 697 animals tested in bacteriology, histopathology and PCR, 345 (49.5%), 94 (13.5%) and 258 (37.0%) came from population A, B and

C respectively. The cross-tabulated distribution of the outcome of the three tests is shown in Table 3.

Table 3: Number of animals in analysis 2 according to (i) their test results for bacteriology (bac), histopathology and PCR and (ii) their population of origin

Histopathology	Population	POS_{bac}/ POS_{PCR}	POS_{bac}/ NEG_{PCR}	NEG_{bac}/ POS_{PCR}	NEG_{bac}/ NEG_{PCR}	Total
POS	A	163	14	32	31	240
	B	21	3	10	13	47
	C	32	0	2	37	71
NEG	A	8	1	9	87	105
	B	0	0	1	46	47
	C	6	0	3	178	187
Total		230	18	57	392	697

POS: positive/NEG: negative.

The test characteristics and infection prevalences in populations A, B and C estimated from this data are presented in Table 4. The model assuming a moderate dependence between bacteriology and PCR (i.e. proportion of maximal conditional covariance equal to 0.4) had the smallest DIC, suggesting this model should be preferred. However, models with a proportion of conditional covariance from 0.3 to 0.5 had very close DIC, suggesting an equivalent fit. Specificity estimates for bacteriology and PCR decreased as the conditional covariance increased, whereas that of histopathology increased. According to the three preferred models, bacteriology and PCR had high estimated specificities. Even if their posterior 95% CI overlapped, the specificity of the bacteriology might be slightly higher than that of PCR (see Figure 2 for the model with the smallest DIC). The estimated specificity of the histopathology was much lower than those of the two other tests. Sensitivity estimates for bacteriology and PCR decreased as the conditional covariance increased, whereas the sensitivity of the histopathology remained almost unchanged. Nevertheless, sensitivity of bacteriology was always lower than that of PCR and of histopathology. Regarding the two latter tests, even if the posterior 95% CI for their sensitivities slightly overlapped, it seemed likely that sensitivity of the histopathology was higher than that of PCR (see Figure 2 for the model with the smallest DIC).

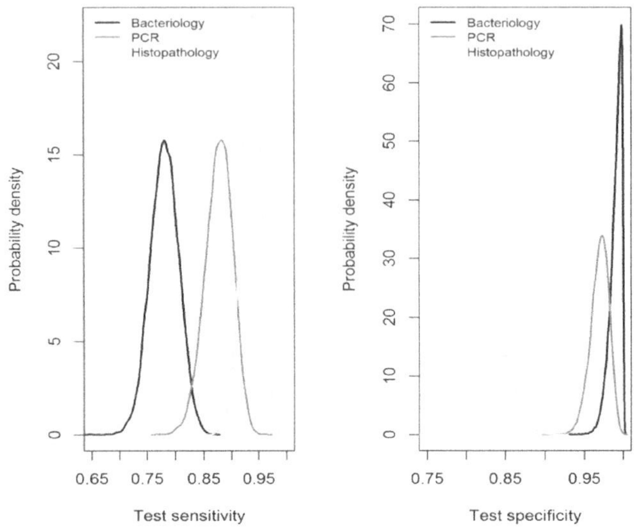

Figure 2: Posterior distributions for sensitivities and specificities of bacteriology, histopathology and PCR.

These estimations are based on the model with a proportion of maximal conditional covariance between bacteriology and PCR equal to 0.4 (model with the smallest DIC).

Table 4: Mean estimates and 95% posterior credible intervals (CI) of the sensitivity (Se) and specificity (Sp) of bacteriology, histopathology and PCR and population specific animal prevalences of the eight models assuming different proportions of maximum conditional covariances (analysis 2)

Conditional covariance[a]	Test characteristics (95% CI)						Prevalences[b] (%)			DIC[c]
	Se_{bact}	Sp_{bact}	Se_{hist}	Sp_{hist}	Se_{PCR}	Sp_{PCR}	A	B	C	
0	82.2 (77.1–86.9)	99.6 (98.6–100.0)	92.9 (89.4–95.9)	80.2 (75.9–84.1)	92.3 (88.6–95.4)	97.5 (95.0–99.4)	65.3 (59.8–70.6)	36.7 (26.8–46.9)	16.0 (11.7–20.9)	116.9
0.1	81.4 (76.3–86.1)	99.6 (98.5–100.0)	93.0 (89.3–96.0)	80.8 (76.5–84.8)	91.3 (87.2–94.8)	97.4 (95.0–99.3)	66.0 (60.3–71.4)	37.1 (27.3–47.5)	16.0 (11.7–21.0)	116.2
0.2	80.5 (75.4–85.2)	99.4 (98.1–100.0)	93.2 (89.6–96.2)	81.5 (77.2–85.6)	90.2 (85.6–94.0)	97.3 (94.9–99.3)	66.8 (61.1–72.2)	37.6 (27.6–48.0)	16.0 (11.6–21.0)	115.4

0.3	79.4 (74.2–84.2)	99.3 (97.6–100.0)	93.4 (89.7–96.5)	82.4 (77.9–86.6)	88.9 (84.0–93.2)	97.2 (94.6–99.2)	67.5 (61.8–73.1)	38.3 (28.1–48.9)	16.1 (11.6–21.2)	114.8
0.4	**78.1 (72.9–82.8)**	**99.1 (97.1–100.0)**	**93.6 (89.9–96.9)**	**83.3 (78.7–87.6)**	**87.7 (82.5–92.3)**	**97.0 (94.3–99.0)**	**68.3 (62.5–73.9)**	**38.9 (28.5–49.8)**	**16.2 (11.5–21.4)**	**114.2**
0.5	76.6 (71.2–81.6)	98.8 (96.6–100.0)	93.9 (90.0–97.4)	84.1 (79.5–88.7)	86.5 (81.1–91.3)	96.7 (93.9–98.9)	69.0 (63.0–74.7)	39.7 (29.2–50.6)	16.3 (11.4–21.7)	114.5
0.6	75.0 (69.4–80.1)	98.5 (96.1–100.0)	94.2 (90.2–97.7)	85.1 (80.3–89.6)	85.3 (79.8–90.4)	96.3 (93.3–98.7)	69.6 (63.7–75.3)	40.6 (29.8–51.8)	16.4 (11.3–22.2)	115.8
0.7	73.5 (67.8–78.8)	98.2 (95.5–99.9)	94.6 (90.5–98.2)	85.9 (80.9–90.4)	84.6 (79.0–89.7)	95.9 (92.6–98.5)	69.9 (64.0–75.6)	41.4 (30.6–52.7)	16.5 (11.4–22.4)	119.3

[a]Proportion of maximum upper limit of conditional covariance.

[b]Analysis 2 was carried out on a subgroup of animals (animals subjected to histopathology which all presented macroscopic bTB-like lesions), hence the differences in animal prevalence estimates between analysis 1 and analysis 2.

[c]Deviance information criterion.

Note: the results of the model with the lowest DIC are in bold.

DISCUSSION

We estimated the sensitivity and specificity of bacteriology, histopathology and PCR for bTB diagnosis in the absence of a gold standard. This is, to our knowledge, the first report of a latent class analysis used for evaluation of bTB diagnostic tests. In this paper, we followed the STRADAS guidelines (Standards for Reporting of Animal Diagnostic Accuracy Studies), that were initially developed to improve the quality of reporting of the design, conduct and results of paratuberculosis test accuracy studies [18]. The analysis showed that PCR is more sensitive than bacteriology and has a good specificity (although possibly lower than that of bacteriology). This makes PCR a useful tool that could potentially become an official bTB diagnostic test within the EU. Histopathology seems to be more sensitive than bacteriology but also less specific, which means that this test cannot be used alone as a confirmation test.

Latent class analysis relies on a number of critical assumptions. First, at least two populations with differing prevalences are required. Here, the population was stratified into three subpopulations (populations A, B and C) based on the incidence of infection within each canton over the 2000–2007 period. Posterior distributions of estimated prevalences in populations A, B and C did not overlap, which suggests that the stratification of the population was successful. According to Toft et al. [19], it is preferable to have a difference in prevalence as large as possible. In our first analysis, prevalences within each population were close (11.1–13.9% in population A, 4.3–8.1% in population B and 2.4–4.2% in population C), which could affect the precision of the estimates as well as the estimates themselves [19]. However, in the second analysis, these potential biases were a priori absent as there was a very large difference in the animal prevalence estimates within population A (62.5–73.9%), B (28.5–49.8%) and C (11.5–21.4%). The second assumption is that of constant test sensitivity and specificity between populations. Johnson et al. [20] stated that a diagnostic test might have increased sensitivity when applied in high prevalence populations due to the greater burden of clinically affected animals and higher quantities of bacteria shed. There might also be problems with constant specificities across populations, e.g. because of geographic differences in cross-reactions to other mycobacteria [2]. Toft et al. [19] demonstrated that a difference in test sensitivity between populations may result in estimates that are biased toward the sensitivity of the test in the population with the highest infection prevalence. Similarly, Johnson et al. [20] showed that, when neither evaluated test is perfectly specific, the results of the Bayesian analysis are all wrong if the assumption of constant sensitivity has failed. However, in our case, bacteriology has an almost perfect specificity, which would allow us to interpret test accuracy estimates for sensitivity as estimates of the average sensitivity across populations if the assumption of constant sensitivity failed [6], [20]. Finally, bacteriology and PCR were considered as conditionally dependent tests. In the first analysis, the eight models had an equivalent DIC and the estimates were very close for models with a proportion of maximal conditional covariance from 0 to 0.3. As in Paul et al. [8], this could mean that the two tests might be conditionally dependent but this dependence does not affect sensitivity and specificity when evaluated against each other. In the second analysis, models with a proportion of maximal conditional covariance from 0.3 to 0.5 had smaller DIC, with a difference of at least 2 from the model assuming bacteriology and PCR as independent (i.e. model with a proportion of maximal conditional covariance equal to 0). Smaller DIC values indicate that these models fit better. It was therefore necessary to take into account the correlation between the two tests.

For all the parameters to be estimated, we used uninformative priors. However, bacteriology is considered to be 100% specific and either an informative prior distribution or a fixed value for this parameter could have been used in the models. We ran the model with the smallest DIC in analysis 1 and the model with the smallest DIC in analysis 2 using either an informative prior distribution (a beta (34.1664, 1.335) distribution whose mode is 99% and 5% percentile 90%) or a fixed value (100%) for the specificity of bacteriology. The estimated test characteristics and infection prevalences were influenced by none of these options (results not shown).

In this study, PCR sensitivity was higher than previously reported. Estimates of sensitivity for this test were broadly the same in the first (mean: 90.9% [87.3–94.3%]) and second analysis (mean: 87.7% [82.5–92.3%]), suggesting that PCR sensitivity was not influenced by the presence or absence of bTB-like lesions on tested animals. On the contrary, Parra et al. [21] reported a sensitivity of PCR from 61.1% for samples with non-visible lesions to 80.6% for chronic lesions when using a manual extraction system and considering bacteriology and/or presence of bTB-like lesions and positive bacterioscopy as the gold standard. In Taylor et al. [22], comparison of RD4 PCR and IS*1081* PCR with the gold standard of bacteriology showed a sensitivity of approximately 50% and 70% respectively. Thacker et al. [23] reported that out of 30 *M. bovis* culture positive tissues, only 20 were positive using IS*6110* real-time PCR and in Proaño-Perez et al. [24], eight animals were found PCR positive over the twelve positive in bacteriology. Only Cardoso et al. [25] reported a frequency of PCR-positive results similar to that of culture-positive results (51.5% versus 54.5%) over 35 lymph nodes samples from animals with macroscopic lesions consistent with *M. bovis* infection. In our study, the sensitivity of PCR was higher than that of bacteriology. It may thus be suspected that bacteriology under French conditions lacks sensitivity. The decontamination procedure of this bacteriology uses sulphuric acid (H_2SO_4), which is very efficient to kill contaminating microorganisms but also toxic for mycobacteria. However, the LDAR 24 and the French National Reference Laboratory produced very satisfactory results in the ring trial organized by the European Union Reference Laboratory for bTB (EURL personal communication). Thus, the assumption of poor sensitivity of French bacteriology is not justified, which is confirmed by our estimates (mean of 79.2% [74.4–84.4%] in the first analysis and of 78.1% [72.9–82.8%] in the second one). These estimates are consistent with

Corner et al. [26] who reported a proportion of infected samples identified by culture between 58.0% and 80.0% depending on the culture media used and the decontamination procedure employed. It is interesting to point out that not all the PCR methods employed in the literature are equivalent (e.g. detected genome sequences are sometimes different as well DNA extraction methods). The good sensitivity reported here could then be attributable to the inherent good quality of our methodology. Besides, unlike previous studies on PCR sensitivity [21]–[25], by applying a latent class analysis, we were able to estimate the characteristics of the three tests without a gold standard. According to Toft et al. [9], when using a single, imperfect test as a reference, estimates of sensitivity and specificity will always be biased and as a result, the new test (here PCR) will always be reported to have properties which are equal to or not as good properties as those of the reference test (here bacteriology). The method we used removed these potential biases. This could also explain the discrepancy between our results and those previously reported. Bacteriology remains, however, necessary for molecular epidemiology studies. Indeed, *M. bovis* full genotyping is only possible on cloned bacterial isolates. Regarding histopathology, our estimates (mean Se of 93.6% [89.9–96.9%] and mean Sp of 83.3% [78.7–87.6%]) are consistent with those obtained by Varello et al. [5]: considering culture as a reference test on suspected lymph nodes samples from 173 cattle positive reactive in ante mortem tests, they reported a relative sensitivity of histopathology of 93.4% (95% CI: 87.4–97.1%) and a relative specificity of 92.3% (95% CI: 81.5–97.9%).

CONCLUSION

This latent class analysis has provided estimates of the characteristics of two currently used tests (bacteriology and histopathology) and a novel one, a PCR method using the LSI VetMAX™ *Mycobacterium tuberculosis* Complex PCR Kit, for bTB diagnosis under field conditions in France. This study showed that PCR had a higher sensitivity than bacteriology while having a good specificity. In addition, PCR allows the confirmation of infection in less than 48 hours. Hence, PCR seems to be a good confirmatory test for bTB surveillance and control programs.

ACKNOWLEDGMENTS

The authors wish to acknowledge the Regional Analysis Laboratory of Côte d'Armor and the Pathologic Anatomy Laboratory of the National Veterinary School in Lyon for performing all the histopathological analyses as well as the French Ministry of Agriculture for authorizing data analysis. They are grateful to Victoria Boschiroli for her helpful comments.

REFERENCES

1. Schiller I, RayWaters W, Vordermeier HM, Jemmi T, Welsh M, et al. (2011) Bovine tuberculosis in Europe from the perspective of an officially tuberculosis free country: trade, surveillance and diagnostics. Vet Microbiol 151: 153–159.
2. Clegg TA, Duignan A, Whelan C, Gormley E, Good M, et al. (2011) Using latent class analysis to estimate the test characteristics of the gamma-interferon test, the single intradermal comparative tuberculin test and a multiplex immunoassay under Irish conditions. Vet Microbiol 151: 68–76.
3. de la Rua-Domenech R, Goodchild AT, Vordermeier HM, Hewinson RG, Christiansen KH, et al. (2006) Ante mortem diagnosis of tuberculosis in cattle: a review of the tuberculin tests, gamma-interferon assay and other ancillary diagnostic techniques. Res Vet Sci 81: 190–210.
4. Karolemeas K, de la Rua-Domenech R, Cooper R, Goodchild AV, Clifton-Hadley RS, et al. (2012) Estimation of the relative sensitivity of the comparative tuberculin skin test in tuberculous cattle herds subjected to depopulation. PLoS One 7: e43217.
5. Varello K, Pezzolato M, Mascarino D, Ingravalle F, Caramelli M, et al. (2008) Comparison of histologic techniques for the diagnosis of bovine tuberculosis in the framework of eradication programs. J Vet Diagn Invest 20: 164–169.
6. Branscum AJ, Gardner IA, Johnson WO (2005) Estimation of diagnostic-test sensitivity and specificity through Bayesian modeling. Prev Vet Med 68: 145–163.
7. Hui SL, Walter SD (1980) Estimating the error rates of diagnostic tests. Biometrics 36: 167–171.
8. Paul S, Toft N, Agerholm JS, Christoffersen AB, Agger JF (2013) Bayesian estimation of sensitivity and specificity of *Coxiella burnetii* antibody ELISA tests in bovine blood and milk. Prev Vet Med 109: 258–263.
9. Toft N, Akerstedt J, Tharaldsen J, Hopp P (2007) Evaluation of three serological tests for diagnosis of Maedi-Visna virus infection using latent class analysis. Vet Microbiol 120: 77–86.
10. Thierry D, Brisson-Noel A, Vincent-Levy-Frebault V, Nguyen S, Guesdon JL, et al. (1990) Characterization of a *Mycobacterium*

tuberculosis insertion sequence, IS6110, and its application in diagnosis. J Clin Microbiol 28: 2668–2673.

11. OIE (2012) Bovine tuberculosis Chap 2.4.7. In: OIE, editor. Manual of diagnostic tests and vaccines for terrestrial animals (mammals, birds and bees). 7 ed. Paris. 674–689.
12. Henault S, Karoui C, Boschiroli ML (2006) A PCR-based method for tuberculosis detection in wildlife. Dev Biol (Basel) 126: 123–132 discussion 325–126.
13. Zhang J, Abadia E, Refregier G, Tafaj S, Boschiroli ML, et al. (2010) Mycobacterium tuberculosis complex CRISPR genotyping: improving efficiency, throughput and discriminative power of 'spoligotyping' with new spacers and a microbead-based hybridization assay. J Med Microbiol 59: 285–294.
14. Enoe C, Georgiadis MP, Johnson WO (2000) Estimation of sensitivity and specificity of diagnostic tests and disease prevalence when the true disease state is unknown. Prev Vet Med 45: 61–81.
15. Gardner IA, Stryhn H, Lind P, Collins MT (2000) Conditional dependence between tests affects the diagnosis and surveillance of animal diseases. Prev Vet Med 45: 107–122.
16. Jones G, Johnson WO, Hanson TE, Christensen R (2010) Identifiability of models for multiple diagnostic testing in the absence of a gold standard. Biometrics 66: 855–863.
17. Lunn D, Spiegelhalter D, Thomas A, Best N (2009) The BUGS project: Evolution, critique, and future directions. Statistics in Medicine 28: 3049–3067.
18. Gardner IA, Nielsen SS, Whittington RJ, Collins MT, Bakker D, et al. (2011) Consensus-based reporting standards for diagnostic test accuracy studies for paratuberculosis in ruminants. Prev Vet Med 101: 18–34.
19. Toft N, Jorgensen E, Hojsgaard S (2005) Diagnosing diagnostic tests: evaluating the assumptions underlying the estimation of sensitivity and specificity in the absence of a gold standard. Prev Vet Med 68: 19–33.
20. Johnson WO, Gardner IA, Metoyer CN, Branscum AJ (2009) On the interpretation of test sensitivity in the two-test two-population problem: assumptions matter. Prev Vet Med 91: 116–121.
21. Parra A, Garcia N, Garcia A, Lacombe A, Moreno F, et al. (2008) Development of a molecular diagnostic test applied to

experimental abattoir surveillance on bovine tuberculosis. Vet Microbiol 127: 315–324.

22. Taylor GM, Worth DR, Palmer S, Jahans K, Hewinson RG (2007) Rapid detection of *Mycobacterium bovis* DNA in cattle lymph nodes with visible lesions using PCR. BMC Vet Res 3: 12.
23. Thacker TC, Harris B, Palmer MV, Waters WR (2011) Improved specificity for detection of *Mycobacterium bovis* in fresh tissues using IS6110 real-time PCR. BMC Vet Res 7: 50.
24. Proaño-Perez F, Benitez-Ortiz W, Desmecht D, Coral M, Ortiz J, et al. (2011) Post-mortem examination and laboratory-based analysis for the diagnosis of bovine tuberculosis among dairy cattle in Ecuador. Prev Vet Med 101: 65–72.
25. Cardoso MA, Cardoso RF, Hirata RD, Hirata MH, Leite CQ, et al. (2009) Direct detection of *Mycobacterium bovis* in bovine lymph nodes by PCR. Zoonoses Public Health 56: 465–470.
26. Corner LA, Gormley E, Pfeiffer DU (2012) Primary isolation of *Mycobacterium bovis* from bovine tissues: conditions for maximising the number of positive cultures. Vet Microbiol 156: 162–171.

CHAPTER 7

The Bovine Paranasal Sinuses: Bacterial Flora, Epithelial Expression of Nitric Oxide and Potential Role in the in-herd Persistence of Respiratory Disease Pathogens

Gerard M. Murray[1], Ro´nan G. O'Neill[2], Alison M. Lee[3] , Ma´ire C. McElroy[2] , Simon J. More[4] , Aisling Monagle[3] , Bernadette Earley[5] , Joseph P. Cassidy[3]

[1] Sligo Regional Veterinary Laboratory, Department of Agriculture, Food and Marine, Doonally, Sligo, Ireland
[2] Central Veterinary Research Laboratory, Department of Agriculture, Food and Marine, Backweston, Celbridge, County Kildare, Ireland
[3] School of Veterinary Medicine, University College Dublin, Belfield, Dublin, Ireland
[4] Centre for Veterinary Epidemiology and Risk Analysis, UCD School of Veterinary Medicine, University College Dublin, Belfield, Dublin, Ireland
[5] Animal and Bioscience Research Department, Animal & Grassland Research and Innovation Centre, Teagasc, Grange, Dunsany, County Meath, Ireland

Citation: Murray GM, O'Neill RG, Lee AM, McElroy MC, More SJ, Monagle A, et al. (2017) The bovine paranasal sinuses: Bacterial flora, epithelial expression of nitric oxide and potential role in the in-herd persistence of respiratory disease pathogens. PLoS ONE 12(3): e0173845. https://doi.org/10.1371/journal.pone.0173845

ABSTRACT

The bovine paranasal sinuses are a group of complex cavernous air-filled spaces, lined by respiratory epithelium, the exact function of which is unclear. While lesions affecting these sinuses are occasionally reported in cattle, their microbial flora has not been defined. Furthermore, given that the various bacterial and viral pathogens causing bovine respiratory disease (BRD) persist within herds, we speculated that the paranasal sinuses may serve as a refuge for such infectious agents. The paranasal sinuses of clinically normal cattle (n = 99) and of cattle submitted for post-mortem examination (PME: n = 34) were examined by microbial culture, PCR and serology to include bacterial and viral pathogens typically associated with BRD: *Mycoplasma bovis*, *Histophilus somni*, *Mannheimia haemolytica* and *Pasteurella multocida*, bovine respiratory syncytial virus (BRSV) and bovine parainfluenza-3 virus (BPIV-3). Overall, the paranasal sinuses were either predominantly sterile or did not contain detectable microbes (83.5%: 94.9% of clinically normal and 50.0% of cattle submitted for PME). Bacteria, including BRD causing pathogens, were identified in relatively small numbers of cattle (<10%). While serology indicated widespread exposure of both clinically normal and cattle submitted for PME to BPIV-3 and BRSV (seroprevalences of 91.6% and 84.7%, respectively), PCR identified BPIV-3 in only one animal. To further explore these findings we investigated the potential role of the antimicrobial molecule nitric oxide (NO) within paranasal sinus epithelium using immunohistochemistry. Expression of the enzyme responsible for NO synthesis, inducible nitric oxide synthase (iNOS), was detected to varying degrees in 76.5% of a sub-sample of animals suggesting production of this compound plays a similar protective role in the bovine sinus as it does in humans.

INTRODUCTION

The bovine paranasal sinuses are a group of complex cavernous air-filled spaces, lined by respiratory epithelium, which develop by evagination of the nasal mucosa into spongy bone between the external and internal laminae of the cranial and facial bones of the bovine skull [1]. The bovine skull possesses several paranasal sinuses—frontal (rostral and caudal), maxillary, lacrimal, palatine, sphenoid and nasal conchal sinuses—some of which communicate with each other (e.g. the maxillary sinus with the palatine and lacrimal sinuses) while others open independently into the nasal meatus. In cattle the paranasal sinuses continue to develop, changing in shape and size, up to seven years of age [2]. There is uncertainty regarding the function of the paranasal sinuses in terrestrial vertebrates, with a role in increasing the thermal and mechanical protection of the brain, without concurrently increasing the weight of the skull, being proposed. However, the most common current hypothesis is that sinuses are functionless structures resulting from the removal of mechanically unnecessary bone—a process sometimes referred to as 'opportunistic pneumatisation' [3].

In cattle, conditions of the paranasal sinuses such as neoplasia [4], sinus cysts [5], or sinusitis [6, 7] are reported relatively infrequently. Many cases of frontal sinusitis occur secondary to dehorning [6], but infection of the paranasal sinuses extending from the nasal mucosa or haematogenous spread from a generalised infection is also possible [8]. Reports investigating the microbial flora present in the bovine paranasal sinuses are lacking.

In human medicine, rhinosinusitis is highly prevalent [9] leading to wide ranging research into its aetiology, pathogenesis and treatment. Since Lundberg [10] showed that nitric oxide (NO) was produced in large quantities in human paranasal sinuses, there has been much speculation that this bioactive signalling molecule plays an important role in the pathogenesis of human rhinosinusitis. NO is considered an effective biocide against a wide spectrum of bacteria, viruses and fungi [11], attributes which have led to a proposed role for NO in the maintenance of sterility in human paranasal sinuses [10]. Studies of the function of this compound in bovine paranasal sinuses have not been carried out.

Bovine respiratory disease (BRD) is a term that encompasses pneumonias in young cattle caused by an array of infectious agents including bacteria and viruses such as BRSV and BPIV-3. Given that many of these pathogens persist within cattle herds as residents of the bovine respiratory tract [12, 13], we speculated that the paranasal sinuses could serve as a refuge facilitating this persistence from year to year.

The aim of this cross-sectional study was thus to describe for the first time, the microbial flora of the paranasal sinuses of cattle and to assess, in particular, if bacterial and viral pathogens typically associated with BRD: *Mycoplasma bovis*, *Histophilus somni*, *Mannheimia haemolytica* and *Pasteurella multocida*, bovine respiratory syncytial virus (BRSV) and bovine parainfluenza-3 virus (BPIV-3), were harbored at this anatomical location. We also used immunohistochemistry (IHC) to examine

NO expression within paranasal sinus epithelium and how this expression correlated with the microbial flora present.

MATERIALS AND METHODS

Study Population

Between the months of August 2014 and January 2015, 99 heads of clinically normal cattle were retrieved from a Department of Agriculture, Food and Marine—regulated slaughterhouse engaged in halal slaughter for human consumption. All cattle were aged between 18 and 24 months and were deemed healthy and fit for human consumption by veterinary ante-mortem and post-mortem examinations. Heads were skinned during the slaughter process and the mandible was removed. All heads were individually identified and transported in an upright manner in separate sealed containers to Sligo Regional Veterinary Laboratory (SRVL) for processing immediately after slaughter. Between January 2015 and August 2016, the heads of 34 animals ranging in age between 1 and 163 months of age, which died on-farm from various causes and were submitted for post-mortem examination (PME) to SRVL, were similarly processed for examination. None of the study animals were euthanized specifically for the purposes of this research.

Sampling of the Paranasal Sinuses

A standard protocol for sampling of the paranasal sinuses was deployed on all 133 heads. Briefly, remaining connective tissue was removed over four sites on each side of the skull, as shown in Fig 1. The selected sites, representing access to the caudal frontal, medial rostral frontal, lateral rostral frontal and maxillary sinuses were disinfected (Anistel, Tristel Solutions Ltd., Cambridgeshire, UK), dried with absorbent paper and sprayed with ethanol. A disinfected 16 millimetre drill bore-piece on an electric drill was used to remove a small circular piece of the skull bone over the sinus of interest. Two sterile cotton swabs (one for routine bacteriology and another for PCR analysis) were inserted in succession through the circular opening at an angle to ensure that the edges of the opening did not come into contact with the swab. The mucosa of the sinus was swabbed in a circular motion for three seconds at each sinus sampling site (n = 8) prior to removal of each swab and transportation to the laboratory for immediate processing. A swab of incised lung, and of tracheal mucosa, taken from a transverse incision at the tracheal bifurcation, for bacterial (n = 57) and viral (n = 133) PCR analysis and heart blood (n = 131) for viral serology were also harvested.

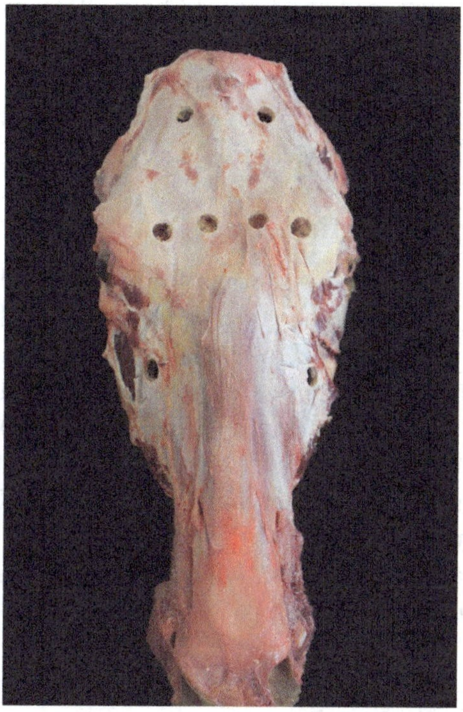

Figure 1: The eight sampling sites on the bovine head used to sample the caudal frontal, medial rostral frontal, lateral rostral frontal and maxillary sinuses on both sides of the bovine skull.

Paranasal sinuses were subsequently exposed in a subset of healthy (n = 5) cattle and those submitted for PME (n = 12) using a circular saw to reflect the dorsal bone of the caudal frontal, medial rostral frontal and lateral rostral frontal sinuses. Sinus mucosa was harvested, fixed in 10% buffered formalin for four days, embedded in paraffin wax and cut with a microtome. General histopathology and immunohistochemistry (IHC) to detect the expression of inducible nitric oxide synthase (iNOS) were also carried out on these sections.

Bacteriology

The time between specimen collection and inoculation was approximately one hour. Bacteriology swabs from each of the eight paranasal sinuses from each animal were cultured on blood agar at 37°C for 48 hours, chocolate agar at 37°C in 5% carbon dioxide for 48 hours, on blood agar at 37°C under anaerobic conditions for 48 hours and on McConkey agar at 37°C for 48 hours. Plates were examined at 24 and 48 hours. Colonies present after 48 hours were sub-cultured onto Colombia blood agar and chocolate agar for a further 48 hours. All isolates were identified by traditional identification methods using a combination of gram staining, catalase, oxidase, urease, indole and API®strip tests (bioMérieux, Marcy l'Etoile, France).

Polymerase Chain Reaction (PCR) Analysis

PCR analysis was performed on swabs from each of the eight paranasal sinuses as well as from the trachea and lungs for the detection of *M. bovis* (n = 67 animals), *H. somni* (n = 67), *M. haemolytica* (n = 56) and *P. multocida* (n = 56) as previously described [14,15,16,17]. Reverse transcriptase PCR analyses to identify BRSV (n = 133) and BPIV-3 (n = 133) were also performed individually on samples taken from the paranasal sinuses, trachea and lungs as previously described [18]. DNA or RNA extracted from laboratory strains of the target organisms using commercial DNA or RNA isolation kits (for *M. bovis*, *H.somni* and respiratory viruses (QIAamp DNA mini kit and QIAamp RNA mini kit, Qiagen Ltd, Manchester, United Kingdom)) or specific commercially manufactured oligonucleotides (for *P. multocida* and *M. haemolytica* (Metabion International AG, Munich, Germany)) were used as positive controls; reaction mixtures without the template DNA or RNA were used as negative controls in all amplifications for each PCR. The β-actin results reflected good sample quality for virology PCR in all samples from 125 of the 133 animals sampled. Six animals recorded unsatisfactory sample quality from a single sinus while two animals recorded unsatisfactory sample quality from two individual sinuses.

Serology

Serum samples (n = 131) were analysed for the presence of immunoglobulin G to BRSV [19] and BPIV-3 [20] by commercially available indirect enzyme-linked immunosorbent assays (ELISA; SVANOVA Biotech, Uppsala, Sweden). The optical density (OD) was measured at 450 nm and was corrected by subtraction of the mean OD value for the control antigen. Percent positivity (PP) was calculated as follows: corrected OD of the sample/corrected OD of the positive control × 100. Serum samples were considered positive if their PP value was greater than, or equal to, 10.

Immunohistochemistry (IHC) for the Detection of Inducible Nitric Oxide Synthase (iNOS)

Histopathological examination and IHC for iNOS detection was performed on 5µm-thick sections of formalin-fixed, paraffin-embedded bovine sinus mucosa from the caudal frontal sinus mounted on glass slides (Superfrost Plus, Fischer Scientific, Dublin, Ireland). Lung tissue from *Mycobacterium bovis*-infected mice, known to express iNOS, was used as a positive control. Slides were dewaxed in two changes of xylene for 10 minutes each, then dehydrated through two changes of alcohol for seven minutes each followed by three washes for five minutes each in phosphate buffered saline (PBS, BP399-4, Fisher Scientific, Dublin, Ireland). Antigen retrieval was carried out by microwaving sections at 700W for 10 minutes (Sanyo Microwave) while immersed in 10mM tri-sodium citrate buffer (S/3320/53, Fischer Scientific, Dublin, Ireland) at pH 6. Slides were then washed in PBS three times for 5 minutes each. Non-specific antigens were blocked by incubating slides in 10% bovine serum albumin (A3294, Sigma, Missouri, USA) for 10 minutes at room temperature. Then slides were incubated with rabbit polyclonal anti-mouse iNOS primary antibody (ab15323, Abcam, Cambridge, UK) at a 1:50 dilution. Rabbit Immunoglobulin Fraction (X 0903, DakoCytomation, Glostrup, Denmark) or PBS alone were used as negative controls. Incubation was carried out at 37°C for one hour, then at room temperature overnight (12 hours approximately). Slides were washed in PBS and incubated with biotinylated anti-rabbit IgG (Vectastain ABC kit, AK-5001, Vector Laboratories Inc., Burlinghame, CA, USA) at room temperature for 30 minutes, followed by washing in PBS and a 30-minute incubation with alkaline phosphatase reagent for 30 minutes at 37°C (Vectastain ABC kit). Slides were washed once more in PBS followed by incubation with Vector Red alkaline phosphatase substrate (Vector Red AP substrate kit, SK-5100, Vector Laboratories Inc.) for 30 minutes at room temperature and then counterstained with haematoxylin. The process was repeated using Rabbit Immunoglobulin Fraction (X 0903, DakoCytomation, Glostrup, Denmark) in place of the primary antibody as a negative control.

RESULTS

Study Population

The study population of both clinically normal animals (n = 99) and cattle submitted for PME (n = 34) comprised of 88 male and 45 female animals.

The average age of all animals sampled was 34.4 months (age range 1–200 months). None of the animals sampled had gross evidence of lesions in their sinuses. Of the animals submitted for PME, pneumonia was the most commonly recorded diagnosis (n = 9). *H. somni* (n = 5) and *M. haemolytica* (n = 1), were the respiratory pathogens identified with greatest frequency in the sinuses of these 9 animals. *E. coli* (n = 2) and *Clostridium spp.* (n = 1) were also identified while 1 animal had sinuses which were sterile or did not contain detectable microbes. Other diagnoses recorded in the study population included peritonitis (n = 3), enteritis (n = 2), black disease (n = 2) and pericarditis (n = 2), among others (S1 Table).

Routine Bacteriology

Routine bacteriology results are presented in Table 1. Of 133 animals sampled, 83.5% had sinuses which were sterile or did not contain detectable microbes (94.9% of clinically normal and 50.0% of animals submitted for PME). Recognised BRD pathogens were isolated at low frequency: *M. haemolytica* (n = 2; both isolated from a single maxillary sinus), *Trueperella pyogenes* (1; maxillary sinus), *P. multocida* (1; caudal frontal sinus) and *Bibersteinia trehalosi* (1; maxillary sinus). With the exception of one *M. haemolytica* isolate, all were isolated from cattle that had been submitted for PME. Anaerobic bacteria were isolated from the paranasal sinuses of two animals—from three (both rostral lateral frotal and one rostral medial frontal sinus) and two sinuses (caudal frontal sinus and rostral lateral frontal sinus), respectively.

Table 1: The number and relative frequency of detection of bacterial and viral pathogens in the paranasal sinuses (caudal frontal, rostral medial frontal, rostral lateral frontal and maxillary sinuses) of clinically normal cattle (n = 99) and of cattle submitted for Post-Mortem Examination (PME) (n = 34)

Polymerase chain reaction (PCR) analyses were performed for BRSV (n = 133), BPIV-3 (n = 133), *M.haemolytica* (n = 56), *P. multocida* (n = 56), *H. somni* (n = 67) and *M. bovis* (n = 67).

	Number of animals based on:			Percentage with detection of bacterial or viral pathogens of:	
Result	Bacteriological culture of sinuses	PCR result from sinuses	Bacteriological culture or PCR result from sinuses	All sinuses sampled	All cattle sampled

	Clinically normal	Submitted for PME	Clini-cally normal	Submitted for PME	Clini-cally nor-mal	Submit-ted for PME	Clini-cally nor-mal	Sub-mitted for PME	Clini-cally normal	Sub-mitted for PME	All
Sterile or not detected	94	17	94	21	94	17	98.4%	79.0%	94.9%	50%	83.5%
Histophilus somni	0	0	2	10	2	10	0.3%	10.3%	2.0%	29.4%	9.0%
E. coli	1	6	N/A	N/A	1	6	0.1%	3.3%	1.0%	17.6%	5.3%
Pasteurella multocida	0	1	2	3	2	4	0.4%	1.5%	2.0%	11.8%	4.5%
Mannheimia haemolytica	1	1	2	1	3	2	0.4%	0.7%	3.0%	5.9%	3.8%
Mycoplasma bovis	N/A	N/A	0	4	0	4	0%	2.6%	0%	11.8%	3.0%
Aerococcus viridians	0	2	N/A	N/A	0	2	0%	1.5%	0%	5.9%	1.5%
Nocardia spp.	1	1	N/A	N/A	1	1	0.1%	0.7%	1.0%	2.9%	1.5%
Clostridium chauvoei	0	1	N/A	N/A	0	1	0%	1.1%	0%	2.9%	0.8%
Other *Clos-tridium spp.*	0	1	N/A	N/A	0	1	0%	0.7%	0%	2.9%	0.8%
Trueperella pyogenes	0	1	N/A	N/A	0	1	0%	0.4%	0%	2.9%	0.8%
Bibersteinia trehalosi	0	1	N/A	N/A	0	1	0%	0.4%	0%	2.9%	0.8%
Streptococ-cus bovis 2	0	1	N/A	N/A	0	1	0%	0.4%	0%	2.9%	0.8%
Niesseria spp.	1	0	N/A	N/A	1	0	0.1%	0%	1.0%	0%	0.8%
Proteus spp.	1	0	N/A	N/A	1	0	0.1%	0%	1.0%	0%	0.8%
Pseudomo-nas spp.	0	1	N/A	N/A	0	1	0%	1.5%	0%	2.9%	0.8%
BPIV-3	N/A	N/A	0	1	0	1	0%	0.4%	0%	2.9%	0.7%
BRSV	N/A	N/A	0	0	0	0	0%	0%	0%	0%	0.0%

Bacterial PCR

Bacterial PCR results are also presented in Table 1. *H. somni* (n = 12), *P. multocida* (6), *M. haemolytica* (5) and *M. bovis* (4) were detected at higher frequency than by bacteriological culture. *H. somni* was the most frequently detected pathogen (9.0%) in the paranasal sinuses of all the animals examined, however, *M. haemolytica* was marginally more frequently detected within the paranasal sinuses of healthy animals (3.0% versus 2.0%).

Of 111 animals recorded with sinuses which were sterile or contained undetectable microbes on bacteriological culture, 40 (32 clinically normal and eight animals submitted for PME) were also negative on PCR for *M. haemolytica*, *P. multocida*, *H. somni* and *M. bovis* nucleic acid, three were negative for *H. somni* and *M. bovis* nucleic acid but were not tested for *P. multocida* and *M. haemolytica*. Nucleic acid from at least one of these four bacterial pathogens was detected in nine animals; *H. somni* (six animals) was the pathogen most frequently detected in these. Lung and tracheal bacterial PCR results are presented in Table 2.

Table 2: The relative frequency of detection of antibodies to BPIV-3 and BRSV in serum and the relative frequency of detection of selected viral and bacterial BRD-causing pathogens in the trachea and lungs of the study population

Test		**Number tested**	**Percentage positive**
Serology (ELISA)	BPIV-3 antibodies	131	91.6%
	BRSV antibodies	131	84.7%
Lung and trachea (PCR)	BPIV-3	131	0.8%
	BRSV	131	0%
	H. somni	57	8.8%
	P. multocida	57	5.3%
	M. haemolytica	57	1.8%
	M. bovis	57	1.8%

Viral Serology and Viral PCR

The seroprevalence of BPIV-3 and BRSV from all animals was 91.6% and 84.7%, respectively (Table 2). Data on the vaccination status of the study population was not available. BPIV-3 nucleic acid was detected by PCR in the maxillary sinus of an eight month old heifer submitted for PME. BPIV-3 nucleic acid was also detected by PCR in the lung of this animal. BRSV was not detected by PCR in any of the study population. Lung viral PCR results are presented in Table 2 indicating the detection of BPIV-3 in the lung and tracheal tissue of one animal while BRSV was not detected in the lung or trachea of any animal examined.

Histopathological Examination and Immunohistochemistry (IHC) for iNOS

No significant changes were observed on histopathological examination of the sampled sinus epithelia (n = 17). Intra-epithelial cytoplasmic iNOS expression was detected in 76.5% (n = 13) of a sub-sample of 17 animals (Fig 2a, 2b and 2c). In six animals staining was evident in 50% or more of the epithelial cells (Fig 2a and 2b) while in three animals, staining was detected in 5% or less of epithelial cells examined. The staining intensity was scored as moderate or strong in 11 of the 13 animals in which positive staining was recorded.

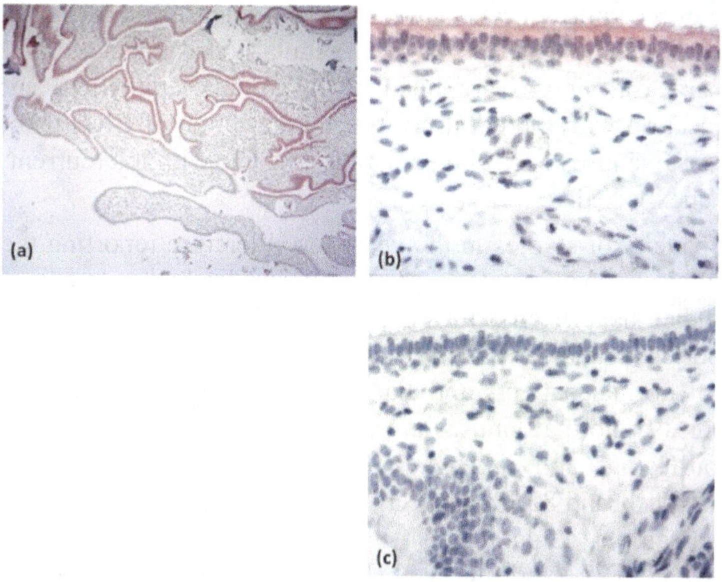

Figure 2: a, b and c: Photomicrographs of bovine paranasal sinus epithelium with positive red immunohistochemical staining of epithelial cells (a and b) denoting expression of inducible nitric oxide synthase (iNOS) and of the negative control (c) (aX10, b X40, c X40).

DISCUSSION

This is the first study to document the microbial flora of the bovine paranasal sinuses and to demonstrate the expression of NO by the sinus epithelium. The fact that bacterial and viral pathogens central to the pathogenesis of BRD were found at very low frequency and/or very low abundance in the sinuses suggests they do not function as a refuge for their persistence within herds.

One of the few previous bacteriological studies of the paranasal sinuses of cattle reported *T. pyogenes* (n = 3) and *P. multocida* (n = 2) in 12 cases of chronic frontal sinusitis in dairy cows [6]. Inflammation was not a feature of any of our sample population although we did identify the pathogens *M. haemolytica* (n = 2), *P. multocida* (n = 1) and *T. pyogenes* (n = 1) at low frequency. In each animal from which one of these BRD pathogens was isolated, only a single paranasal sinus was positive and, with the exception of one *M. haemolytica* isolate, these findings were in animals which had been submitted for PME as a result of various diseases. The presence of recognised BRD pathogens as commensals in the nasal mucosa and lungs of healthy animals has been regularly documented [21, 22] and, as the paranasal sinuses are lined by respiratory epithelium and are in communication with the nasal meatus, the detection of BRD pathogens in the paranasal sinuses of healthy animals, or those with intercurrent disease, was not unexpected.

The paucity of studies in the veterinary literature reporting the normal paranasal sinus flora in cattle is not replicated in human medicine where many studies have been published, albeit with conflicting results. Brook [23] reported primarily anaerobic bacterial isolates from the maxillary sinuses of normal patients and Jiang and colleagues [24] concluded that endoscopically normal sinuses are not sterile; Sobin and colleagues [25] and Abou-Hamad and colleagues [26] reported sterile human maxillary sinuses in 100% and 82.1% respectively of their study populations. A study of a larger sample (42 patients/84 sinuses) of healthy human frontal sinuses recorded that 85.7% were sterile [27]. All of these studies employed routine bacteriological examination, however, the advent of culture—independent bacterial sequencing techniques has increased the sensitivity of analyses and allowed the detection of low abundance or non-culturable bacteria which has led to the conclusion that diverse populations of bacteria inhabit the paranasal sinuses of both healthy and sick human patients [28].

We found that bovine paranasal sinuses were either predominantly sterile or contained undetectable microbes with a greater frequency of identification of bacterial pathogens in animals submitted for PME. We used a combination of bacteriological culture and PCR to detect bacterial species and the majority of detections were achieved by PCR. This likely further reflects the presence of low bacterial loads in these animals. Hauser et al [29], in a study which compared DNA-based molecular techniques with bacteriological culture of the paranasal sinuses of 54 human patients with chronic rhinosinusitis, concluded that clinical culture was poorly representative of the actual bacteria present and cited the inability of culture media to replicate sinus conditions, the selection for fast growing bacteria over pathogens, the inability to detect low abundance bacteria, poor sample handling and simple misidentification among the reasons for the poor correlation between both methods. In contrast, Abou-Hamad and colleagues [26] in their survey of the paranasal sinuses of healthy humans concluded that the use of PCR rather than routine bacteriology would overestimate the presence of bacteria. While we employed PCR on a proportion of our population and did not detect selected pathogens in most samples tested, it remains possible that other bacteria are present at low abundance in the bovine paranasal sinuses. The next generation sequencing techniques and phylogenetic oligonucleotide arrays employed in recent studies of human paranasal sinuses [28] detect bacteria that are not readily identified using conventional culture or PCR. The role of prior antimicrobial administration in exerting bias on our results can be largely discounted because even though some animals submitted for PME may have received antimicrobials prior to death, those sourced from the slaughterhouse would have been subject to strict withdrawal periods prior to slaughter if antimicrobials had been administered. Our findings suggest that BRD pathogens are present at very low frequency, and possibly low abundance, in the bovine paranasal sinuses. Lima and colleagues [30] recently evaluated the upper respiratory tract microbiota of dairy calves and reported an association between bacterial abundance, rather than bacterial community structure, and subsequent BRD morbidity risk. Future research examining the microbiome of the bovine paranasal sinus could possibly determine if a similar association exists between the bacterial community structure or bacterial abundance in the bovine paranasal sinuses and subsequent risk of BRD morbidity.

The selection of the four sampling sites employed in this study on each side of the skull was based on considerations of the ease of access for sampling as well as the interconnection of some of the paranasal sinuses sampled with others which are not as accessible (e.g. the maxillary sinus communicates with the palatine sinus and the lacrimal sinus). Cotton swabs were used for sampling based on the findings of human paranasal sinus bacteriology from cross-sectional studies where various sampling methods have been employed. Although some studies have employed sinus lavage with sterile saline or mucosal biopsy with positive results [26, 31, 32], Jiang and colleagues [24] reported that cotton swabs have a higher isolation rate in both patients with chronic maxillary sinusitis and those without this diagnosis than mucosal biopsies of endoscopically normal maxillary sinuses.

Nitric oxide (NO) is a free radical lipophylic gas which has been detected in the exhaled breath of humans and other animals, but not in cattle [33, 34]. In humans NO is an endogenous mediator of many physiological processes including the regulation of blood flow, neurotransmission, haemostasis and chronic inflammation [35]. NO in paranasal sinus gas in humans has been found to be markedly higher than that detected in exhaled air [10, 36] and the paranasal sinuses have been proposed as a reservoir for NO [37]. Lewandowski and colleagues [38] proposed that paranasal sinuses might be a key anatomic site for the production of nasal NO based on the findings of very low NO levels in the expired air of baboons, the only mammal known to lack paranasal sinuses. NO has broad spectrum antibacterial, antiviral and antifungal properties [39] and a role for NO in human paranasal sinus host defence and in the maintenance of sterility of the human paranasal sinuses under normal conditions has been proposed by Lundberg [10]; NO also has a role in human respiratory tract mucocillary clearance [40, 41]. Defects in NO production are believed to play a role in the pathogenesis of sinusitis in humans [42], with Lindberg and colleagues [43] reporting that patients with chronic sinusitis recorded lower levels of NO production in the upper airways when compared to healthy patients.

NO is synthesised from L-arginine and oxygen by the enzyme nitric oxide synthase (NOS) in a wide variety of cell types, including epithelial, endothelial, nervous and inflammatory cells. Three isoforms of NOS have been identified, two of which are constitutive, calcium-dependent and found in endothelial and neuronal cells, and one which is inducible (iNOS), calcium-independent and present in respiratory epithelium. Under normal conditions, iNOS is expressed only weakly in cells or not at all [35]. Lundberg and colleagues [10] demonstrated the presence of iNOS, using immunohistochemistry, and iNOS messenger RNA, using in situ hybridisation, in human paranasal sinus epithelial cells; staining was strongest in the apical part of the cell. Lundberg and colleagues [44] reported that the NOS present in the paranasal sinus mucosa of humans was identical to iNOS and was calcium-independent but its regulation in the paranasal sinus appeared to be fundamentally different from iNOS found in inflamed tissue or activated white blood cells as it was constantly expressed, was not downregulated by glucocorticoids and behaved similarly to constitutive

NOS. The presence of iNOS in the paranasal sinus epithelium of cattle in our study in the absence of infectious agents or inflammation is further evidence of constitutive expression and is a plausible explanation for the absence of a substantial or consistent microbial flora.

Our findings may be relevant to developments in the treatment or prevention of BRD. Regev-Shoshani and colleagues [39,45,46] have used an intranasal NO-releasing solution as an alternative to the prophylactic use of antimicrobials in cattle. These authors reported that nasal and muzzle infusion of cattle on feedlot arrival with NO releasing solution at a flow rate of 160 ppm NO in a 3 L/min gas flow for 5 seconds was not inferior to antimicrobial prophylaxis in cattle at low/medium risk of developing BRD.

While serology indicated widespread exposure of both clinically normal and cattle submitted for PME to BPIV-3 and BRSV (seroprevalences of 91.6% and 84.7%, respectively), we cannot exclude the possibility that these results reflect vaccinal antibody titres in some of the study population. However, recent reports of respiratory disease diagnoses in weanling cattle examined post-mortem in Ireland [22] recorded low frequency of respiratory vaccination in the study population and the authors suspect a similar low frequency of vaccination in the study population of the current study. BPIV-3 was detected by PCR in the sinus of only one animal while BRSV was not detected in the sinus of any animal suggesting that the persistence of BRD-causing viruses is unlikely in the sinuses. Sequencing of BRSV from outbreaks has indicated that while viruses within outbreaks appear identical, they vary spatially and temporally between outbreaks implying the ongoing introduction of novel strains to cattle populations [47,48]. Similar uncertainty exists regarding the maintenance of BPIV-3 virus infection within herds [49, 50, 51, 52]. We interpreted the detection of BPIV-3 in a single sinus of one animal, which also had BPIV-3 in its lungs, as evidence of active infection rather than persistence. Although infection with either of these BRD-causing viruses has seasonality [18], our study populations were sampled over a number of months rather than at a specific time period to avoid the possibility that seasonally low circulation of virus would decrease the likelihood of their detection.

CONCLUSIONS

This novel study has found that the microbial flora of the paranasal sinuses of cattle is very limited and that in a majority of animals these sites are sterile or contain undetectable microbes. The research further suggests this finding is related to the expression of the antimicrobial compound NO by sinus epithelium. The fact that bacterial and viral pathogens central to the pathogenesis of BRD were found at very low frequency and/or very low abundance in the sinuses suggests these locations do not function as a refuge for the persistence of these infectious agents within herds.

ACKNOWLEDGMENTS

The authors wish to acknowledge the assistance of the staff of Sligo Regional Veterinary Laboratory (RVL) in data management, sampling and bacteriological analysis. The assistance provided by the staff of Virology division, Central Veterinary Research Laboratory (CVRL), Department of Agriculture, Food and Marine, Backweston Campus, Co. Kildare, Athlone RVL and Pathology division (CVRL) was also gratefully received. We are also grateful for the cooperation of Mr. Philip Breslin, veterinary inspector, Department of Agriculture, Food and Marine and also wish to acknowledge the contribution of Ms. Sheila Worrall and Mr. Joseph Brady, Veterinary Pathology Department, UCD for their technical assistance in the development of the iNOS IHC and Mr. Brian Cloak, UCD, for assistance with photography.

REFERENCES

1. Budras KD, Habel RE, Mulling CKW, Greenough PR. Bovine Anatomy: An Illustrated Text. 2nd ed, Schluetersche, Germany; 2011.
2. Dyce KM, Sack WO, Wensing CJG. Textbook of Veterinary Anatomy. 3rd ed Philadelphia: Saunders; 2002.
3. Farke AA. Evolution and functional morphology of the frontal sinuses in Bovidae (Mammalia: Artiodactyla), and implications for the evolution of cranial pneumaticity. Zool J Linnean Soc. 2010; 159: 988–1014.
4. Crocker CB, Rings DM. Lymphosarcoma of the frontal sinus and nasal passage in a cow. J Am Vet Med Assoc. 1998; 213: 1472–1474.
5. Mundell LD, Smith BP, Hoffman RL. Maxillary sinus cysts in two cattle. J Am Vet Med Assoc. 1996; 209: 127–129.
6. Ward JL, Rebhun WC. Chronic frontal sinusitis in dairy cattle: 12 cases (1978–1989). J Am Vet Med Assoc. 1992; 201: 326–328.
7. Crilly JP, Hopker A. Case Report: diagnosis and treatment of maxillary sinusitis in an Aberdeen Angus cow. Livestock. 2013; 18: 230–235.
8. Mugera GM. Veterinary pathology in the tropics: for students and practitioners. 1st ed Nairobi University Press; New Age International; 2000
9. Anand VK. Epidemiology and economic impact of rhinosinusitis. Ann Otol Rhinol Laryngol Suppl. 2004; 193: 3–5.
10. Lundberg JO, Farkas-Szallasi T, Weitzberg E, Rinder J, Lidholm J, Anggåard A, et al. High nitric oxide production in human paranasal sinuses. Nat Med. 1995; 4: 370–373.
11. De Groote MA, Fang FC. NO inhibitions: antimicrobial properties of nitric oxide. Clin Infect Dis. 1995; 21: 162–165.
12. Griffin D, Chengappa MM, Kuszak J, McVey DS. Bacterial pathogens of the bovine respiratory disease complex. Vet Clin North Am Food Anim Pract. 2010; 26:381–394. 10.1016/j.cvfa.2010.04.004
13. Guterbock WM. The impact of BRD: the current dairy experience. Anim Health Res Rev. 2014; 15:130–134. 10.1017/S1466252314000140
14. Rodríguez F, Bryson DG, Ball HJ, Forster F. Pathological and immunohistochemical studies of natural and experimental Mycoplasma bovis pneumonia in calves. J Comp Path. 1996; 115: 151–162.
15. Bryson DG, Ball HJ, McAliskey M, McConnell W, McCullough

SJ. Pathological, immunocytochemical and microbiological findings in calf pneumonias associated with Haemophilus somnus infection. J Comp Pathol. 1990; 103: 433–445.

16. Klima CL, Alexander TW, Hendrick S, McAllister TA. Characterization of *Mannheimia haemolytica* isolated from feedlot cattle that were healthy or treated for bovine respiratory disease. Can J Vet Res. 2014; 78: 38–45.
17. Liu D, Lawrence ML, Austin FW. Specific PCR identification of *Pasteurella multocida* based on putative transcriptional regulator genes, J Micro Meth. 2004; 58:263–267.
18. O'Neill R, Mooney J, Connaghan E, Furphy C, Graham DA. Patterns of detection of respiratory viruses in nasal swabs from calves in Ireland: a retrospective study. Vet Rec. 2014; 175: 351–356. 10.1136/vr.102574
19. Elvander M, Edwards S, Näslund K, Linde N. Evaluation and application of an indirect ELISA for the detection of antibodies to bovine respiratory syncytial virus in milk, bulk milk, and serum. J Vet Diagn Invest. 1995; 7:177–182. 10.1177/104063879500700202
20. Graham DA, Mawhinney KA, McShane J, Connor TJ, Adair BM, Merza M. Standardization of enzyme-linked immunosorbent assays (ELISAs) for quantitative estimation of antibodies specific for infectious bovine rhinotracheitis virus, respiratory syncytial virus, parainfluenza-3 virus, and bovine viral diarrhea virus. J Vet Diagn Invest. 1997; 9:24–31. 10.1177/104063879700900105
21. Caswell JL. Failure of respiratory defenses in the pathogenesis of bacterial pneumonia of cattle. Vet Path. 2014; 51: 393–409.
22. Murray GM, More SJ, Sammin D, Casey MJ, McElroy MC, O Neill RG, et al. Pathogens, patterns of pneumonia, and epidemiologic risk factors associated with respiratory disease in recently weaned cattle. J Vet Diag Invest. 2017; 29: 20–34.
23. Brook I. Aerobic and anaerobic bacterial flora of normal maxillary sinuses. Laryngoscope. 1981; 91: 372–376.
24. Jiang RS, Liang KL, Jang JW, Hsu CY. Bacteriology of endoscopically normal maxillary sinuses. J Laryngol Otol. 1999; 113: 825–828.
25. Sobin J, Engquist S, Nord CE. Bacteriology of the maxillary sinus in healthy volunteers. Scand J Infect Dis. 1992; 24: 633–635.
26. Abou-Hamad W, Matar N, Elias M, Nasr M, Sarkis-Karam D, Hokayem

N, et al. Bacterial flora in normal adult maxillary sinuses. Am J Rhinol. 2009; 23: 261–263.

27. Albu S, Florian IS. Bacteriology of normal frontal sinuses. Am J Otolaryngol Head Neck Med Surg. 2013; 34: 327–330.
28. Abreu NA, Nagalingam NA, Song Y, Roediger FC, Pletcher SD, Goldberg AN, et al. Sinus microbiome diversity depletion and Corynebacterium tuberculostearicum enrichment mediates rhinosinusitis. Sci Transl Med. 2012; 4: 151re24.
29. Hauser LJ, Feazel LM, Ir D, Fang R, Wagner BD, Robertson CE, et al. Sinus culture poorly predicts resident microbiota. Int Forum Allergy Rhinol. 2015; 5: 3–9. 10.1002/alr.21428
30. Lima SF, Teixeira AG, Higgins CH, Lima FS, Bicalho RC. The upper respiratory tract microbiome and its potential role in bovine respiratory disease and otitis media. Sci Rep. 2016; 6: 29050 10.1038/srep29050
31. Su WY, Liu C, Hung SY, Tsai WF. Bacteriological study in chronic maxillary sinusitis. Laryngoscope. 1983; 93:931–934.
32. Meltzer EO, Hamilos DL, Hadley JA, Lanza DC, Marple BF, Nicklas RA, et al.(2004): Rhinosinusitis: Establishing definitions for clinical research and patient care. J Allergy Clin Immunol. 2004; 114: 155–212. 10.1016/j.jaci.2004.09.029
33. Gustafsson LE, Leone AM, Persson MG, Wiklund NP, Moncada S. Endogenous nitric oxide is present in the exhaled air of rabbits, guinea pigs and humans. Biochem Biophys Res Commun. 1991; 181: 852–857.
34. Schedin U, Roken BO, Nyman G, Frostell C, Gustafsson LE. Endogenous nitric oxide in the airways of different animal species. Acta Anaesthesiol Scand. 1997; 41: 1133–1141.
35. Serrano C, Valero A, Picado C. Nasal nitric oxide. Arch Bronconeumol. 2004; 40: 222–230.
36. Lundberg JO, Rinder J, Weitzberg E, Lundberg JM, Alving K. Nasally exhaled nitric oxide in humans originates mainly in the paranasal sinuses. Acta Physiol Scand. 1994; 152:431–432. 10.1111/j.1748-1716.1994.tb09826.x
37. Andersson JA, Cervin A, Lindberg S, Uddman R, Cardell LO. The paranasal sinuses as reservoirs for nitric oxide. Acta Otolaryngol. 2002; 122: 861–865.
38. Lewandowski K, Busch T, Lohbrunner H, Rensing S, Keske U,

Gerlach H, et al. Low nitric oxide concentrations in exhaled gas and nasal airways of mammals without paranasal sinuses. J Appl Physiol. 1998; 85: 405–10.

39. Regev-Shoshani G, Crowe A, Miller C. A nitric oxide releasing solution as a potential treatment for fungi associated with tinea pedis. J App Microbiol. 2013a; 114: 536–544.
40. Lindberg S, Cervin A, Runer T. Low levels of nasal nitric oxide (NO) correlate to impaired mucociliary function in the upper airways. Acta Otolaryngol. 1997; 117:728–734.
41. Kim JW, Min YG, Rhee CS, Lee CH, Koh YY, Rhyoo C, et al. Regulation of mucociliary motility by nitric oxide and expression of nitric oxide synthase in the human sinus epithelial cells. Laryngoscope. 2001; 111: 246–250. 10.1097/00005537-200102000-00011
42. Naraghi M, Deroee AF, Ebrahimkhani M, Kiani S, Dehpour A. Nitric oxide: a new concept in chronic sinusitis pathogenesis. Am J Otolaryngol. 2007; 28: 334–337. 10.1016/j.amjoto.2006.10.014
43. Lindberg S, Cervin A, Runer T. Nitric oxide (NO) production in the upper airways is decreased in chronic sinusitis. Acta Otolaryngol. 2009; 117: 113–117.
44. Lundberg JO, Weitzberg E, Rinder J, Rudehill A, Jansson O, Wiklund NP, et al. Calcium-independent and steroid-resistant nitric oxide synthase activity in human paranasal sinus mucosa. Eur Resp J. 1996; 9: 1344–1347.
45. Regev-Shoshani G, Church JS, Cook NJ, Schaefer AL, Miller CC.Prophylactic nitric oxide treatment reduces incidence of bovine respiratory disease complex in beef cattle arriving at a feedlot. Res Vet Sci. 2013b; 95: 606–611.
46. Regev-Shoshani G, McMullin B, Nation N, Church JS, Dorin C, Miller C. Non-inferiority of nitric oxide releasing intranasal spray compared to sub-therapeutic antibiotics to reduce incidence of undifferentiated fever and bovine respiratory disease complex in low to moderate risk beef cattle arriving at a commercial feedlot. Prev Vet Med, 2015; In Press.
47. Larsen LE, Tjornehoj K, Viuff B. Extensive sequence divergence among bovine respiratory syncytial viruses isolated during recurrent outbreaks in closed herds. J Clin Microbiol. 2000; 38: 4222–4227.

48. Bidokhti MR, Tråvén M, Ohlson A, Zarnegar B, Baule C, Belák S, et al. Phylogenetic analysis of bovine respiratory syncytial viruses from recent outbreaks in feedlot and dairy cattle herds. Arch Virol. 2012; 157: 601–607. 10.1007/s00705-011-1209-3
49. Allan EM, Pirie HM, Selman IE, Snodgrass DR. Some characteristics of a natural infection by parainfluenza-3 virus in a group of calves. Res Vet Sci.1978; 24: 339–46.
50. Frank GH, Marshall RG. Relationship of serum and nasal secretionneutralizing antibodies in protection of calves against parainfluenza-3 virus. Am J Vet Res 1971; 32: 1707–1713.
51. Frank GH, Marshall RG. Parainfluenza-3 virus infection of cattle. J Am Vet Med Assoc 1973; 163: 858–860.
52. Castrucci G, Cilli V, Frigeri F, Ferrari M, Ranucci S, Rampichini L. Reactivation of Bovid herpesvirus 1 and 2 and parainfluenza-3 virus in calves latently infected. Comp Immunol Microbiol Infect Dis.1983; 6:193–199.

CHAPTER 8

An Unusual Case of Bovine Anthrax in the Canton of Jura, Switzerland in 2017

Stefanie Gobeli Brawand[1], Sonja Kittl[1], Martina Dettwiler[2], Andreas Thomann[1], Simon Feyer[1], José Cachim[3], Grégoire Theubet[4], Nicole Liechti[5,6], Matthias Wittwer[5], Nadia Schürch[5], Simone Oberhänsli[6], Andreas Heinimann[7] and Jörg Jores[1]

[1]Institute of Veterinary Bacteriology, University of Bern, Bern, Switzerland
[2] Institute of Animal Pathology, University of Bern, Bern, Switzerland
[3] Animal Health, General Management of Agriculture, Viticulture and Veterinary Affairs (DGAV), Canton of Vaud, Switzerland
[4] Vétérinaires Mont-Terri Sàrl, Courgenay, Switzerland. 5 Spiez Laboratory, Federal Office for Civil Protection, Spiez, Switzerland
[6] Interfaculty Bioinformatics Unit, University of Bern, Bern, Switzerland
[7] Institute of Geography and Centre for Development and Environment, University of Bern, Bern, Switzerland

Citation: Gobeli Brawand, S., Kittl, S., Dettwiler, M. et al. An unusual case of bovine anthrax in the canton of Jura, Switzerland in 2017. BMC Vet Res 15, 265 (2019). https://doi.org/10.1186/s12917-019-1996-4

ABSTRACT

Background

Anthrax caused by *Bacillus anthracis* is a zoonotic disease mainly affecting herbivores. The last Swiss outbreak was over 20 years ago. We describe a recent anthrax outbreak involving two cows from the same herd. One cow was designated as a peracute clinical case with sudden death and typical lung lesions, while the other cow presented with protracted fever and abortion.

Case Presentation

On April 29th 2017, a 3.5-year-old Montbéliard dairy cow was found dead while out at pasture with haemorrhage from the nose. The veterinarian suspected pneumonia and performed a necropsy on site. Subsequently, a lung and liver sample were sent to the laboratory. Unexpectedly, *Bacillus anthracis* was isolated, a pathogen not found in Switzerland for decades. Several days later, a second cow from the same farm showed signs of abortion after protracted fever. Since these symptoms are not typical for anthrax, and the bacteria could not be demonstrated in blood samples from this animal, a necropsy was performed under appropriate biosafety measures. Subsequently, *Bacillus anthracis* could be isolated from the placenta and the sublumbal lymph nodes but not from the blood, liver, spleen and kidney. The outbreak strain (17OD930) was shown to belong to the lineage B.Br. CNEVA, the same as Swiss strains from previous outbreaks in the region. We speculate that the disease came from a temporarily opened cave system that is connected to an old carcass burial site and was flushed by heavy rainfall preceding the outbreak.

Conclusion

Even in countries like Switzerland, where anthrax is very rare, new cases can occur after unusual weather conditions or ground disturbance. It is important for public officials to be aware of this risk to avoid possible spread.

BACKGROUND

Anthrax is caused by the Gram-positive bacterium *Bacillus anthracis*. It is primarily a disease of herbivores with high mortality [1]. *B. anthracis* forms spores outside the warm-blooded host after contact with oxygen and is therefore able to survive for decades in the environment. Bovines are usually infected through ingestion, or possibly inhalation, of spores when grazing on contaminated pastures [1]. After an incubation time of 1–14 days, normally, sudden death without any specific preceding symptoms is observed [1]. If the course of the disease is more protracted, as is more often the case in pigs or horses, fever and development of oedemas can be seen [1, 2]. Animals found dead on pasture, especially with haemorrhage from the mouth, nostrils or anus, should alert a veterinarian to the possibility of contracting anthrax [2].

Burial of the carcasses of diseased animals is contraindicated as it bears the risk of spore release, e.g. by physical soil disturbance at burial sites. In countries like Switzerland, where anthrax has not been observed in decades, new outbreaks can nevertheless occur if ancient burial sites are disturbed and thus spores released. Such outbreaks have also been described in Sweden [3].

Furthermore, anthrax is a zoonosis transmitted mainly through direct or indirect contact with infected animals or by occupational exposure [1]. The last major anthrax outbreak in Switzerland occurred in 1985 in the canton of Grisons and caused the death of 11 cattle and 2 goats, but no human cases were reported from this outbreak [4]. However, there was a large outbreak in a wool factory in the North-eastern part of Switzerland lasting from 1978 to 1981, involving 25 workers who developed cutaneous or respiratory anthrax. The source of this outbreak was imported goat hair from Pakistan [5]. The last two cases of bovine anthrax in Switzerland date back to 1997 and 1993 and occurred in the municipality Muotathal (canton of Schwyz) and in the municipality Develier (canton of Jura), respectively (Figs. 1, 2).

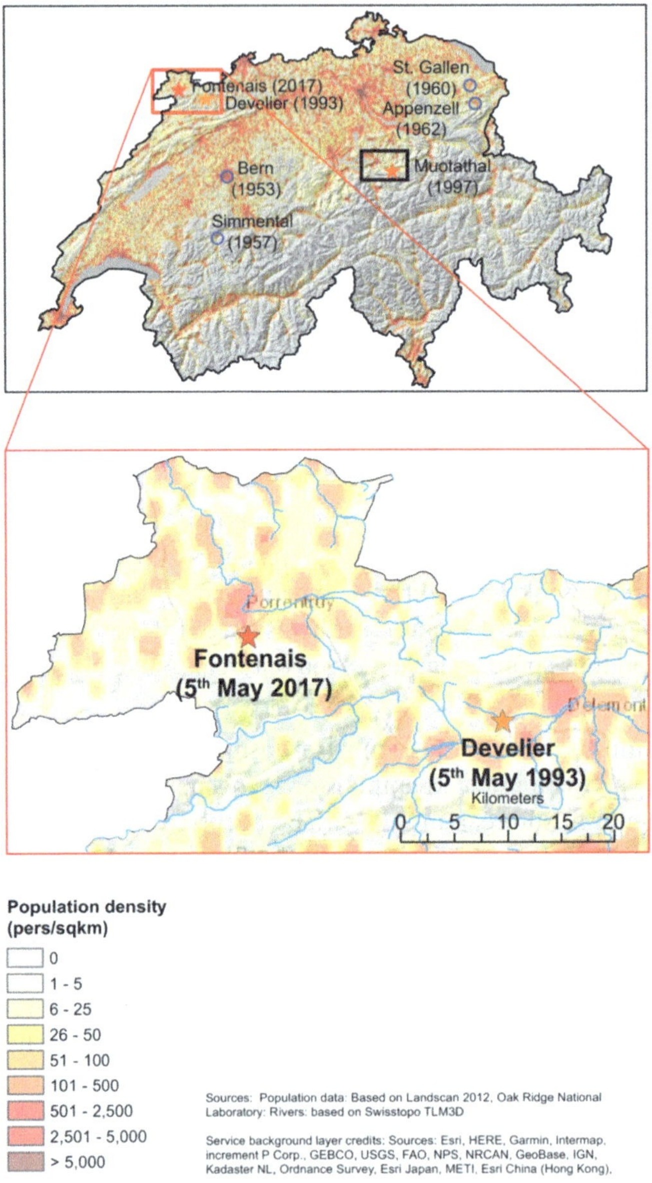

Figure 1: Population density in the region of the most recent bovine anthrax cases in the canton of Jura, Switzerland.

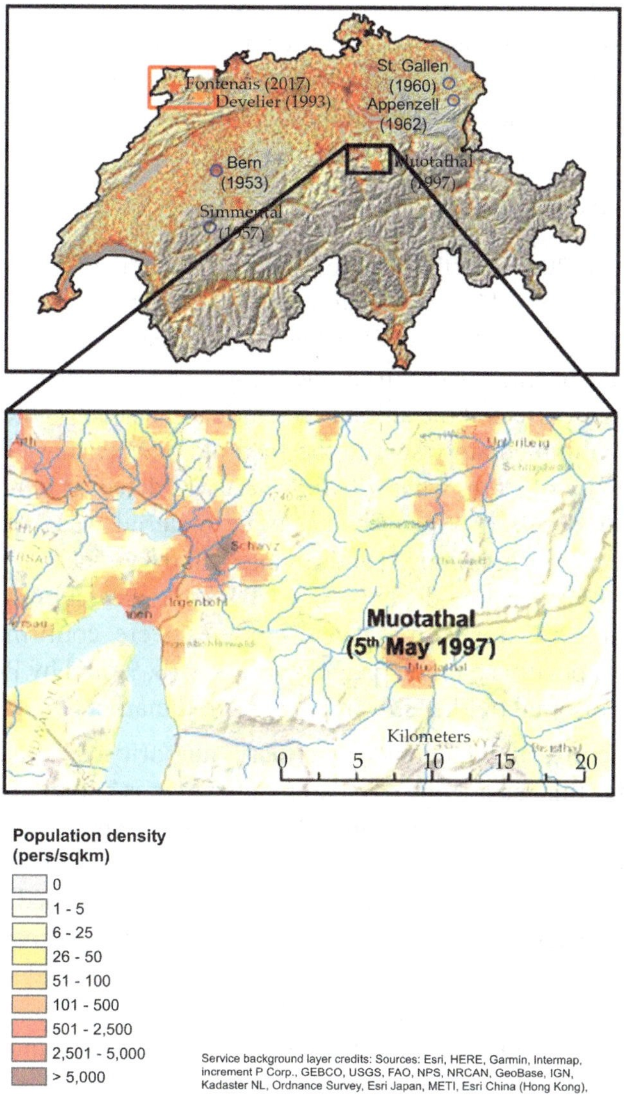

Figure 2: Population density in the region of the bovine anthrax cases in the Muotathal, canton of Schwyz, Switzerland.

We describe here an outbreak of bovine anthrax in Switzerland affecting two cows. The second case, where the animal showed protracted fever and signs of abortion, is an uncommon presentation of anthrax in cattle. Additionally, the suspected source of the outbreak – a temporary cave opening leading to a cave labyrinth connected to an old carcass dumping site – highlights the possibility of anthrax recurring in countries where it has not been observed for decades. As anthrax is a potentially fatal zoonosis, awareness among veterinarians and farmers is important, especially in regions where the disease is thought to be only of historical interest [6].

CASE PRESENTATION

On April 29th 2017, in the municipality of Fontenais in the canton of Jura (Fig. 1), in only 17 km air-line distance from the location of the 1993 outbreak, a 3.5-year-old Montbéliard dairy cow was found dead at pasture, showing dark haemorrhage from the nose. The veterinarian suspected pneumonia and dissected the animal on a trailer in a shed on the farm. Liver and lung were immediately sent to the Institute of Veterinary Bacteriology at the University of Bern, where *B. anthracis* was detected in lung tissue by microscopy and culture. Specifically, tissue material was Giemsa-stained for microscopic examination [7] and streaked onto trypticase soy agar plates containing 5% sheep blood (TSA-SB) (BD Becton Dickinson) following incubation at 37 °C for 18 h. Typical colonies were confirmed to be *B. anthracis* by a positive gamma phage lysis assay [8, 9] and by PCR, specific for the chromosomal marker *sap* and the plasmid markers *pag* and *cap* [10].

Consequently, the cantonal veterinary authority ordered a second-degree ban for this farm. Nine days later, nine bovines belonging to the same herd developed fever ranging from 39.5 to 40.2 °C. Venous blood samples from several affected animals were submitted for bacteriological analysis, where Giemsa-stain as well as a bacterial culture, performed as described above, were negative for *B. anthracis*. While eight animals recovered without any treatment, one animal, a pregnant 4.5-year-old cow, showed haemorrhagic vaginal discharge about 3 days after onset of fever. This cow was treated with a single dose of penicillin (Benzylpenicillinum procainum 20 mg/kg body weight intramuscularly; Procacillin®, MSD Animal Health GmbH, Luzern, Switzerland). The cantonal veterinary authorities asked for a necropsy of this animal under appropriate biosafety measures at the Institute of Animal Pathology (ITPA) of the University of Bern, as repeated blood samples taken before antibiotic treatment had tested negative, and clinical signs were unusual for an infection with *B. anthracis*. The diseased animal was transported to the Institute of Animal Pathology and euthanised 16 h after treatment with penicillin. Necropsy performed under special preventive measures [11] revealed macroscopically inconspicuous liver, spleen and kidneys, while the placenta showed haemorrhagic placentomes, and sublumbal lymph nodes were enlarged and necrotic. Giemsa-stained smears of the placenta and sublumbal lymph nodes revealed a massive yield of encapsulated rods, and culture followed by PCR confirmed the presence of *B. anthracis*. Resistance testing showed that the isolated strain was sensitive to penicillin.

In order to identify the potential source of infection, several people, including the farmer and his family, the cantonal veterinarian and a speleologist were interviewed on May 22, 2017. Based on the topographical, geographical and hydrological features, an outbreak hypothesis and a sampling plan were developed. The sampling scheme was based on the assumption that due to rainfall, anthrax spores were flushed out from a nearby cavity known as "trou des gez" and came up to the surface on the pasture where the outbreak occurred. A speleologist had dug up soil in this area while opening a small, historically known cave next to an old tree. According to his testimony, water spilt out and flowed over the grass area, whereupon he closed the cave again. This happened a few days prior to the onset of the outbreak. Soil, grass and water samples were taken from the farmland and in the cavity "trou des gez", known to be a former burial site for many types of carcasses. In order to screen the farm infrastructure for contamination, swabs were taken from the trailer on which the first diseased cow had been necropsied and from the box in the barn where the second animal began to abort.

The sample analysis was performed at Spiez Laboratory using a modified protocol [12]. Briefly, 5 g of soil/grass was homogenised in PBS/Tween and mixed for several hours to dislodge spores. To settle solid particles, the samples were centrifuged at low speed (500 x *g*) for 30 s. To inactivate vegetative cells, the supernatant was heated at 80 °C for 30 min and subsequently centrifuged at 5,000 x *g* for 15 min to pellet spores and residual particles. To allow germination, the pellet was dissolved in 9 ml tryptic soy broth (TSB, BioMérieux) and incubated for 30 min at 37 °C. For PCR analysis, 1 mL was centrifuged at 5,000 x *g* for 15 min and the pellet was dissolved in AVL buffer (Qiagen). Prior to DNA extraction, the suspension was heated at 100 °C for 15 min followed by sterile filtration (0.45 µm, Millipore). Water samples were filtered through 0.45 µl filters and the filters, as well as the swabs, were eluted in PBS/Tween. Further steps were analogous to the soil samples. Real-time PCR analysis was performed as previously described. For PCR positive samples, serial dilutions of the TSB suspensions were streaked directly on cereus ID agar (Merck) for cultivation under BSL-3 (biosafety level 3) conditions. PCR analysis of the swab samples showed that *B. anthracis* spores were present on the trailer and in the shed where the first diseased cow was necropsied as well as in the box of the second affected cow. An isolate was retrieved from the trailer, whereas the soil and grass samples tested negative with the internal process controls confirming accurate sample preparation.

Daily rainfall patterns in the fourteen days preceding this case and the previous bovine anthrax cases in Muotathal and Develier were analysed and compared with the long-term daily mean from the years 1961 to 2016 using the RhiresD product [13]. The analysis revealed that in the 14-day-period before each case precipitation amounts were indeed higher than the long-term historic mean (+43% in Develier, +75% in Muotathal and+60% in Fontenais; Fig. 3, Fig. 4), substantiating the role of heavy rains ahead of soil disturbance.

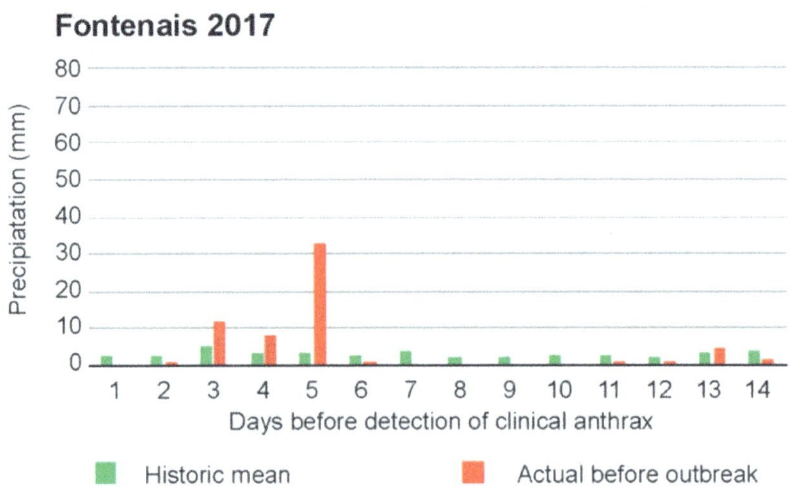

Figure 3: Daily rainfall amount preceding the most recent bovine anthrax case in Fontenais, Canton of Jura, Switzerland.

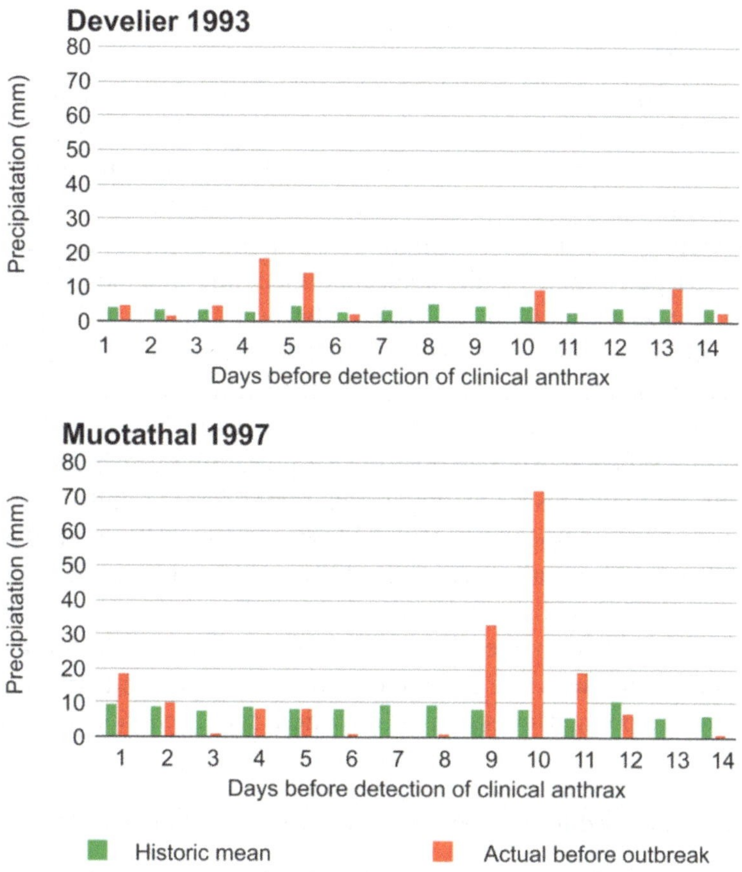

Figure 4: Daily rainfall amount preceding the bovine anthrax case in Muotathal, Canton of Schwyz, and Develier, canton of Jura, Switzerland.

To assess the potential risk to humans the population density around the outbreak area was determined based on Landscan 2012, Oak Ridge National Laboratory (population data) and on Swisstopo TLM3D (rivers). High population densities ranging between 500 and 2,500 persons living per km2 in a perimeter of 5 km around the 2017 outbreak location and the other two most recent Swiss bovine anthrax outbreak locations could be identified (Figs. 1, 2).

In order to sequence and subsequently compare the *B. anthracis* strain isolated in 2017 to other isolates from Switzerland and neighbouring countries, DNA was extracted following BSL-3 safety procedures. The strain was grown on TSA-SB at 37 °C for only 6 h to avoid production of spores. Extraction was performed using the QIAmp® DNA Blood Mini Kit (QIAGEN) using the manufacturer's instructions for isolation of DNA for Gram-positive bacteria. The protocol was adapted as follows: bacteria were not pelleted, but a loop full was directly suspended in enzyme solution and was incubated at 56 °C overnight. The eluate was run through a 0.2 µm filter before transferring it to a BSL-2 lab, where the DNA was additionally concentrated by precipitation. Illumina paired-end sequencing (read length 250 bp) was performed by the Lausanne Genomic Technologies Facility (Lausanne, Switzerland) on a MiSeq platform. Approximately 13 Mio paired reads were obtained resulting in 600x coverage. The reads were quality controlled using fastqc [14] and then trimmed with trimmomatic 0.33 [15]. Reads were mapped to the Ames Ancestor (NC_007530.2) using the Burrows-Wheeler Aligner bwa 0.7.13 [16], followed by SNP (single nucleotide polymorphism) calling using samtools (1.3 and 0.1.19) [17]. For comparison, previously sequenced Swiss strains [18], as well as other published strains from all three clades [18], were also mapped to the same reference and SNPs called as described above. For the Swiss strains, reads were downloaded from the Sequence Read Archive while for the other strains, the assemblies were downloaded from GenBank and artificial reads created for mapping (for strain information and accession numbers see Table 1).

Table 1: Strains used for phylogenetic analyses sorted by clade [18]

Strain ID	Country of isolation	Year of isolation	Host species	GenBank Accession No	clade
9080G	Georgia	1998	soil	NZ_CM002398.1	A
A0193	USA	–	bovine	NZ_ABKF00000000.1	A
A0389	Indonesia	–	–	NZ_ABLB00000000.1	A
A0488	UK	1935	bovine	NZ_ABJC00000000.1	A
A1039	Bolivia	1999	bovine	NZ_LAKZ00000000.1	A
A1075	Chile	–	bovine	NZ_LBFE00000000.1	A
Ames Ancestor	USA	–	–	NC_007530.2	A
Australia94	Australia	1994	bovine	NZ_AAES00000000.1	A
Canada_bison	Canada	–	–	NZ_CP010322.1	A
JF3853	Switzerland	1952	bovine	ERR899845	A
PAK1	Pakistan	1978	ovine	NZ_CP009325.1	A
Turkey32	Turkey	1991	human	NZ_CP009315.1	A

Vollum	UK	1963	bovine	NZ_CP007666.1	A
17OD930	Switzerland	2017	bovine	SRP144421 (this study)	B
A0442	South Africa	–	–	NZ_ABKG00000000.1	B
A0465	France	1997	bovine	NZ_ABLH00000000.1	B
ANS-ES00–82	France	2000	bovine	NZ_JHDS00000000.2	B
BA1035	South Africa	–	human	NZ_CP009700.1	B
BF1	Germany	2009	bovine	AMDT00000000.1	B
CNE-VA-9066	France	1992	bovine	NZ_AAEN00000000.1	B
HYU01	Korea	2009	soil	NZ_CP008846.1	B
JF3852	Switzerland	1953	bovine	ERR899844	B
JF3854	Switzerland	1957	bovine	ERR899846	B
JF3887	Switzerland	1960	bovine	ERR899847	B
JF3888	Switzerland	1962	bovine	ERR899848	B
K3	South Africa	–	human	NZ_CP009331.1	B
KrugerB	South Africa	–	–	NZ_AAEQ00000000.1	B
RA3	France	1998	bovine	NZ_CP009697.1	B
SVA11	Sweden	2011	bovine	NZ_CP006742.1	B
Zimbabwe89	Zimbabwe	–	–	NZ_JMPU00000000.1	B
2002013094	USA	1956	soil	NZ_CP009902.1	C

For the phylogenetic analyses, gap regions in the alignment were removed using MEGA7 [19] and the program modeltest-ng 0.1.3 (https://github.com/ddarriba/modeltest) was used to determine the best nucleotide substitution model. The program was run setting the topology parameter to maximum likelihood and otherwise using default parameters. The best model according to Bayesian information criteria was then chosen for phylogenetic tree construction with PhyML 3.3.20180214 [20, 21] and the tree was visualised in MEGA7. This whole genome SNP analysis revealed that *B. anthracis* strain 17OD930 clusters together with Swiss strains from prior outbreaks, as well as German and French strains (all of the B.Br. CNEVA lineage [18]). However, 17OD930 is not specifically closely related to any of the Swiss strains as compared to other strains in the same cluster. This indicates that it is not directly linked with earlier outbreaks strains that have been sequence characterised here (Fig. 5).

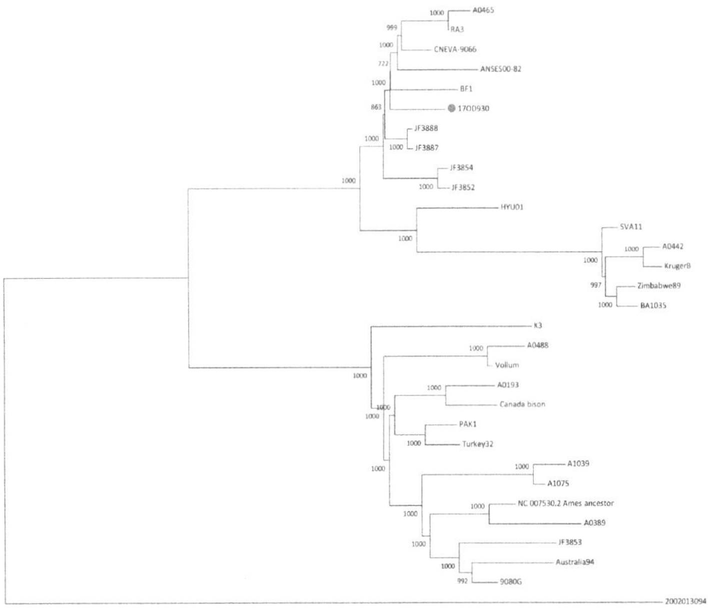

Figure 5: Whole genome SNP analysis comparing the 2017 outbreak strain to known strains of *B. anthracis*. The tree was constructed using PhyML with 1,000 replicates. The scale indicates the number of substitutions per site. The Clade C strain (2002013094) is used as outgroup.

Additionally, de novo assembly was performed using SPAdes 3.10.1 [22]. Fourteen contigs were obtained for the chromosome and their locations determined by mapping to the Ames Ancestor (NC_007530.2). Gaps were bridged by read-mapping to the Ames ancestor and the RA3 strain (NZ_CP009697.1). Ambiguities were resolved by Sanger sequencing. The closed genome sequence was then submitted to GenBank (GenBank accession No: CP029323).

The direct comparison of both isolates from this recent outbreak (retrieved from the trailer where the first cow was necropsied, and the isolate from the second cow) shows as expected identical sequencing data which confirms that both cows were infected by the same genotype, i.e. strain.

While the above-mentioned analyses were conducted, the cantonal authorities decreed measures to contain the outbreak. As stated before, the cantonal veterinary office immediately ordered a so-called second-degree ban upon being informed of the positive anthrax result. The ban included restrictions on the movement of animals and persons living on the farm.

Temperature measurements of all animals were conducted twice daily. Two neighbouring sheep farms were instructed to observe their animals for clinical signs. The cantonal office of public health together with the cantonal veterinary office surveyed the health of the involved humans. Nasal swabs analysed by PCR were negative. The federal office of public health, as well as the federal food safety and veterinary office, were also informed and the latter announced the case to the OIE. The public was informed by means of a press release. Epidemiological investigations to discern the source of infection were conducted as described, and the small recently human-made cave opening extending to an old tree on the pasture was firmly closed again. Decontamination of the barn, its annexes, the quarantine box and the milking room, as well as the shed and trailer, was conducted by the cantonal veterinary service in collaboration with the cantonal fire brigade using 10% formalin solution followed by 1% peracetic acid and hydrogen peroxide for final disinfection. The manure heap was treated by self-heating up to 70 °C for four days.

DISCUSSION AND CONCLUSIONS

Anthrax in cattle manifests as a severe disease with a fulminant course ending in sudden death in most cases. A protracted course followed by abortion as described here is unusual in this species [2]. The disease process fits the unusual findings of the necropsy, which revealed a macroscopically and bacteriologically inconspicuous spleen. Usually, this organ shows a dramatic macroscopic change in cases of infection with *B. anthracis* [2]. This case confirms, that anthrax spores are still present in Switzerland and infections may take unexpected clinical courses. In this unusual anthrax outbreak the responsible veterinarian did not anticipate an infection with *B. anthracis* and therefore opened the carcass which poses a risk of contracting anthrax. The veterinarian was consequently treated with antibiotics and did not develop disease. The very low prevalence of anthrax in Switzerland contributes to a limited awareness of the disease among practising veterinarians. It is very important to include anthrax infection as a differential diagnosis when the sudden death of cattle at pasture occurs and to treat carcasses of such animals accordingly to minimise the risk of infection and spread of the pathogen. Furthermore, this outbreak clearly showed that different animals from the same herd infected with anthrax can present very different clinical outcomes including unusual placentitis. In addition, population density around the sites of the last bovine anthrax cases was shown to be high, underlining the possible risk these bovine anthrax cases pose to the human population. Informing the public properly and swiftly helps to prevent human infections in the area of the outbreak. Medical doctors should also keep anthrax on their list of differential diagnosis in regions that experienced anthrax outbreaks even a long time ago.

Although *B. anthracis* is usually susceptible to penicillin [1], *B. anthracis* could still be detected in the organs of the diseased animal. The isolate tested sensitive for penicillin and no antibiotic resistance genes or point mutations were detected that could substantiate the isolation of live bacteria after antibiotic treatment. A possible explanation may be that the interval of 16 h between treatment and euthanasia was too short for the antibiotic to eliminate the bacteria or that the drug did not pass the barrier between bloodstream and placenta because the process of abortion was already too advanced.

Although numerous environmental samples were taken, the source of the infection could not be determined. Nevertheless, it is plausible that the water pouring from the recent cave opening next to an old tree contaminated a limited area of the pasture where the cows gathered for milking at the farm. Grazing and animal movement through this highly frequented area of the pasture may have dispersed and diminished the spore concentration. The rather atypical manifestation of the disease in the lung is in accordance with a superficial deposition of the spores on the vegetation by water flow and subsequent inhalation during grazing.

Data from rainfall analysis could not be evaluated statistically, however, there were some rainy days prior to the outbreak, which may have led to the flush out of anthrax spores from the described cave labyrinth in the karst landscape. Unfortunately, no isolates of past anthrax outbreaks in the Canton of Jura were available for genetic comparison. The strains from other former Swiss outbreaks did not show any closer relation to our strain than isolates from surrounding countries and have therefore no correlation with the described outbreak.

ACKNOWLEDGEMENTS

We thank Isabelle Brodard for gDNA extraction for sequencing. We acknowledge J. Gigandet for performing the field necropsy on the first cow. Calculations were performed on UBELIX (http://www.id.unibe.ch/hpc), the HPC cluster at the University of Bern.

REFERENCES

1. Ashford D, Beyer W, Cherkasskiy B, Doganay M, Dragon D, Hugh-Jones M, Kaufmann A, Leuenberger R, Turner A, Wilson W. Anthrax in humans and animals. 4th ed. Geneva: WHO, FAO, OIE; 2008.
2. Quinn PJ, Markey BK, Leonard FC, FitzPatrick ES, Fanning S, Hartigan PJ. Veterinary microbiology and microbial diseases. 2nd ed. Oxford: Wiley-Blackwell; 2011.
3. Agren J, Finn M, Bengtsson B, Segerman B. Microevolution during an Anthrax outbreak leading to clonal heterogeneity and penicillin resistance. PLoS One. 2014;9(2):e89112.
4. Kuoni E, Zindel W. Anthrax 1985 in Graubunden. Schweiz Arch Tierheilkd. 1986;128(5):261–7.
5. Pfisterer RM. An anthrax epidemic in Switzerland. Clinical, diagnostic and epidemiological aspects of a mostly forgotten disease. Schweiz Med Wochenschr. 1991;121(22):813–25.
6. Elvander M, Persson B, Sternberg Lewerin S. Historical cases of anthrax in Sweden 1916-1961. Transbound Emerg Dis. 2017;64(3):892–8.
7. Carter GR, Cole JR. Diagnostic procedures in veterinary bacteriology and mycology. 5th ed. San Diego: Academic Press; 1990.
8. Abshire TG, Brown JE, Ezzell JW. Production and validation of the use of gamma phage for identification of bacillus anthracis. J Clin Microbiol. 2005;43(9):4780–8.
9. Brown ER, Cherry WB. Specific identification of *Bacillus anthracis* by means of a variant bacteriophage. J Infect Dis. 1955;96(1):34–9.
10. Ryu C, Lee K, Yoo C, Seong WK, Oh HB. Sensitive and rapid quantitative detection of anthrax spores isolated from soil samples by real-time PCR. Microbiol Immunol. 2003;47(10):693–9.
11. Dettwiler M, Mehinagic K, Gobeli Brawand S, Thomann A, Feyer S, Husser L, Theubet G, Gigandet J, Rottenberg S, Posthaus H. Bacillus anthracis as a cause of bovine abortion - a necropsy case requiring special biosafety measures. Schweiz Arch Tierheilkd. 2018;160(9):547–52.
12. Turner WC, Kausrud KL, Beyer W, Easterday WR, Barandongo ZR, Blaschke E, Cloete CC, Lazak J, Van Ert MN, Ganz HH, et al. Lethal exposure: an integrated approach to pathogen transmission via environmental reservoirs. Sci Rep. 2016;6:27311.
13. MeteoSwiss [https://www.meteoswiss.admin.ch/content/dam/

meteoswiss/de/service-und-publikationen/produkt/raeumliche-daten-niederschlag/doc/ProdDoc_RhiresD.pdf].

14. Babraham Bioinformatics [http://www.bioinformatics.babraham.ac.uk/projects/fastqc/]. Accessed 19 Dec 2017.
15. Bolger AM, Lohse M, Usadel B. Trimmomatic: a flexible trimmer for Illumina sequence data. Bioinformatics. 2014;30(15):2114–20.
16. Li H, Durbin R. Fast and accurate short read alignment with burrows-wheeler transform. Bioinformatics. 2009;25(14):1754–60.
17. Li H, Handsaker B, Wysoker A, Fennell T, Ruan J, Homer N, Marth G, Abecasis G, Durbin R, Genome Project Data Processing S. The sequence alignment/map format and SAMtools. Bioinformatics. 2009;25(16):2078–9.
18. Derzelle S, Aguilar-Bultet L, Frey J. Whole genome SNP analysis of bovine B. anthracis strains from Switzerland reflects strict regional separation of Simmental and Swiss Brown breeds in the past. Vet Microbiol. 2016;196:1–8.
19. Kumar S, Nei M, Dudley J, Tamura K. MEGA: a biologist-centric software for evolutionary analysis of DNA and protein sequences. Brief Bioinform. 2008;9(4):299–306.
20. Guindon S, Dufayard JF, Lefort V, Anisimova M, Hordijk W, Gascuel O. New algorithms and methods to estimate maximum-likelihood phylogenies: assessing the performance of PhyML 3.0. Syst Biol. 2010;59(3):307–21.
21. Guindon S, Gascuel O. A simple, fast, and accurate algorithm to estimate large phylogenies by maximum likelihood. Syst Biol. 2003;52(5):696–704.
22. Bankevich A, Nurk S, Antipov D, Gurevich AA, Dvorkin M, Kulikov AS, Lesin VM, Nikolenko SI, Pham S, Prjibelski AD, et al. SPAdes: a new genome assembly algorithm and its applications to single-cell sequencing. J Comput Biol. 2012;19(5):455–77.

SECTION 3:
SWINE

CHAPTER 9

Effect of Spatial Separation of Pigs on Spread of Streptococcus suis Serotype 9

Niels Dekker[1,2], Annemarie Bouma[1] , Ineke Daemen[1] , Don Klinkenberg[1] , Leo van Leengoed[1] , Jaap A. Wagenaar[2] , Arjan Stegeman[1]

[1]Faculty of Veterinary Medicine, Department of Farm Animal Health, Utrecht University, Utrecht, The Netherlands

[2] Faculty of Veterinary Medicine, Department of Infectious Diseases and Immunology, Utrecht University, Utrecht, The Netherlands

ABSTRACT

The spread of an infectious agent in a population can be reduced by interfering in the infectiousness or susceptibility of individuals, and/or in their contact structure. The aim of this study was to quantify the effect of prevention of direct contact between infectious and susceptible pigs on the transmission of *Streptococcus suis (S. suis)*. In three replicate experiments, *S. suis*-free pigs were housed in boxes either in pairs (25 pairs) or alone (15 pigs). The distance between the boxes was ±1 m. At 7 weeks of age, one

Citation: Dekker N, Bouma A, Daemen I, Klinkenberg D, van Leengoed L, Wagenaar JA, et al. (2013) Effect of Spatial Separation of Pigs on Spread of Streptococcus suis Serotype 9. PLoS ONE 8(4): e61339. https://doi.org/10.1371/journal.pone.0061339

pig of each pair was inoculated intranasally with *S. suis* serotype 9; the other pigs were exposed to *S. suis* by either direct (pairs) or indirect contact (individually housed pigs). Tonsillar brush and saliva swab samples from all pigs were collected regularly for 4 weeks post inoculation to monitor colonization with *S. suis*. All inoculated pigs became infected, and their pen mates became colonized within 2 days. Thirteen indirectly exposed pigs became positive within 7–25 days after exposure. The rate of direct transmission β_{dir} was estimated to be 3.58 per pig per day (95% CI: 2.29–5.60). The rate of indirect transmission increased in time, depending on the cumulative number of days pigs tested positive for the presence of *S. suis*. The estimate β'_{ind} was 0.001 (95% CI: 0.0006–0.0017) new infections per pig per day for each day that an infected pig was tested positive for *S. suis*. We conclude that prevention of direct contact reduces the rate at which susceptible pigs become colonized. Simulation studies using these parameters showed, however, that such intervention measure would not limit *S. suis* serotype 9 spread in a commercial pig farm to a relevant extent, implying that spatial separation of groups op pigs within a compartment would not be effective on a farm.

INTRODUCTION

Infectious diseases are a large problem in animal husbandry and various control measures are implemented to reduce their impact, either by blocking transmission of the infection or by minimizing disease upon infection. Transmission of an infection depends on the infectiousness of infected individuals, the susceptibility of uninfected ones, and their mutual interaction [1]–[4]. Direct contact between infected and susceptible animals is considered to be the most important risk factor for spread of infections [5]–[8]. Consequently, interference in the contact structure between infected and susceptible animals, either by reducing the frequency or the intensity of the contact, might contribute to reduction of transmission of infections within a farm [2], [5], [6].

Separation of animals by adjustments of the housing or management systems has been applied on dairy and sheep farms to prevent transmission of pathogens from dams to offspring [5], [9]. Implementation of this measure is also considered for pig farms to improve the health status [10]–[13]. Measures to reduce contacts between pigs, however, do not fit well in current pig management practices for various reasons. One is that it is current management practice to mix pigs to create homogeneous groups; another reason is that individual housing is not acceptable for welfare reasons. Therefore, before considering measures to reduce infection transmission by separation of pigs or groups of pigs, it should be demonstrated that such measures have a substantial effect.

Evaluation of such separation measures can be performed at different scales, e.g. at region scale, with herds as units, at herd scale, with compartments as units, or at compartment scale, with individual pigs as units. The dynamics of spread of a pathogen among individuals within a compartment highly determines the dynamics at a higher scale e.g. between compartments or farms. Consequently, the effectiveness of a measure with respect to reducing the spread of a pathogen at a higher scale depends on the effect at a lower scale [14]. As a first step, we therefore focus on the lowest scale, i.e. individual pigs in a compartment.

One of the infectious agents on pig farms is *Streptococcus suis* (*S. suis*), which may cause meningitis, arthritis or septicaemia in piglets. The prevalence of infected pigs differs between farms, and varies with age [15]–[26]; mortality up to 20% has been reported [27], [28].Several serotypes of *S. suis* are circulating, of which serotypes 2 and 9 are most often isolated from clinical cases [28]–[33]. Disease caused by *S. suis* is a major determinant for abundant use of antibiotics in pig farming [34]. Prudent use of antimicrobial therapy is propagated and therefore alternative control measures should be seriously considered.

Many routes have been suggested for pig-to-pig transmission of *S. suis*. As *S. suis* is frequently isolated from the pigs upper respiratory tract, the direct oro-nasal route is generally assumed an important one [27], [28]. Other, indirect routes are also possible (e.g. airborne) [35].

The aim of our study was to determine whether the spread of *S. suis* within a compartment could be reduced by prevention of direct contact between pigs. We quantified the transmission rate of *S. suis* serotype 9 in an experimental set up between pigs housed in pairs and between pigs housed in pens placed at a distance of approximately 1 m. Such a design has been considered as robust in quantifying transmission [36], [37], and has been used for several other bacterial pathogens [38]–[40]. The parameter estimates derived from our experiments were used in a simulation model to study the effect of separation of pigs in a hypothetical conventional farm. Our experimental results show that the transmission rate is influenced by contact structure between pigs, but the simulation showed that separation most likely does not restrict spread of *S. suis* in a conventional stable compartment.

MATERIALS AND METHODS

Ethics Statement

The experiments were approved by the Animal Care and Ethics Committee of Utrecht University, in accordance with the Dutch law on experimental animals (approval number DEC 2008.II.08.072). To reduce the number of animals, we carried out experiments to study the effect of vaccination on transmission of *S. suis* as well. The results of that research objective have been published elsewhere [41].

Inoculum

S. suis serotype 9 strain 7997 was used as inoculum (provided by H. Smith, Central Veterinary Institute, Wageningen UR, Lelystad, The Netherlands). The strain was isolated from a clinical case in a commercial farm in The Netherlands. The strain contains genes encoding for several (putative) virulence factors, like suilysine (SLY) and variant (higher molecule weight) muramidase released protein (MRP*) [42], and belongs to the clonal complex (CC) that includes the vast majority of invasive *S. suis* serotype 9 strains found in the Netherlands, i.e. CC16 [43]. After overnight culture from a −80°C stock on agar plates at 37°C and 5% CO_2, one colony was suspended in 10 mL Todd-Hewitt broth (TH) (BioTrading, The Netherlands), and incubated for 3–4 h at 37°C until an optical density of 0.5–0.6 at 600 nm. After overnight storage at 4°C this suspension was diluted tenfold in TH and cultured for 2 h at 37°C resulting in an optical density of 0.5–0.6 at 600 nm. Ten mL of this suspension was washed twice and suspended in 10 mL physiologic saline solution. The bacterial concentration of the final suspension was $2–3\times10^8$ colony forming units (CFU) per mL.

Pigs

We used Landrace×Yorkshire pigs from sows housed at the animal facilities of the department. Piglets were caesarean derived and colostrum deprived. The first 4 weeks of life the piglets were housed in isolators, and thereafter in ground floor pens (12–15 pigs/pen). The pigs were fed with milk replacers during the first 4 weeks, and then with gamma-irradiated pelleted concentrates (Sloten B.V., The Netherlands; Trouw Nutrition Nederland B.V., The Netherlands). The feed contained *Enterococcus faecium* (DSM 7134), *Bacillus licheniformis* (DSM 5749) and *Bacillus subtilis* (DSM 5750).

Experimental Design

Three replicate experiments were conducted sequentially (for details: see Table 1 and Figure 1). Pigs were housed either in pairs or alone in boxes. Pair-wise housed pigs were used to measure direct transmission, the pigs housed alone to determine indirect transmission. The distance between boxes was 80–100 cm. This distance was sufficient to prevent direct contact between pigs in different pens, and might be applicable between groups in commercial farms. The boxes had multiplex walls (height: 80 cm), iron grid floor with rubber lying area, with a total area of 1.2 m^2 per box.

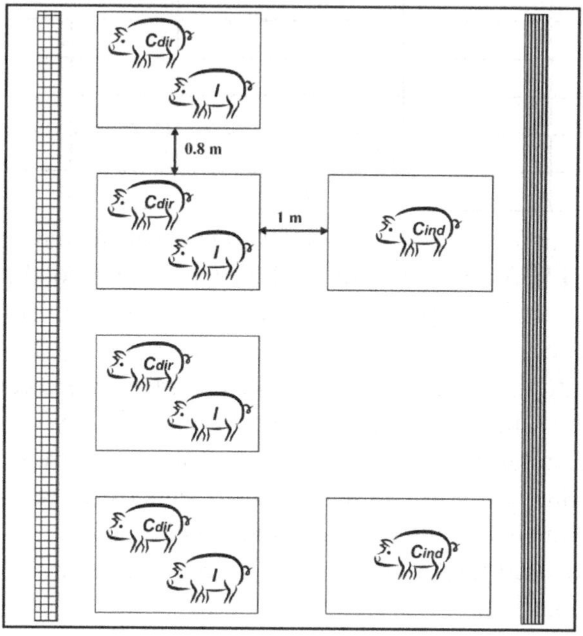

Figure 1: Design of stable compartment in animal experiments.

Experimental design for evaluation of the effect of prevention of direct contact by spatial separation on transmission of *S. suis* serotype 9 among pigs. Inoculated pigs (*I*) were intranasally infected with *S. suis* serotype 9 two days before placing in these boxes. Stable compartments contained 2 to 4 boxes with pair-wise housed pigs (C_{dir}+*I*) and 1 to 3 boxes with pigs housed individually (C_{ind}). C_{dir} pigs had direct contact with *I* pigs; C_{ind} pigs were only indirectly exposed to *S. suis*. The air inlet in the compartment is marked with ; the air outlet with .

Table 1: Design of animal experiments to evaluate the effect of spatial separation on *S. suis* **serotype 9 transmission**

	stable compartment		individually housed	pair-wise housed	
experiment		N pigs	N indirect contacts	N direct contacts	N inoculated pigs
I	A	7	1	3	3
	B	5	1	2	2
II	A	10	2	4	4
	B	10	2	4	4
	C	6	2	2	2
III	A	10	2	4	4
	B	10	2	4	4
	C	7	3	2	2

At the age of 7 weeks, one pig in each pair was inoculated intranasally with 1×10^9 CFU *S. suis* in 5 mL saline while sedated. Inoculation was performed in another stable compartment to prevent infection of contacts due to spread of the inoculum. Two days later the inoculated pigs were placed back into their original boxes. The pen mates were considered to be directly exposed, the pigs housed alone to be indirectly exposed. In total 25 inoculated (referred to as *I*), 25 direct contact pigs (referred to as C_{dir}) and 15 individually housed, indirect contact pigs (referred to as C_{ind}) were used. Compartments contained 2–4 boxes with pairs and 1–3 boxes with individually housed pigs, and per experiment 2–3 separate compartments were used (Table 1 and Figure 1).

Rectal temperatures and clinical observations were recorded daily. Neurologic signs and signs caused by lesions of the locomotor apparatus were scored separately. Pigs that showed severe clinical signs were euthanized. The remaining pigs were euthanized at 26–28 days post inoculation (DPI).

Biosecurity protocols were applied to prevent contact infections by animal handling. Animal handlers wore sterile gloves, boots and coveralls which were changed before handling between each box. C_{ind} pigs were sampled first, and C_{dir} pigs were sampled before *I* pigs.

To study the effect of vaccination on transmission of *S. suis* 13 of the 25 pairs of pigs (see Table S1 for details) were vaccinated at 4 and 7 weeks of age with a formalin inactivated whole bacterin vaccine, containing the same *S. suis* strain as used for inoculation (homologous vaccination). The results showed that both level of *S. suis* colonization and within-pen transmission were not affected by vaccination [36].

Sampling

Saliva and tonsil brushing samples were taken at 5 days and 3 h before inoculation, and 2–7, 9, 12, 15, 19, 22 DPI, and at the end of the experiment. We used a sterile steel wedge to open the mouth; saliva was sampled by turning round a swab (Cultiplast®) under the tongue for 5 s. Both palatine tonsillar areas were brushed for 3 s each with a sterile toothbrush. The brush and swab heads were put in separate sterile tubes containing 10 mL saline solution and transported to the laboratory. All pigs were necropsied and macroscopically affected organs were, if present, sampled for bacteriological examination. Palatine tonsillar tissue was collected.

Laboratory Tests

Quantitative bacteriological examination was performed on all samples. Serial dilutions of swab or tonsil brushing samples (10^0–10^{-4}) were plated on selective agar plates, containing Columbia agar, 6% sheep blood, 0.2 μg/mL crystal violet and colistin/oxolinic acid (BioTrading, The Netherlands) [41], [44]. After incubation for 18–24 h at 37°C and 5% CO_2, plates with 10–200 colonies were selected. Suspected *S. suis* colonies were counted and per plate two *S. suis* suspected colonies were subcultured and tested for amylase activity [45]. Isolates that showed amylase activity were stored at −20°C in 0.5 mL TE buffer pH 7.5 (10 mM) until further processing. DNA-isolation of these isolates was performed with InstaGene™ Matrix (Biorad, The Netherlands), according to the manufacturers instructions. A real time polymerase chain reaction (RT-PCR) on the *cps9H* gene [41], [46] was performed on suspected isolates from the first two and the last saliva and tonsil samples in which amylase positive colonies were detected. All colonies that were visually classified as *S. suis* tested amylase-positive, and were all positive in the *cps9H*-PCR. Tonsil tissue samples that were obtained at necropsy were submerged in boiling water (5 s), crunched in a Stomacher®

macerator (Laméris, The Netherlands) for 10 min, and diluted, plated and confirmed as described above for samples, except for the dilutions that were plated (10^{-1} to 10^{-5}).

Statistical Analysis of Culture Data and Clinical Data

Mean *S. suis* ^{10}logCFU levels of the first 3 samples taken after onset of colonization in individual pigs were compared among the 3 groups (C_{dir}, C_{ind} and I) using a Kruskal-Wallis one-way ANOVA. The same procedure was performed for mean levels over the whole period after pigs tested positive, and for *S. suis* counts (in ^{10}logCFU/g tissue) in tonsillar tissue. For samples where no *S. suis* was detected, the detection limit of the test procedure (i.e. 2×10^3 CFU/sample) was used for the calculations. It did not affect the results if we had used a much lower value, i.e. 2 CFU/sample, instead for analysis.

Clinical signs were compared between the groups with a Kruskal-Wallis test applied on the percentage of days a pig showed signs. If the result of the Kruskal-Wallis test was significant, the Mann-Whitney U test was performed to determine differences between groups. The Bonferroni approach was used to correct for multiple comparisons. Differences in outcome variables were considered significant if $P<0.05$. Analysis of culture and clinical data was performed in SPSS version 16.0.2.

Statistical Analysis of Transmission

To analyze transmission of *S. suis*, a stochastic SI (susceptible-infectious) model was applied. A pig was considered to be infected if at least one sample tested positive. The transmission rates β_{dir} and β_{ind} represent the rates at which infectious animals transmit the infection to other animals within and among pens, respectively. The parameter β_{ind} includes the process of *S. suis* transmission without direct contact between hosts, and thus exclusively via the environment. The parameter β_{dir} includes both transmission via direct contact and transmission via the environment, as these two routes cannot be distinguished in a pair wise setting.

According to the standard SI model, susceptible pigs become infectious at rate $\beta_{dir} * I_w$ (or $\beta_{ind} * I_b$) in which I_w and I_b are the numbers of infectious animals present within the same pen, and in the whole compartment, respectively. The standard model assumes that susceptible animals are infected at a constant rate if I_w (or I_b) is constant, which was more or less constant during our experiment. For within-pen transmission the data allowed the standard model to be used. For between-pen transmission, however, we did not see new infections in C_{ind} until one week after exposure to inoculated pigs, and counted 13 cases in the two weeks thereafter. This implies that for between-pen transmission the assumption of constant infection rate per unit of time was not met, and the standard SI model could not be used for analysis. For between-pen transmission, we therefore fitted an alternative model in which susceptible animals are infected at rate $\beta'_{ind} * \Sigma I$, in which ΣI is the cumulative number of shedding days of all infectious animals in that stable compartment (at time *t*). In this model a shedding day was defined as a day that *S. suis* was found in the tonsil and/or nasal swab sample taken at that day. This model reflects a gradual build-up of infectivity in the environment. The alternative transmission rate β'_{ind} (unit: 1/day^2) is now interpreted as the number of new infections an animal will cause *each day* in a susceptible population, *for each day* that it has already been infectious. In the analysis it was assumed that from the moment that pigs died or were euthanized, they did not add any infectivity to the cumulative infectivity anymore. The infectivity they had added before death was still included in the total infectivity in that compartment for the remaining experimental period.

Additionally, as the bacterial load in infected pigs might influence the transmission rate, models were tested which included cumulative colonization or shedding levels in tonsillar brushing or saliva samples instead of shedding days. The 'cumulative colonization or shedding level' at a particular day was calculated by adding the CFU/sample up to that day, using interpolation for all days on which no sample was taken. We used Akaike's Information Criterion (AIC) to test if this alternative model fitted the data better than the original model [47]. The transmission rate calculations were performed in statistical program R version 2.13.0.

As mentioned, the experiment was designed to serve a dual purpose to reduce the number of experimental animals. As vaccinated and non-vaccinated pigs were distributed over all but one compartments, it was not possible to distinguish between infectivity of vaccinated and unvaccinated pairs. In addition, the bacterial load in swab or brushing samples did not differ significantly between vaccinated and unvaccinated pigs. It was therefore assumed that all colonized pigs contributed equally to transmission of *S. suis*. Therefore, vaccination was not included as explanatory variable in the statistical analysis to determine the effect of spatial separation. Power analysis was based on the research question about vaccination [41].

Simulation Study

The effect of separation of groups of pigs on the spread of *S. suis* in a conventional farm compartment was studied using a simulation model. A situation with separated litters (scenario A) was compared with one of mixing litters (scenario B). In scenario A, the compartment contained 10 pens each consisting of 10 pigs. In scenario B, 10 litters of 10 pigs were mixed to form two groups of 50 pigs housed in two pens. The model was independent of distances between pens. Simulations were started with one infected pig. The estimate for β'_{ind} was divided by two to compensate for the in comparison to a compartment extra space needed to house 100 pigs. Per situation 10.000 iterations were performed. Simulation was done with statistical program R version 2.13.0.

RESULTS

Colonization

No *S. suis* was detected in samples taken before inoculation. All inoculated pigs tested positive for *S. suis* at 2 DPI and all C_{dir} pigs tested positive within two days after exposure. The first C_{ind} pigs tested positive on day 7 post exposure and their number gradually increased to 13 (Figure 2). Once positive, pigs remained positive during the remaining experimental period.

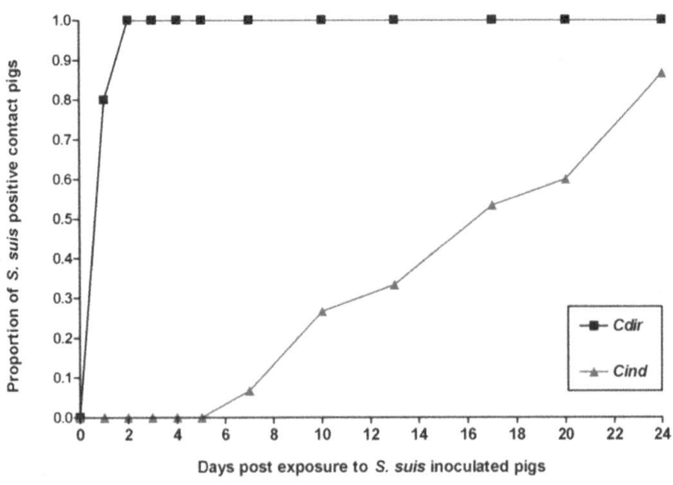

Figure 2: Transmission of *S.*suis to pigs with or without direct contact with infectious ones.

Diagram showing the proportion of *S. suis* positive pigs over time for directly contact exposed pigs (C_{dir};N=25) and pigs housed at a distance (C_{ind}; N=15) after *S. suis* serotype 9 inoculated pigs entered the stable compartments in the transmission experiments. Two C_{ind} pigs remained negative.

Mean levels of *S. suis* are shown in Figure 3. No differences were observed between the animal categories (C_{dir}, C_{ind} and *I)* in mean levels in tonsil and saliva samples (*P*-values are 0.25 and 0.12, respectively, for the first 3 positive samples and 0.19 and 0.66, respectively, for the whole sampling period). No differences between these categories were observed in the load of *S. suis* in their tonsil tissue samples (*P*=0.99). Based on these results, the assumption of homogeneity of infectivity was considered reasonable.

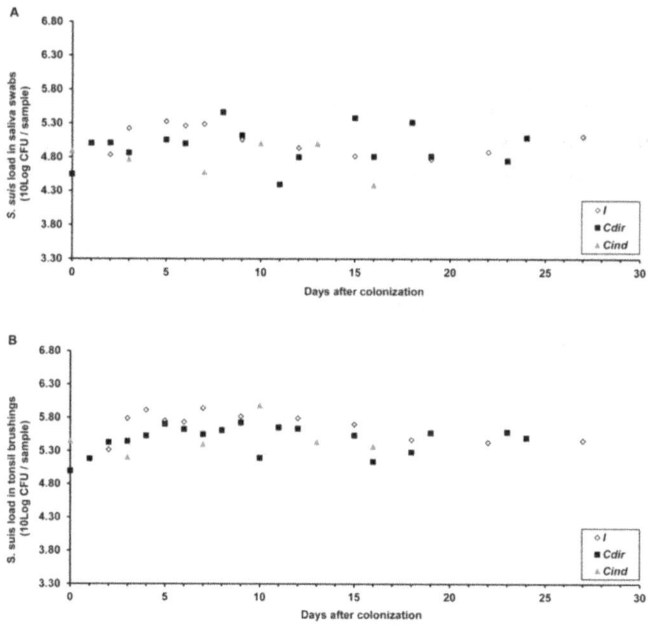

Figure 3: Mean *S.* suis loads in saliva and tonsil samples.

Mean loads of *S. suis* serotype 9 colony forming units (CFU) in saliva (panel A) and tonsil brushing samples (panel B) shown separately for pigs inoculated with (*I*), directly (C_{dir}) or indirectly contact exposed (C_{ind}) to *S. suis*. Time is expressed as days after a pig was found positive firstly. As sampling was not conducted daily, and because of the delay in *S. suis* transmission to mainly C_{ind} pigs the numbers and time points of observations differ between the groups. No significant differences were observed in mean levels between C_{dir}, C_{ind} and I pigs for the first 3 positive samples and for mean levels over the whole sampling period.

Estimation of Transmission Rates

The rate for within-pen transmission β_{dir} was estimated at 3.58 (95% CI: 2.29–5.60) new infections an animal will cause in a susceptible population, and for between-pen transmission β'_{ind} was 0.001 (95% CI: 0.0006–0.0017) new infections an animal will cause each day in a susceptible population, for each day that it has already been infectious. Spatial separation reduced the transmission on average 36 times (range 20–895). Addition of cumulative *S. suis* levels did not change the fit of the model for estimating indirect transmission.

Simulation Study

In both scenarios, all pigs were *S. suis* positive before 14 days (Figure 4). It was estimated that if *S. suis* free litters were separated from infected ones (scenario A), a prevalence of 50% or 90% would be reached at on average day 8 (2.5th–97.5th percentile: 6–10 days) and day 11 (2.5th–97.5th percentile: 9–13 days). When pigs were mixed (scenario B), this was reached after 4 (2.5th–97.5th percentile: 3–5 days) and 7 days (2.5th–97.5th percentile: 5–9 days), respectively.

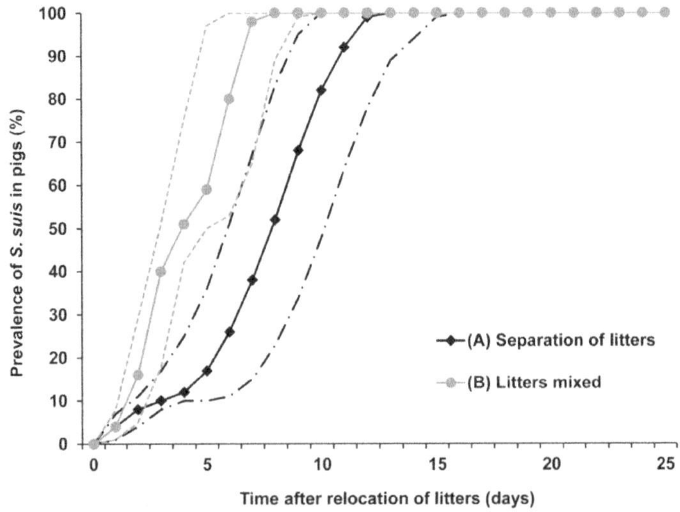

Figure 4: *S.* suis spread within a simulated hypothetical stable compartment under different weaning scenarios.

Course of percentage of *S. suis* positive pigs over time after relocation of pigs at weaning for two simulated situations based on a hypothetical conventional pig farm setting. In scenario A, 10 litters of 10 pigs each were introduced in a stable compartment and kept spatially separated. In scenario B, 10 litters were mixed to form two groups of 50 pigs. In both scenarios it was assumed that that at relocation one of the 100 pigs was infected with *S. suis* and that direct contact between pigs in different pens was impossible. Per scenario 10.000 iterations were performed. The solid lines represent the median values, and the dashed lines the 2.5th and 97.5th percentiles of these simulations. The black lines represent scenario A, and the grey lines scenario B.

Clinical Signs and Mortality

Lameness was the only clinical sign observed in C pigs; it occurred in 4 out of 25 C_{dir} and 3 out of 15 C_{ind} pigs (P=0.99). Signs were observed more often in I than in C_{dir} (P=0.01) or C_{ind} pigs ($P < 0.001$). Out of the 25 I pigs, in 20 pigs fever was recorded on one or more days, in 10 pigs lameness signs, in 3 pigs neurologic signs and 4 pigs died before the end of the experiment.

DISCUSSION

The objective of this study was to quantify the effect of separation of pigs on the transmission rate of *S. suis* serotype 9. All directly and indirectly exposed pigs became colonized after exposure to inoculated ones, but the rate at which it occurred was approximately 36 times lower for indirectly compared to directly exposed pigs. This finding suggests that this intervention measure could contribute to the reduction of *S. suis* spread in a farm. However, the simulation study using these parameters showed that the cumulative incidence in piglets did not differ between 'separated' and 'mixed' groups to a relevant level. This implies that it is unlikely that separation of pigs as simulated in this simulation exercise will contribute to reduction of the prevalence of *S. suis* serotype 9 infections in a farm compartment, also taking into account that on conventional farms less stringent hygienic measures are implemented than in this experiment.

Separation sharply reduced the transmission rate in the experiment, but had only a small effect on colonization time in a farm compartment. This counterintuitive finding is caused by the large number of pigs present in such a compartment. Although each of the indirectly exposed pigs only has a small probability of getting infected, the probability that at least one of them will contract the infection is considerable and, moreover, strongly increases as the number of infectious pigs in the 'source' pen increases. Once indirect transmission has taken place, direct transmission further increases the infectious load and consequently increases exposure of the remaining uninfected pigs and the probability of further indirect transmission.

In total 15 C_{ind} pigs were colonized, most likely via contaminated dust particles, as has been described for *S. suis* serotype 2 [35], [48]. We consider other indirect routes of *S. suis* spread, e.g. via animal caretakers [20], [21], [49], less likely because of the stringent biosecurity measures that were applied.

A model assuming increasing infectivity fitted the data of indirect transmission better than a model assuming fixed infectivity. The increasing infectivity is likely to have occurred due to accumulation of *S. suis* in the environment [28], a phenomenon which has also been observed in studies with other pathogens [40], [50]–[55]. In studies with *Campylobacter jejuni* and *Escherichia coli* in broilers, for example, a similar pattern of increasing probability of infection was observed [40], [50]. The authors of these studies concluded that the model fitted the data better when assuming increasing infectivity [40], [50]. We also monitored the levels of the pathogen in the samples, as an association between these levels and an increasing infectivity could be suggested. The levels remained rather similar during the experiment, and addition of cumulative levels of *S. suis* in saliva or tonsil samples did not further improve the fit of the model.

Transmission of *S. suis* is not only dependent on the type of contacts between pigs, but also on the infectivity and susceptibility of individual pigs, which might be affected by (changes in) host behaviour. This may, for example, occur after mixing pigs from different litters leading to fighting [56]–[58], which would change the frequency and/or intensity of contacts between pigs [11], [12], [56]–[58]. Moreover, the stress induced by mixing [11], [12], [56]–[58] could further increase both susceptibility, and infectivity by increasing colonization and shedding of already infected hosts, as described for other bacterial pathogens [59]–[61]. If and to which extent the transmission changes after mixing is, however, not known. Moreover, even with adjustments the conclusions about the relevance of the effect of spatial separation would not change, as in the optimal 'separation' scenario the spread of *S. suis* serotype 9 was already so rapid that all pigs in a compartment were infected within about two weeks.

In the experiment stringent hygienic measures were applied between pens at compartment scale, and the same was assumed in the simulation study. In pig farms it is probably nearly impossible to perform such strict measures between pens in the same compartment, but they might be implemented at herd scale between compartments. As the rate of airborne spread is probably lower between compartments than between pens, spread of *S. suis* strains between different compartments, which are physically separated, might be reduced.

ACKNOWLEDGMENTS

The animal care personnel is acknowledged for the excellent assistance during the animal experiments.

REFERENCES

1. Bailey NTJ (1975) General epidemics. In: Bailey NTJ, editor. The mathematical theory of infectious diseases and its applications.New York: Hafner press.81–133.
2. Koopman JS, Longini IM Jr (1994) The ecological effects of individual exposures and nonlinear disease dynamics in populations. American Journal of Public Health 84: 836–842.
3. Diekmann O, Heesterbeek JAP (2000) Mathematical epidemiology of infectious diseases. model building, analysis and interpretation. Chichester: John Wiley & Son. 303 p.
4. De Jong MC, Kimman TG (1994) Experimental quantification of vaccine-induced reduction in virus transmission. Vaccine 12: 761–766.
5. Dore E, Pare J, Cote G, Buczinski S, Labrecque O, et al. (2012) Risk factors associated with transmission of mycobacterium avium subsp. paratuberculosis to calves within dairy herd: A systematic review. Journal of Veterinary Internal Medicine/American College of Veterinary Internal Medicine 26: 32–45.
6. Eble P, de Koeijer A, Bouma A, Stegeman A, Dekker A (2006) Quantification of within- and between-pen transmission of foot-and-mouth disease virus in pigs. Veterinary Research 37: 647–654.
7. van Roermund HJ, Eble PL, de Jong MC, Dekker A (2010) No between-pen transmission of foot-and-mouth disease virus in vaccinated pigs. Vaccine 28: 4452–4461.
8. Torremorell M, Allerson M, Corzo C, Diaz A, Gramer M (2012) Transmission of influenza A virus in pigs. Transboundary and Emerging Diseases 59 (Suppl. 1)68–84.
9. Reina R, Berriatua E, Lujan L, Juste R, Sanchez A, et al. (2009) Prevention strategies against small ruminant lentiviruses: An update. Veterinary Journal (London, England: 1997) 182: 31–37.
10. Amass SF, Baysinger A (2006) Swine disease transmission and prevention. In: Straw BE, Zimmerman JJ, D'Allaire S, Taylor DJ, editors. Diseases of Swine.Ames, Iowa, USA: Blackwell Publishing. pp. 1075–1098.
11. Ekkel ED, Savenije B, Schouten WG, Tielen MJ (1996) Health, welfare, and productivity of pigs housed under specific-stress-free conditions in comparison with two-site systems. Journal of Animal Science 74:

2081–2087.

12. Horvath G, Jaczkone Jarabin I, Visnyei L (2000) Effects of regrouping, feeding and drinking methods on weight gain of weaned piglets. DTW. Deutsche Tierarztliche Wochenschrift 107: 364–367.
13. Cleveland-Nielsen A, Nielsen EO, Ersboll AK (2002) Chronic pleuritis in danish slaughter pig herds. Preventive Veterinary Medicine 55: 121–135.
14. Van Nes A, De Jong MC, Buijtels JA, Verheijden JH (1998) Implications derived from a mathematical model for eradication of pseudorabies virus. Preventive Veterinary Medicine 33: 39–58.
15. Swildens B, Nielen M, Wisselink HJ, Verheijden JH, Stegeman JA (2007) Elimination of strains of streptococcus suis serotype 2 from the tonsils of carrier sows by combined medication and vaccination. The Veterinary Record 160: 619–621.
16. Torremorell M, Calsamiglia M, Pijoan C (1998) Colonization of suckling pigs by streptococcus suis with particular reference to pathogenic serotype 2 strains. Canadian Journal of Veterinary Research=Revue Canadienne De Recherche Veterinaire 62: 21–26.
17. Clifton-Hadley FA, Alexander TJ, Upton I, Duffus WP (1984) Further studies on the subclinical carrier state of streptococcus suis type 2 in pigs. The Veterinary Record 114: 513–518.
18. Van Leengoed LA, Vecht U, Verheyen ER (1987) Streptococcus suis type 2 infections in pigs in the netherlands (part two). The Veterinary Quarterly 9: 111–117.
19. Brisebois LM, Charlebois R, Higgins R, Nadeau M (1990) Prevalence of streptococcus suis in four to eight week old clinically healthy piglets. Canadian Journal of Veterinary Research=Revue Canadienne De Recherche Veterinaire 54: 174–177.
20. Robertson ID, Blackmore DK, Hampson DJ, Fu ZF (1991) A longitudinal study of natural infection of piglets with streptococcus suis types 1 and 2. Epidemiology and Infection 107: 119–126.
21. Dee SA, Carlson AR, Winkelman NL, Corey MM (1993) Effect of management practices on the streptococcus suis carrier rate in nursery swine. Journal of the American Veterinary Medical Association 203: 295–299.
22. Cloutier G, D'Allaire S, Martinez G, Surprenant C, Lacouture S, et al. (2003) Epidemiology of streptococcus suis serotype 5 infection in a pig

herd with and without clinical disease. Veterinary Microbiology 97: 135–151.

23. Clifton-Hadley FA, Alexander TJ, Enright MR, Guise J (1984) Monitoring herds for streptococcus suis type 2 by sampling tonsils of slaughter pigs. The Veterinary Record 115: 562–564.
24. Flores JL, Higgins R, D'Allaire S, Charette R, Boudreau M, et al. (1993) Distribution of the different capsular types of streptococcus suis in nineteen swine nurseries. The Canadian Veterinary Journal.La Revue Veterinaire Canadienne 34: 170–171.
25. Luque I, Blume V, Borge C, Vela AI, Perea JA, et al. (2009) Genetic analysis of streptococcus suis isolates recovered from diseased and healthy carrier pigs at different stages of production on a pig farm. Vet J. Vet J. 186(3) pp. 396–398.
26. Robertson ID, Blackmore DK (1989) Prevalence of streptococcus suis types 1 and 2 in domestic pigs in australia and new zealand. The Veterinary Record 124: 391–394.
27. Staats JJ, Feder I, Okwumabua O, Chengappa MM (1997) Streptococcus suis: Past and present. Veterinary Research Communications 21: 381–407.
28. Higgins R, Gottschalk M (2006) Streptococcal diseases. In: Straw BE, Zimmerman JJ, D'Allaire S, Taylor DJ, editors. Diseases of Swine. Ames, Iowa, USA: Blackwell Publishing. pp. 769–783.
29. Gottschalk M, Segura M (2000) The pathogenesis of the meningitis caused by streptococcus suis: The unresolved questions. Veterinary Microbiology 76: 259–272.
30. Allgaier A, Goethe R, Wisselink HJ, Smith HE, Valentin-Weigand P (2001) Relatedness of streptococcus suis isolates of various serotypes and clinical backgrounds as evaluated by macrorestriction analysis and expression of potential virulence traits. Journal of Clinical Microbiology 39: 445–453.
31. Wisselink HJ, Smith HE, Stockhofe-Zurwieden N, Peperkamp K, Vecht U (2000) Distribution of capsular types and production of muramidase-released protein (MRP) and extracellular factor (EF) of streptococcus suis strains isolated from diseased pigs in seven european countries. Veterinary Microbiology 74: 237–248.
32. Silva LM, Baums CG, Rehm T, Wisselink HJ, Goethe R, et al. (2006) Virulence-associated gene profiling of streptococcus suis

isolates by PCR. Veterinary Microbiology 115: 117–127.
33. Tarradas C, Perea A, Vela AI, Goyache J, Dominguez L, et al. (2004) Distribution of serotypes of streptococcus suis isolated from diseased pigs in spain. The Veterinary Record 154: 665–666.
34. Callens B, Persoons D, Maes D, Laanen M, Postma M, et al. (2012) Prophylactic and metaphylactic antimicrobial use in belgian fattening pig herds. Preventive Veterinary Medicine 106: 53–62.
35. Berthelot-Herault F, Gottschalk M, Labbe A, Cariolet R, Kobisch M (2001) Experimental airborne transmission of streptococcus suis capsular type 2 in pigs. Veterinary Microbiology 82: 69–80.
36. Velthuis AG, de Jong MC, de Bree J, Nodelijk G, van Boven M (2002) Quantification of transmission in one-to-one experiments. Epidemiology & Infection 128: 193–204.
37. Velthuis AG, Bouma A, Katsma WE, Nodelijk G, De Jong MC (2007) Design and analysis of small-scale transmission experiments with animals. Epidemiology & Infection 135: 202–17.
38. Feberwee A, Mekkes DR, Klinkenberg D, Vernooij JC, Gielkens AL, et al. (2005) An experimental model to quantify horizontal transmission of mycoplasma gallisepticum. Avian Pathology 34: 355–361.
39. van Bunnik BA, Katsma WE, Wagenaar JA, Jacobs-Reitsma WF, de Jong MC (2012) Acidification of drinking water inhibits indirect transmission, but not direct transmission of campylobacter between broilers. Preventive Veterinary Medicine 105: 315–319.
40. Bunnik B (2012) Unexpected delay in transmission between spatially separated hosts. Book of Abstracts of the 13th International Symposium on Veterinary Epidemiology and Economics: 103. Available: http://www.isvee13.org/programme/scientific-programme. Accessed 2012 Nov 23.
41. Dekker CN, Bouma A, Daemen AJ, van Leengoed LA, Jonker FH, et al. (2012) Homologous whole bacterin vaccination is not able to reduce streptococcus suis serotype 9 strain 7997 transmission among pigs or colonization. Vaccine 30: 1379–1387.
42. de Greeff A, Wisselink HJ, de Bree FM, Schultsz C, Baums CG, et al. (2011) Genetic diversity of streptococcus suis isolates as determined by comparative genome hybridization. BMC Microbiology 11: 161.
43. Schultsz C, Jansen E, Keijzers W, Rothkamp A, Duim B, et al. (2012) Differences in the population structure of invasive streptococcus

suis strains isolated from pigs and from humans in the netherlands. PLoS One 7(5) e33854: 10.1371/journal.pone.0033854.

44. Wisselink HJ, Reek FH, Vecht U, Stockhofe-Zurwieden N, Smits MA, et al. (1999) Detection of virulent strains of streptococcus suis type 2 and highly virulent strains of streptococcus suis type 1 in tonsillar specimens of pigs by PCR. Veterinary Microbiology 67: 143–157.
45. Devriese LA, Ceyssens K, Hommez J, Kilpper-Balz R, Schleifer KH (1991) Characteristics of different streptococcus suis ecovars and description of a simplified identification method. Veterinary Microbiology 26: 141–150.
46. Smith HE, Veenbergen V, van der Velde J, Damman M, Wisselink HJ, et al. (1999) The cps genes of streptococcus suis serotypes 1, 2, and 9: Development of rapid serotype-specific PCR assays. Journal of Clinical Microbiology 37: 3146–3152.
47. Dohoo I, Martin W, Stryhn H (2007) Veterinary epidemiologic research. Prince Edward Island: AVC Inc.706 p.
48. Clifton-Hadley FA, Enright MR (1984) Factors affecting the survival of streptococcus suis type 2. The Veterinary Record 114: 584–586.
49. Enright MR, Alexander TJ, Clifton-Hadley FA (1987) Role of houseflies (musca domestica) in the epidemiology of streptococcus suis type 2. The Veterinary Record 121: 132–133.
50. Bunnik B (2012) An experimental and modelling study on the mechanisms of indirect transmission. Abstracts of the 13th International Symposium on Veterinary Epidemiology and Economics: 376 . Available: http://www.isvee13.org/programme/scientific-programme. Accessed 2012 Nov 23. [Google Scholar]
51. Courcoul A, Hogerwerf L, Klinkenberg D, Nielen M, Vergu E, et al.. (2011) Modelling effectiveness of herd level vaccination against Q fever in dairy cattle. Veterinary Research 42: : 68. Available: http://www.veterinaryresearch.org/content/42/1/68.Accessed 23 November 2012.
52. Weesendorp E, Loeffen W, Stegeman A, de Vos C (2011) Time-dependent infection probability of classical swine fever via excretions and secretions. Preventive Veterinary Medicine 98: 152–164.
53. Lurette A, Belloc C, Touzeau S, Hoch T, Ezanno P, et al.. (2008) Modelling salmonella spread within a farrow-to-finish pig herd. Veterinary Research 39: : 49. Available: http://dx.doi.org/10.1051/

vetres:2008026. Accessed 23 November 2012.

54. Marce C, Ezanno P, Seegers H, Pfeiffer DU, Fourichon C (2011) Within-herd contact structure and transmission of mycobacterium avium subspecies paratuberculosis in a persistently infected dairy cattle herd. Preventive Veterinary Medicine 100: 116–125.

55. Marce C, Ezanno P, Weber MF, Seegers H, Pfeiffer DU, et al. (2010) Invited review: Modeling within-herd transmission of mycobacterium avium subspecies paratuberculosis in dairy cattle: A review. J Dairy Sci 93: 4455–4470.

56. Friend TH, Knabe DA, Tanksley TD Jr (1983) Behavior and performance of pigs grouped by three different methods at weaning. Journal of Animal Science 57: 1406–1411.

57. Hessel EF, Reiners K, Van den Weghe HF (2006) Socializing piglets before weaning: Effects on behavior of lactating sows, pre- and postweaning behavior, and performance of piglets. Journal of Animal Science 84: 2847–2855.

58. Jensen P (1994) Fighting between unacquainted pigs - effects of age and of individual reaction pattern. Applied Animal Behaviour Science 41: 37–52. [Google Scholar]

59. Lyte M, Vulchanova L, Brown DR (2011) Stress at the intestinal surface: Catecholamines and mucosa-bacteria interactions. Cell and Tissue Research 343: 23–32.

60. Rostagno MH (2009) Can stress in farm animals increase food safety risk? Foodborne Pathogens and Disease 6: 767–776.

61. Naresh R, Hampson DJ (2011) Exposure to norepinephrine enhances brachyspira pilosicoli growth, attraction to mucin and attachment to caco-2 cells. Microbiology (Reading, England) 157: 543–547.

CHAPTER 10

Association between Transmission Rate and Disease Severity for Actinobacillus Pleuropneumoniae Infection in Pigs

Tijs J Tobias[1], Annemarie Bouma[1] , Angeline JJM Daemen[1] , Jaap A Wagenaar[2,3], Arjan Stegeman[1] and Don Klinkenberg[1]

[1]Department of Farm Animal Health, Faculty of Veterinary Medicine, Utrecht University, Yalelaan 7, Utrecht 3584 CL, The Netherlands
[2] Department of Infectious Diseases and Immunology, Faculty of Veterinary Medicine, Utrecht University, Yalelaan 1, Utrecht 3584 CL, The Netherlands
[3] Central Veterinary Institute of Wageningen UR, Lelystad 8200 AB, The Netherlands

ABSTRACT

A better understanding of the variation in infectivity and its relation with clinical signs may help to improve measures to control and prevent (clinical) outbreaks of diseases. Here we investigated the role of disease severity on infectivity and transmission of *Actinobacillus pleuropneumoniae*, a bacterium causing respiratory problems in pig farms. We carried out transmission

Citation: Tobias, T.J., Bouma, A., Daemen, A.J. et al. Association between transmission rate and disease severity for Actinobacillus pleuropneumoniae infection in pigs. Vet Res 44, 2 (2013). https://doi.org/10.1186/1297-9716-44-2

experiments with 10 pairs of caesarean-derived, colostrum-deprived pigs. In each pair, one pig was inoculated intranasally with 5×10^6 CFUs of *A. pleuropneumoniae* strain 1536 and housed together with a contact pig. Clinical signs were scored and the course of infection was observed by bacterial examination and qPCR analysis of tonsillar brush and nasal swab samples. In 6 out of 10 pairs transmission to contact pigs was observed, but disease scores in contact infected pigs were low compared to the score in inoculated pigs. Whereas disease score was positively associated with bacterial load in inoculated pigs and bacterial load with the transmission rate, the disease score had a negative association with transmission. These findings indicate that in pigs with equal bacterial load, those with higher clinical scores transmit *A. pleuropneumoniae* less efficiently. Finally, the correlation between disease score in inoculated pigs and in positive contact pigs was low. Although translation of experimental work towards farm level has limitations, our results suggest that clinical outbreaks of *A. pleuropneumoniae* are unlikely to be caused only by spread of the pathogen by clinically diseased pigs, but may rather be the result of development of clinical signs in already infected pigs.

INTRODUCTION

The course of an infection can vary widely between individual hosts, affecting clinical signs of the infected host as well as the ability to transmit the infection to another member of the population. Variation in infectivity between individuals has shown to be of importance for the course of an epidemic [1-3] and can be associated with variation in clinical signs [4-6]. For some diseases, effective control measures should therefore target the severely diseased individual to reduce the size of an outbreak [7].

Current control measures for bacterial diseases in livestock farming often rely on an intervention with antimicrobials for the prevention or treatment of disease. As the use of antimicrobials is under debate, proper insight in the course of outbreaks and associations between disease and pathogen dispersion during outbreaks is crucial to design alternative control measures.

A. pleuropneumoniae is such a pathogen, causing clinical outbreaks and commonly controlled by antimicrobials, with high variation in infectivity [8]. Worldwide, *A. pleuropneumoniae* is highly prevalent among pig farms [9] and outbreaks of severe respiratory disease, with fibro hemorrhagic necrotizing pneumonia and fibrinous pleurisy [10], are observed occasionally. Variation in outbreaks with respect to size [11] and severity [12,13] are related to factors such as strain and serotype differences [14-16], immunity of the host [17,18] and co-infections [19,20].

Currently, the association between disease and pathogen dispersion and thereby the clinical course of *A. pleuropneumoniae* outbreaks is poorly understood. It could be that already subclinically infected pigs develop clinical signs at the same time because of the presence of a risk factor (trigger) that may be either infectious or non infectious [21]. Alternatively, it may be that an outbreak starts with only a few diseased pigs, which rapidly spread a clinical form of the infection on the farm, a hypothesis that is supported by a study that describes a rapid increase of prevalence during and after outbreaks [22]. The second hypothesis suggests a strong association between clinical signs and infectiousness, as well as a strong association between clinical signs of infected pigs within a transmission chain.

The aim of this study is to assess for *A. pleuropneumoniae*, the association between severity of clinical disease, infectiousness and transmission. In an experimental setting, the variation in disease between inoculated pigs was used to relate disease severity to the excreted bacterial load, the rate of transmission to a contact exposed pig and disease severity in that contact exposed pig.

MATERIAL AND METHODS

Experimental Design

To study *A. pleuropneumoniae* transmission, ten pair-wise transmission trials, split up in two separate and identically performed runs, were conducted. Per run twelve Caesarean-derived, colostrum-deprived (CD/CD) piglets (Landrace × York × Pietrain) were reared in incubators until 21 days of age as described before [23]. CD/CD pigs were used as a result of pilot studies that showed interference of *Pseudomonas spp.* and other *Pasteurellaceae spp.* for detection and quantification of *A. pleuropneumoniae* Colony Forming Units (CFUs) in samples obtained from inoculated SPF piglets (free of *A. pleuropneumoniae)*. At day 21 pigs were moved to an experimental unit with eight pens, length × width: 1.1 × 1.1 meter, with 1/3 slatted floor and walls of 0.8 meter high, as in Dekker et al. [23]. The experimental unit was equipped with a laminar ventilation air flow. The room temperature was set constant at 25°C and air speed at 2.4 m/s.

Pigs were ear tagged and randomly assigned to pen 1, 2 or 3 for an eight day habituation period until inoculation. The habituation period was included for acclimatisation to the conditions in the experimental unit, to improve feed intake as well as to train the animals for clinical examination. Because trained pigs are likely to respond less stressfully to physical examinations, measurement errors were thus minimised.

After the habituation period pigs were randomly assigned a status and a pair number. Pairs consisted of I-pigs, that were inoculated, and C-pigs, that were contact exposed. Before inoculation, all C-pigs were moved to their designated pens. The I-pigs were housed in pen 1 for inoculation and moved to their pen six hours after inoculation. Per run two S-pigs were assigned as sentinel and housed together in a separate pen to monitor indirect transmission of *A. pleuropneumoniae*.

Table 1: Cumulative clinical disease score, necropsy and bacteriology results for *A. pleuropneumoniae* infection per pig

Pair	Pig status	Run/ pen	Survival until day	Clinical score		Pathology				
				RHS4	RHS20	Pleurisy score	Pneumonia score	Bacteriology		
								Lung sample	Nose tonsil	Tonsil
1	I	1/4	4	31.13	86.23	20.7	6.6	+	+	+
	C		21	3.10	5.11	0	0	ND	-	-
2	I	1/5	4	13.68	82.74	19.3	1.0	+	+	+
	C		21	2.48	6.72	0	0.26	-	-	-
3	I	1/6	2	66.88	93.38	33.9	33.2	+	+	+
	C		21	2.35	6.26	0	1.5	-	-	-
4	I	1/7	21	6.80	6.23	0	0	ND	-	+
	C		21	7.90	8.90	0	0	ND	-	+
5	I	1/8	21	16.18	12.87	5.9	0.3	+	+	+
	C		21	5.58	8.77	3.2	1.6	-	-	± dub

6	I	2/4	7	23.70	72.13	6.9	4.7	+	+	+
	C		21	7.43	6.42	0	0	ND	-	+
7	I	2/5	21	17.65	10.35	7.1	8.6	-	-	+
	C		21	8.70	5.55	0	0	ND	-	+
8	I	2/6	21	24.40	12.70	0	0	+	+	+
	C		21	4.03	4.20	0	0	ND	-	-
9	I	2/7	21	30.70	20.09	9.2	15.0	+	+	+
	C		21	7.13	7.55	0	0	ND	-	-
10	I	2/8	4	46.13	89.23	28.1	26.6	+	+	+
	C		21	5.60	7.82	0	0	ND	-	-
11	S	1/3	11*	0.31	ND	0	0	ND	-	-
	S		21	0.89	5.39	0	0	-	-	-
12	S	2/3	21	0.68	4.05	0	0	-	-	-
	S		21	0.56	5.27	0	0	-	-	-

+=*A. pleuropneumoniae* confirmed, ± dub=dubious growth, ND=Not determined, * euthanized due to lameness.

Experiments were approved by the Animal experiments committee of Utrecht University (AEC) (approval number DEC2010.II.02.25). When pigs had body temperature of>40°C or showed eminent signs of pain they were treated with Fentanyl (B. Braun Melsungen AG, Melsungen, Germany). Fentanyl is a potent analgesic, but does not bear anti-inflammatory capacities. Pigs were euthanized when the results of the daily welfare assessment exceeded the criteria accorded by the AEC.

Inoculation

Inoculation was performed intranasally with a six hour culture of *A. pleuropneumoniae* reference strain 1536. *A. pleuropneumoniae* was cultured on Heart Infusion agar with 5% sheep erythrocytes and 0.2% β-NAD (Nicotinamide adenine dinucleotide) (HIS-V) overnight at 37°C and 5%

CO_2. The next day 1 colony was suspended in 1 mL of saline and 50 μL was plated on a new HIS-V plate and incubated. After six hours, the plate was washed with 5 mL of saline and diluted to 2.5×10^6 CFUs/mL guided by optical density measurements. Before inoculation pigs were sedated with Azaperon (Stressnil®, Janssen Animal Health, Brussels, Belgium) and subsequently pigs were inoculated by dripping 1 mL of the inoculum in each nostril during inhalation (total 5×10^6 CFUs). Before and after inoculation the concentration of the inoculum was determined by plating of serial dilution series. Inoculation dose and method of administration were chosen based on the results of two pilot studies where sufficient variation in disease severity was obtained, when 10^6 Colony Forming Units (CFUs) *A. pleuropneumoniae* serotype 2 (strain 1536) was intranasally administered in CD/CD pigs.

Samples

Before examination, restraining or sampling, all personnel had to change boots, coverall and gloves for each pen. To minimise the risk of transmission due to sampling or examination the contact pig was always sampled before the inoculated pig.

Tonsil and nasal samples were collected on post inoculation day −1, 1, 2, 4, 6, 8, 11, 13, 15, 18 and 21, or before euthanasia of severely diseased animals. Nasal swabs were obtained with a small cotton swab (Applimed SA, Italy) and tonsil scrapings were obtained by brushing the tonsils for 10 s with a soft toothbrush. On day −1 and day 21 the pigs were bled.

Bacteriology

Tonsil brush samples were submerged in 10 mL and nose swabs in 1 mL saline and thoroughly mixed for 20 min before selective bacteriologic examination (SBE). Subsequently, tenfold serial dilution series were made of tonsil brush (10^{-1}, 10^{-2}, 10^{-3}) and nose swab samples (10^0, 10^{-1}, 10^{-2}). Per dilution 50 μL was plated on a selective agar plate with Clindamycin, Gentamicin, Vancomycin and Amphotericin (CGVA plate) [24] and incubated at 37°C and 5% CO_2. *A. pleuropneumoniae* suspected colonies were counted after overnight incubation. Per sample two suspected colonies were confirmed as *A. pleuropneumoniae* when positively tested for satellite growth, Christie–Atkins–Munch–Petersen (CAMP) reaction, urease and mannitol fermentation. Bacterial counts were back calculated to whole sample constituents in CFUs. Additional analyses were performed with an apxIVA qPCR [25].

Clinical Disease Score

A clinical score (CS) was obtained for each pig daily from day −8 to day 21 post inoculation by the same examiner. CS was calculated as the average score (on a 0 – 4 scale) for eight different clinical parameters, scored as described by Hoeltig [26]: behaviour, locomotion score, vomiting, body temperature, respiratory breathing type, respiratory sounds, breathing frequency and coughing. Unlike in Hoeltig et al., cyanosis was not included, because it was never observed. Neither was feed intake, which could only be observed per pair. Clinical scores were obtained twice a day from day 0 – 5 and once a day thereafter, while bacterial samples were collected less frequently. To include all information of clinical observations, average clinical scores (AvgCS) for the days of sampling were derived by averaging all observed CS from one bacterial sampling moment to the next. For example on day 11 the AvgCS was calculated based on the CS of day 9 until 11. To calculate a cumulative respiratory health score at day 4 and 20 (RHS4, RHS20), like in Hoeltig's study, CS was summed until day 4 or 20 and five points per day were added for each remaining day after an animal had died or was euthanized. Subsequently, the summed score was expressed as a percentage of the maximum obtainable score.

Serology

Serology was applied to confirm the cause of clinical symptoms in I-pigs, as well as to detect possible false negative culture or qPCR results in C-pigs. Analysis of serology was performed at the Animal Health Service (Deventer, the Netherlands). Serum samples of day −1 and at euthanasia were analysed by Complement Fixation Test (CFT) titration [27] and a commercial App serotype 2 ELISA (Biovet, St Hyacinthe, Canada). CFT results > 80 in final sample or a distinct (> 0.2) increase of optical density were considered indicative for seroconversion and infection. Additionally, serum samples obtained at euthanasia were analyzed by an ApxIV ELISA (Idexx, Maine, USA). Serum samples of pigs that died before day 14 were not analysed by CFT or ApxIV ELISA, because no seroconversion was expected.

Pathology

At necropsy, all pigs were examined by a veterinary pathologist (at the Veterinary Pathology Diagnostic Centre, Utrecht University). Tonsils were removed and homogenised and macroscopic lung and pleurisy lesions were assessed. A lesion score per lung was calculated, as described by Hannan et al. [28]. Lung lesions and tonsil homogenates were sampled by bacteriologic culture.

Statistical Analyses

The effects of time and pig status (I or C) on AvgCS and the $\log_{10}+1$ of CFUs count in tonsil samples after day 0 were evaluated using mixed effect models with pig number as random effect, to account for repeated observations. Spearman's rank correlation test was used to investigate the correlation between ranks of cumulative clinical scores (RHS4 or RHS20), CFT titres and pathology scores in individual pigs.

To study the effect of disease severity (AvgCS) on transmission from the I-pig to the C-pig (transmission chain), three analyses were performed. First, correlation between AvgCS and $\log_{10}$ CFUs found in tonsil and nasal samples was tested by Spearman's Partial rank correlation using pig number as controlled variable.

Second, the effect of disease severity and bacterial load on the rate of transmission was evaluated. The change from a negative to at least two consecutive positive samples in the contact pig was considered indicative for transmission. Presence or absence of transmission between two samplings (0/1) was used as response variable in a Generalized Linear Model with a binomial error distribution, a complementary log-log link function [8] and ln (time) between samplings as offset. AvgCS and $\log_{10}$ CFUs +1 counts in tonsil and nasal samples were evaluated as model terms and effect estimates are provided with 95% profile confidence intervals. A sensitivity analysis was carried out with respect to the use of SBE as indicator for infectiousness and transmission, by using qPCR results as an additional test (see Additional file 2). The qPCR may be more sensitive than SBE, but specificity for indicating infected and infectious pigs rather than non-viable bacteria may be lower.

Finally, the association between disease severity scores within the transmission chain was analysed. The correlation between CS (AvgCS) of the I-pig, from day of inoculation onwards, with CS (AvgCS) of the C-pig, from the first day of a positive culture and onwards, within pairs, was evaluated using Spearman's Partial rank correlation analysis. Pair number was used as controlled variable.

Statistical analyses were performed using statistical software R version 2.11.1 [29] and additional packages pcor.R and lme4. Corrected Akaike Information Criterion for finite sample sizes (AICc) was used to select the models fitting the data best [30].

RESULTS

Bacteriology Results

Inoculation of I-pigs was successful as demonstrated by at least two or more positive results in SBE in nasal or tonsillar samples in I-pigs upon inoculation. Sequentially taken tonsillar samples of I-pigs were almost all positive and more or less constant after day 2, results of nasal samples showed decreasing CFUs of *A. pleuropneumoniae* over time (Table 2 and Figure 1) and at day 21 only one of five surviving I-pigs had a positive nasal swab sample in SBE.

Table 2: Results of tonsil and nasal sample selective bacterial examination for *A. pleuropneumoniae* in time

Pair	Status	Days post inoculation											Serology			Days to trans-mis-sion based on SBE
		-1	1	2	4	6	8	11	13	15	18	21	CFT	App S2 ELISA	ApxIV ELISA	
1	I	-	tn	tn	tn	†							ND	Neg	ND	
	C	-	-	tn	-	-	-	-	t	-	-	t	<40	Neg	Neg	∞
2	I	-	n	tn	tn	†							ND	Neg	ND	
	C	-	-	-	-	-	-	-	-	-	-	-	<40	Neg	Neg	∞
3	I	-	tn	tn	†								ND	Neg	ND	
	C	-	-	-	t	t	t	t	t	t	t	t	<40	Neg	Neg	4
4	I	-	tn	tn	tn	tn	t	t	t	t	t	t	640	Neg	Neg	
	C	-	-	-	t	t	t	t	t	t	t	t	<40	Neg	Neg	4

5	I	-	tn	tn	tn	tn	tn	tn	tn	tn	tn	tn	320	Neg	D	
	C	-	-	-	-	t	t	t	t	t	t	t	80	Neg	Neg	6
6	I	-	tn	tn	tn	tn	†						ND	Neg	ND	
	C	-	-	-	t	t	t	t	t	t	t	t	40	Neg	Neg	4
7	I	-	tn	tn	tn	tn	t	tn	tn	t	t	t	1280	Neg	P	
	C	-	-	-	t	t	t	t	t	t	t	t	<40	Neg	Neg	4
8	I	-	n	t	t	-	t	-	-	t	-	-	160	Neg	D	
	C	-	-	-	-	-	-	-	-	-	-	-	<40	Neg	Neg	∞
9	I	-	tn	tn	tn	tn	tn	tn	tn	tn	tn	t	1280	Neg	Neg	
	C	-	-	-	t	t	t	t	t	t	t	t	<40	Neg	Neg	4
10	I	-	n	tn	tn	†							ND	Neg	ND	
	C	-	-	-	-	-	-	-	-	-	-	-	<40	Neg	Neg	∞
11	S	-	-	-	-	-	-	†*					ND	Neg	ND	
	S	-	-	-	-	-	-	-	-	-	-	-	<40	Neg	Neg	
12	S	-	-	-	-	-	-	-	-	-	-	-	<40	Neg	Neg	
	S	-	-	-	-	-	-	-	-	-	-	-	<40	Neg	Neg	

Positive tonsil sample results are defined by t, positive nasal sample results are defined by n and negative samples by -. tn represents a positive tonsil and positive nasal sample result. †=euthanized or died before that day. * euthanized due to lameness. ND=not determined, Neg=Negative, D=Dubious, P=Positive.

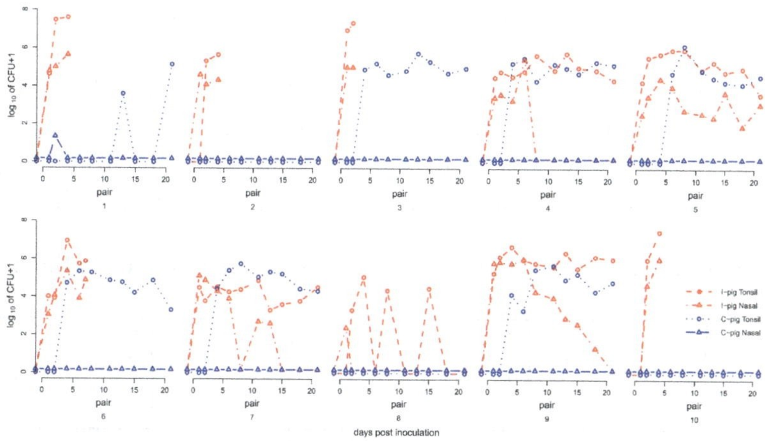

Figure 1: Course of *A. pleuropneumoniae* CFUs counts in tonsil and nasal samples in time. Tonsil sample counts are presented as *A. pleuropneumoniae* $\log_{10}$ (CFUs+1) and nasal samples are presented as $\log_{10}$ (CFUs+1.5) to separate them graphically from tonsil samples.

Transmission occurred in six out of ten pairs as at least two or more positive samples in C-pigs tested positive (Figure 1). In one C-pig (pair 1) two non-consecutive samples were positive in culture, starting long after the I-pig had died and we assume that transmission did not occur in that pair. Only one nasal sample of the C-pigs showed a positive result in SBE (Figure 1).

Evaluation of mixed effect models showed that pig status significantly affected the CFUs count in tonsil samples. The median CFUs count in positive tonsil samples in I-pigs was $10^{5.0}$ (range: $10^{3.3}$ – $10^{7.6}$) CFUs and in transmission positive C-pigs: $10^{4.9}$ (range $10^{3.3}$ – $10^{6.1}$) CFUs. All tonsil and nasal samples from the S-pigs showed negative results in SBE.

Quantitative results from qPCR analysis and SBE were highly correlated ($r=0.89$; $P<0.001$ for all samples), as reported before [25]. As expected, more samples were tested positive with qPCR (204/454) than with SBE (153/454), though with low levels of DNA. Based on qPCR results, transmission occurred in eight pairs. Results and conclusions from qPCR analyses are presented.

Clinical Disease

The CS in I-pigs varied between pigs ranging from 0 to 2.25 (median 0.66), see Table 3 and Figure 2. Clinical disease was more severe than expected from the pilot studies and five I-pigs died or were euthanized between days two and seven after inoculation. Each disease parameter under observation was affected at least once in I-pigs, except for cyanosis which was never observed in any of the pigs. Median CS was 0.3 (range 0.0 – 1.4) in all C-pigs and sentinel pigs (range 0.0 - 0.6). One of the sentinel pigs was euthanized on day 11 because of lameness, due to a sterile fissure in the right femur.

Table 3: Observed median and range of disease severity scores of the pigs

Status	CS	AvgCS	RHS4	RHS20
I-pigs	0.7 (0.0 – 2.3)	0.7 (0.1 – 2.0)	24.0 (6.8 – 66.9)	46.1 (6.3 – 93.4)
C-pigs +	0.3 (0.0 – 1.4)	0.3 (0.0 – 0.8)	7.3 (2.3 – 8.7)	34.9 (27.7 – 44.5)
C-pigs -	0.3 (0.0 – 0.9)	0.2 (0.0 – 0.7)	3.6 (2.5 – 5.6)	31.1 (21.0 – 33.6)
S-pig	0.3 (0.0 – 0.6)	0.2 (0.0 – 0.5)	3.1 (1.5 – 4.4)	5.3 (4.1 – 9.3)

I-pigs=inoculated pigs, C-pigs+= C-pigs in transmission positive pairs, C-pigs –=C-pigs in pairs with no observed transmission.

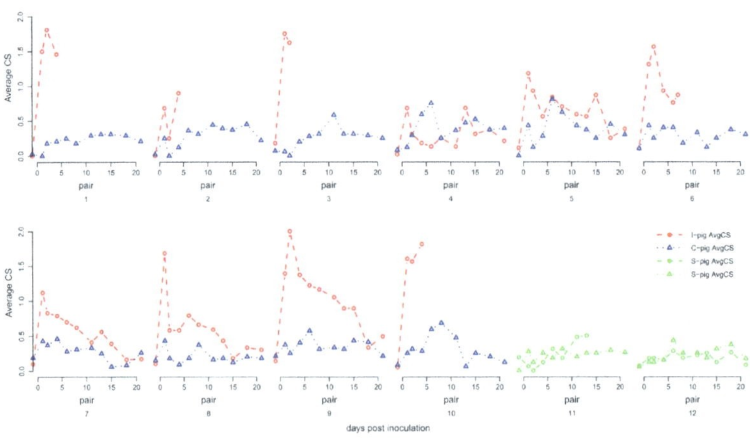

Figure 2: Average Clinical Score (AvgCS) of piglets per pair in time. The course of Average Clinical Score of piglets in time.

Statistical analyses demonstrated that CS and AvgCS were higher in I-pigs and *A. pleuropneumoniae* positive C-pigs compared to negative C-pigs or S-pigs. In addition CS and AvgCS were also significantly higher in I-pigs than in positive C-pigs. AvgCS decreased significantly over time in I-pigs, but not in positive C-pigs. Finally, *A. pleuropneumoniae* negative C-pigs and S-pigs did not differ in the score of CS nor AvgCS.

Pathology

In 8 out of 10 I-pigs macroscopically observable lung or pleurisy lesions were observed (Table 1). Pleurisy and pneumonia scores were highly correlated within the same animal (Spearman's rank correlation $r=0.89$; $P<0.001$). Three C-pigs had small pleurisy lesions and in one of them a small pneumonic lesion was observed. In sentinel pigs no pneumonia or pleurisy was observed. No associations were found between pathologic lesion scores of the pigs from the same pair.

Most tissue samples of macroscopic lung and pleurisy lesions were positive for *A. pleuropneumoniae*. No other bacteria were cultured from these lesions (Table 1). Tonsils of I-pigs were all positive for *A. pleuropneumoniae*, while not all positive C-pigs had a positive tonsil at necropsy.

Serology

All 5 surviving I-pigs seroconverted in CFT (Table 1), with a median titre of 640. Nine C-pigs had negative CFT results and one positive C-pig had a dubious CFT result of 80. The two I-pigs that did not show any sign of pneumonia or pleurisy at necropsy had positive serum samples nevertheless. Results of apxIV ELISA were mostly in accordance with CFT results, except for the I-pigs of pair 4 and 9 (negative in ApxIV Elisa). With the App serotype 2 ELISA, none of the samples tested positive, nor did they show a distinct increase in OD.

Association between Clinical Disease, Pathology and Serology Results

Spearman's correlation analyses showed significant correlations between cumulative RHS4 or RHS20 with pleurisy lesion scores ($r=0.77$; $P<0.001$, respectively $r=0.87$ $P<0.001$) and pneumonia scores ($r>0.59$; $P<0.01$, respectively $r=0.75$ $P<0.001$). Comparable results were found for the association between CFT results and pleurisies ($r=0.53$; $P=0.04$) and pneumonia ($r=0.75$; $P=0.001$) scores.

Disease Severity and Bacterial Load

Partial correlation analysis showed that AvgCS was significantly correlated with bacterial quantities in tonsil ($r=0.54$, $P<0.001$) and nasal samples ($r=0.63$ ($P<0.001$) of I pigs, but not in tonsil samples of C-pigs ($r=0.06$, $P=0.69$) (Figure 3). This means that 29% and 39% of the variance in AvgCS in I-pigs could be explained by the variance in number of *A. pleuropneumoniae* found in tonsil or nasal samples respectively. Since only one positive nasal swab was found in C-pigs, no associations were made to AvgCS. Evaluation of mixed model with time as fixed effect and pig number as random effect, showed that pig status (I- or positive sampled C-pigs) is not needed to explain the height of the bacterial tonsil load by AvgCS. This means that bacterial tonsil load is equally affected by AvgCS in I-pigs and transmission positive C-pigs.

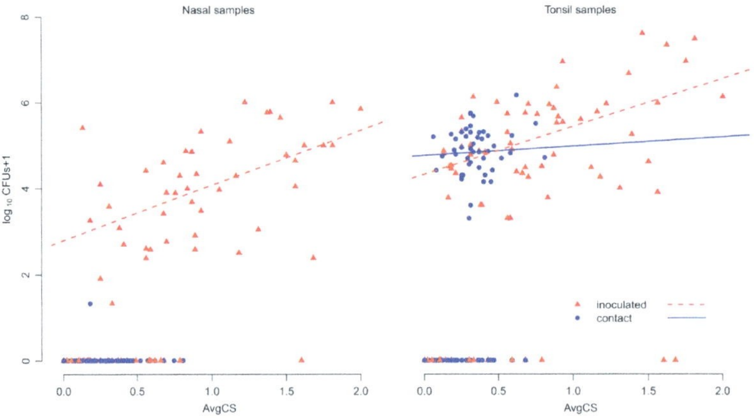

Figure 3: Association between bacterial load and average disease score. Association between $\log_{10}$ (CFUs +1) of *A. pleuropneumoniae* 1536 in nasal and tonsil samples and average clinical score (AvgCS) for I- and C-pigs. Regression lines are based on positive samples.

Effects on Transmission Rate

Transmission was either observed on day 4 or 6 or not at all. The best fitting model (lowest AICc), evaluating the effects on the transmission rate, included both the $\log_{10}+1$ of nasal CFUs and average clinical score (Table 4). Based on the confidence intervals transmission was, negatively affected by AvgCS and positively by nasal bacterial load. Both effects are much smaller and not significant if included in the model individually. The results were not sensitive to the use of SBE rather than qPCR to define transmission and infectiousness.

Table 4: Effects of disease severity and bacterial counts on the transmission rate of *A. pleuropneumoniae* 1536

Model	Intercept	Log_{10}(CFUs+1) Tonsil	Log_{10}(CFUs+1) Nasal	AvgCS	AICc
1	−2.3 (−3.3; -1.5)	X	X	X	28.10
2	−4.0 (−7.8; -1.9)	0.35 (−0.04; 0.93)	X	X	27.39
3	−3.9 (−7.7; -2.0)	X	0.44 (−0.01; 1.17)	X	26.76
4	−2.1 (−4.0; -0.6)	X	X	−0.22 (−2.19; 1.40)	30.27
5	−4.1 (−8.0; -2.0)	0.11 (−0.44; 0.91)	0.34 (−0.36; 1.27)	X	29.01
6	−4.8 (−11.3; -1.6)	0.82 (0.10; 2.24)	X	−1.95 (−5.08; 0.16)	26.93
7	−4.3 (−11.5; -1.4)	X	1.07 (0.25; 3.06)	−2.48 (−5.80; -0.16)	24. 69
8	−5.0 (−12.6; -1.5)	0.31 (−0.21; 1.67)	0.88 (−0.10; 2.92)	−2.67 (−5.95; -0.30)	26.60

Model evaluation for estimation of the effects on the transmission rate, using a Generalized Linear Model with complementary log-log link. Effect estimators are given with the 95% confidence interval.

Association of Disease Severity within Pairs

In transmission positive pairs a significant association of CS between the I- and C-pig was found, r=0.34 (P=0.00). Based on AvgCS, r=0.35 (P=0.06).

DISCUSSION

In this study we investigated the association between transmission of *A. pleuropneumoniae*, bacterial load in the oropharynx and nasal cavity and severity of clinical signs. An association was to be expected if clinical outbreaks are caused by rapid spread of the bacteria by clinically affected pigs. The average clinical score was positively associated with the bacterial load - the amount of bacteria - in the oropharynx and nasal cavity which was, in turn, positively associated with the transmission rate of *A. pleuropneumoniae*. However, corrected for bacterial load, the clinical score was negatively associated with the transmission rate. This means that in pigs with a similar bacterial load, pigs with higher clinical scores transmitted the bacteria less efficiently. Both effects of clinical disease (positive and negative) resulted in a much smaller net effect, not even significantly from zero (Table4, model 4). The association found between daily clinical score within the transmission chain was r=0.34. These findings suggest that it is unlikely that clinical outbreaks of *A. pleuropneumoniae* are caused by rapid transmission of *A. pleuropneumoniae* by clinically affected pigs only. It implies that other causes and mechanisms may cause the occurrence of outbreaks.

Our conclusions were based on the results of bacterial examination, because SBE detects viable bacteria and is, to our opinion, representative for colonization. Additional analyses performed with qPCR (Additional file 2) suggested that transmission may have occurred in two additional pairs, but the number of genomic copies found in those C-pigs was low. Although sensitivity in the SBE is not 100%, we considered these samples containing non-viable bacteria and therefore concluded that these pigs were not infected. Therefore we included the qPCR results as additional data only. Most importantly however, the analysis of transmission based on the qPCR results would lead to the same conclusion about the effect of disease severity and nasal bacterial load on transmission.

Transmission of pathogens is dependent on the susceptibility and the infectiousness of the uninfected and infected individuals, respectively and the contact rate [31]. CD/CD pigs were used and randomly assigned to pairs and treatment. Therefore susceptibility of the pigs was considered to be similar at the start of the experiment. Severity of disease was positively associated with the bacterial load, and bacterial load with the transmission rate, but the severity of disease was negatively associated with the transmission rate. This result could be explained by the effect of disease severity on the contact rate, as *A. pleuropneumoniae* is assumed to spread mainly by direct contact. If clinical signs affect the frequency and/or intensity of contact between pigs, the rate of transmission may be lower when pigs are showing severe signs compared to sub-clinically infected pigs. On the other hand, in similarly affected animals the transmission rate will be mostly associated with the number of bacteria isolated, as has been shown by others [8].

An association between clinical score in I-pigs and positive C-pigs was demonstrated, but the score of inoculated pigs was significantly higher than for contact positive ones. This could be explained by the absence of lung lesions in five of six positive contact pigs. In positive C-pigs higher clinical scores were observed than in negative C-pigs, which was not reported in other studies [8,32]. This may be explained as follows. First, *A. pleuropneumoniae* might have caused only minor pathology in pig tonsils, as previously described for gnotobiotic pigs [33], which may have induced only mild clinical signs. We did not, however, investigate the morphologic changes in pig tonsils. Second, the scoring method we used was more detailed than in the other transmission studies and as signs were only mild a less detailed method might have missed these. It should be mentioned, however, that our scoring method was not applied blindly as the observer was aware of the infection status of each pig, which may have resulted in observation bias.

In previously performed studies with *A. pleuropneumoniae*, infectious pigs were induced by endobronchial application of the inoculum or by exposure to other infected pigs [24], resulting in a more uniform expression of clinical signs. In our study, the inoculum was applied intranasally, as we aimed at inducing variation in clinical signs rather than uniformity. It is known that variation in signs can also be affected by using different doses or different inoculation routes [15,34], but using these methods of challenge was not suitable here as a possible observed difference in transmission could then also be due to inoculation method rather than the clinical score.

In our study bacterial counts in tonsil and nasal sample from inoculated pigs correlated well, but the number of bacteria in nasal samples decreased over time and most nasal samples were found negative in contact infected pigs. Our findings suggest that for transmission studies on farms it is more appropriate to take tonsil samples instead of nasal samples to detect a colonised pig, as the nasal samples may be negative in pigs with low number of bacteria. In experimental studies nasal samples may be useful, as they are easier to collect and, moreover, may reflect the infectiousness of the pigs more appropriately [8].

In this study CD/CD pigs were used, because of diagnostic limitations of bacterial examination in SPF pigs. In the field specific maternally derived antibodies [18], and possibly cross immunity for other *Pasteurellaceae*, can be protective for developing clinical signs. Besides, competition for colonization by other bacteria, especially other *Pasteurellaceae*, or protective effects of milk or colostrum on colonization are plausible under field conditions as well. Therefore extrapolation of our estimates to the field should be done with extreme caution and our conclusion on the relation between clinical signs and transmission of *A. pleuropneumoniae* needs confirmation under more natural circumstances.

Nevertheless our observation that clinical signs reduce transmission of *A. pleuropneumoniae*, e.g. by reducing the contact rate, could have significant consequences for effectiveness of interventions. Isolation of severely affected individuals has been shown to be effective to reduce the size of an outbreak, e.g. for *Salmonella*[7] in cattle. In that case the development of clinical signs usually coincides with a raise in infectiousness and diseased animals are responsible for most of the transmission during the outbreak. The latter was exactly one of the hypotheses for the course of outbreaks of *A. pleuropneumoniae*. If they become diseased, *A. pleuropneumoniae* infected pigs become diseased within a few days after infection and this study has shown that diseased pigs shed more bacteria, so isolation theoretically might be effective. However, most infected pigs do not become diseased, so most infectious pigs will not be noticed. Moreover the results of this study have shown a negative impact of disease on transmission. Thus, while during an outbreak of pleuropneumonia, caused by *A. pleuropneumoniae*, isolation of severely affected pigs may be beneficial for their wellbeing, the effect on the course of the outbreak is likely to be limited.

In conclusion the results of this experiment do not support the hypothesis that outbreaks start with only a few diseased pigs that rapidly spread a clinical form of the infection. It is therefore more likely that an outbreak occurs due to the development of clinical signs at the same time in already infected pigs due to some (non-) infectious trigger.

AUTHORS' CONTRIBUTIONS

TJT performed the animal experiment, was responsible for laboratory analyses, retrieving results and drafting and finalizing the manuscript. DK designed the study and assisted in statistical analysis and preparation of the manuscript. AB was involved in design of the study, laboratory analyses and preparation of the manuscript. AJJMD assisted with laboratory analyses and preparation of the inoculum. JAW and JAS were involved in design of the study and preparation and drafting the manuscript. All authors read and approved the final manuscript.

ACKNOWLEDGEMENTS

Laura Scheske and animal care personnel are acknowledged for excellent technical assistance. Niels Dekker is acknowledged for his help in raising CD/CD pigs. Barbara Tempelmans Plat (Animal Health Service, Deventer, The Netherlands) is acknowledged for serology testing.

REFERENCES

1. Woolhouse MEJ, Dye C, Etard J, Smith T, Charlwood JD, Garnett GP, Hagan P, Hii JLK, Ndhlovu PD, Quinnell RJ, Watts CH, Chandiwana SK, Anderson RM. Heterogeneities in the transmission of infectious agents: implications for the design of control programs. Proc Natl Acad Sci USA. 1997;94:338–342. doi: 10.1073/pnas.94.1.338.
2. Klinkenberg D, Nishiura H. The correlation between infectivity and incubation period of measles, estimated from households with two cases. J Theor Biol. 2011;284:52–60. doi: 10.1016/j.jtbi.2011.06.015.
3. Lloyd-Smith JO, Schreiber SJ, Kopp PE, Getz WM. Superspreading and the effect of individual variation on disease emergence. Nature. 2005;438:355–359. doi: 10.1038/nature04153.
4. Weesendorp E, Backer J, Stegeman A, Loeffen W. Transmission of classical swine fever virus depends on the clinical course of infection which is associated with high and low levels of virus excretion. Vet Microbiol. 2011;147:262–273. doi: 10.1016/j.vetmic.2010.06.032.
5. Cloutier G, D'Allaire S, Martinez G, Surprenant C, Lacouture S, Gottschalk M. Epidemiology of Streptococcus suis serotype 5 infection in a pig herd with and without clinical disease. Vet Microbiol. 2003;97:135–151. doi: 10.1016/j.vetmic.2003.09.018.
6. Geenen PL, Döpfer D, van der Meulen J, de Jong MC. Transmission of F+E. coli in groups of early weaned piglets. Epidemiol Infect. 2005;133:459–468. doi: 10.1017/S0950268804003292.
7. Lanzas C, Brien S, Ivanek R, Lo Y, Chapagain PP, Ray KA, Ayscue P, Warnick LD, Gröhn YT. The effect of heterogeneous infectious period and contagiousness on the dynamics of Salmonella transmission in dairy cattle. Epidemiol Infect. 2008;136:1496–1510. doi: 10.1017/S0950268807000209.
8. Velthuis AG, De Jong MC, Stockhofe N, Vermeulen TM, Kamp EM. TransmissionofActinobacilluspleuropneumoniaeinpigsischaracterized by variation in infectivity. Epidemiol Infect. 2002;129:203–214.
9. Gottschalk M, Taylor DJ. In: Diseases of Swine. 9. Straw BE, Zimmerman JJ, D'Allaire S, Taylor DJ, editor. Ames, Iowa, USA: Blackwell Publishing; 2006. Actinobacillus pleuropneumoniae; pp. 563–576.
10. Bertram TA. Pathobiology of acute pulmonary lesions in swine infected with Haemophilus (Actinobacillus) pleuropneumoniae. Can

Vet J. 1988;29:574–577.

11. Fenwick B, Henry S. Porcine pleuropneumonia. J Am Vet Med Assoc. 1994;204:1334–1340. [PubMed]
12. Shope RE. Porcine contagious pleuropneumonia. I. Experimental transmission, etiology, and pathology. J Exp Med. 1964;119:357–368. doi: 10.1084/jem.119.3.357.
13. Tobias TJ, Raymakers RJ, van Nes A, van Leengoed LA. Outbreak of respiratory distress resembling influenza caused by Actinobacillus pleuropneumoniae in pigs. Vet Rec. 2009;164:402–403. doi: 10.1136/vr.164.13.402.
14. Rosendal S, Boyd DA, Gilbride KA. Comparative virulence of porcine Haemophilus bacteria. Can J Comp Med. 1985;49:68–74.
15. Jacobsen MJ, Nielsen JP, Nielsen R. Comparison of virulence of different Actinobacillus pleuropneumoniae serotypes and biotypes using an aerosol infection model. Vet Microbiol. 1996;49:159–168. doi: 10.1016/0378-1135(95)00184-0.
16. Broes A, Martineau GP, Gottschalk M. Dealing with unexpected Actinobacillus pleuropneumoniae serological results. J Swine Health Prod. 2007;15:264–269.
17. Cruijsen T, van Leengoed LA, Ham-Hoffies M, Verheijden JH. Convalescent pigs are protected completely against infection with a homologous Actinobacillus pleuropneumoniae strain but incompletely against a heterologous-serotype strain. Infect Immun. 1995;63:2341–2343.
18. Cruijsen T, van Leengoed LA, Kamp EM, Bartelse A, Korevaar A, Verheijden JH. Susceptibility to Actinobacillus pleuropneumoniae infection in pigs from an endemically infected herd is related to the presence of toxin-neutralizing antibodies. Vet Microbiol. 1995;47:219–228. doi: 10.1016/0378-1135(95)00109-3.
19. Marois C, Gottschalk M, Morvan H, Fablet C, Madec F, Kobisch M. Experimental infection of SPF pigs with Actinobacillus pleuropneumoniae serotype 9 alone or in association with Mycoplasma hyopneumoniae. Vet Microbiol. 2009;135:283–291. doi: 10.1016/j.vetmic.2008.09.061.
20. Pol JM, van Leengoed LA, Stockhofe N, Kok G, Wensvoort G. Dual infections of PRRSV/influenza or PRRSV/Actinobacillus

pleuropneumoniae in the respiratory tract. Vet Microbiol. 1997;55:259–264. doi: 10.1016/S0378-1135(96)01323-5.

21. Maes D, Chiers K, Haesebrouck F, Laevens H, Verdonck M, de Kruif A. Herd factors associated with the seroprevalences of Actinobacillus pleuropneumoniae serovars 2, 3 and 9 in slaughter pigs from farrow-to-finish pig herds. Vet Res. 2001;32:409–419. doi: 10.1051/vetres:2001133.
22. Loeffen WLA, Kamp EM, Stockhofe-Zurwieden N, Van Nieuwstadt APKMI, Bongers JH, Hunneman WA, Elbers ARW, Baars J, Nell T, Van Zijderveld FG. Survey of infectious agents involved in acute respiratory disease in finishing pigs. Vet Rec. 1999;145:123–129. doi: 10.1136/vr.145.5.123.
23. Dekker CN, Bouma A, Daemen AJ, van Leengoed LA, Jonker FH, Wagenaar JA, Stegeman JA. Homologous whole bacterin vaccination is not able to reduce Streptococcus suis serotype 9 strain 7997 transmission among pigs or colonization. Vaccine. 2012;30:1379–1387. doi: 10.1016/j.vaccine.2011.12.035.
24. Velthuis AG, De Jong MC, Kamp EM, Stockhofe N, Verheijden JH. Design and analysis of an Actinobacillus pleuropneumoniae transmission experiment. Prev Vet Med. 2003;60:53–68. doi: 10.1016/S0167-5877(03)00082-5.
25. Tobias TJ, Bouma A, Klinkenberg D, Daemen AJJM, Stegeman JA, Wagenaar JA, Duim B. Detection of Actinobacillus pleuropneumoniae in pigs by real-time quantitative PCR for the apxIVA gene. Vet J. 2012;193:557–560. doi: 10.1016/j.tvjl.2012.02.004.
26. Hoeltig D, Hennig-Pauka I, Beyerbach M, Thies K, Rehm T, Gerlach GF, Waldmann KH. Comparison of the diagnostic significance of clinical, radiographic and ultrasonographic results after an experimental aerosol infection of pigs with Actinobacillus pleuropneumoniae. Berl Munch Tierarztl Wochenschr. 2008;121:422–431. in German.
27. Nielsen R. Serological and immunological studies of pleuropneumonia of swine caused by Haemophilus parahaemolyticus. Acta Vet Scand. 1974;15:80–89.
28. Hannan PCT, Bhogal BS, Fish JP. Tylosin tartrate and tiamutilin effects on experimental piglet pneumonia induced with pneumonic pig lung homogenate containing mycoplasmas, bacteria and viruses. Res Vet Sci. 1982;33:76–88.

29. R Development Core Team. R: A Language and Environment for Statistical Computing, R Foundation for Statistical. Austria: Vienna; 2010.
30. Burnham KP, Anderson DR. Model Selection and Multimodel Inference, A Practical Information-Theoretic Approach. New York, USA: Springer; 2002.
31. Stegeman JA, Bouma A, de Jong MC. Epidemiological models to assist the management of highly pathogenic avian influenza. Rev Sci Tech. 2011;30:571–579.
32. Nielsen R, Mandrup M. Pleuropneumonia in swine caused by Haemophilus parahaemolyticus. A study of the epidemiology of the infection. Nord Vet Med. 1977;29:465–473.
33. Chiers K, Haesebrouck F, van Overbeke I, Charlier G, Ducatelle R. Early in vivo interactions of Actinobacillus pleuropneumoniae with tonsils of pigs. Vet Microbiol. 1999;68:301–306. doi: 10.1016/S0378-1135(99)00048-6.
34. van Leengoed LA, Kamp EM. Endobronchial inoculation of various doses of Haemophilus (Actinobacillus) pleuropneumoniae in pigs. Am J Vet Res. 1989;50:2054–2059.

CHAPTER 11

Development and Validation of a Real-Time PCR for Chlamydia suis Diagnosis in Swine and Humans

Kristien De Puysseleyr[1]., Leentje De Puysseleyr[1]., Julie Geldhof[1] , Eric Cox[2] , Daisy Vanrompay[1]

[1] Department of Molecular Biotechnology, Faculty of Bioscience Engineering, University of Ghent, Ghent, Belgium
[2]Department of Virology, Parasitology and Immunology, Faculty of Veterinary Medicine, University of Ghent, Ghent, Belgium

ABSTRACT

Pigs are the natural host for *Chlamydia suis*, a pathogen which is phylogenetically highly related to the human pathogen *C. trachomatis*. *Chlamydia suis* infections are generally treated with tetracyclines. In 1998, tetracyline resistant *C. suis* strains emerged on U.S. pig farms and they are currently present in the Belgian, Cypriote, German, Israeli, Italian and Swiss pig industry. Infections with tetracycline resistant *C. suis* strains are mainly associated with severe reproductive failure leading to marked economical

Citation: De Puysseleyr K, De Puysseleyr L, Geldhof J, Cox E, Vanrompay D (2014) Development and Validation of a Real-Time PCR for Chlamydia suis Diagnosis in Swine and Humans. PLoS ONE 9(5): e96704. https://doi.org/10.1371/journal.pone.0096704.

loss. We developed a sensitive and specific TaqMan probe-based *C. suis* real-time PCR for examining clinical samples of both pigs and humans. The analytical sensitivity of the real-time PCR is 10 rDNA copies/reaction without cross-amplifying DNA of other *Chlamydia* species. The PCR was successfully validated using conjunctival, pharyngeal and stool samples of slaughterhouse employees, as well as porcine samples from two farms with evidence of reproductive failure and one farm without clinical disease. *Chlamydia suis* was only detected in diseased pigs and in the eyes of humans. Positive humans had no clinical complaints. PCR results were confirmed by culture in McCoy cells. In addition, *Chlamydia suis* isolates were also examined by the *tet*(C) PCR, designed for demonstrating the tetracycline resistance gene *tet*(C). The *tet*(C) gene was only present in porcine *C. suis* isolates.

INTRODUCTION

Chlamydiaceae are obligate intracellular Gram-negative bacteria that cause a broad range of diseases in both humans and animals. Pigs can become infected by *Chlamydia (C.) pecorum*, *C. abortus*, *C. psittaci* and *C. suis* [1]. Seroprevalence rates of *Chlamydiaceae* in German, Swiss, Italian and Belgian pigs are high (40–96%) [2], [3], [4], [5]. As far as known, the pig is the only host for *C. suis,* which is phylogenetically closely related to the human pathogenic agent *C. trachomatis* [1]. *Chlamydia suis* infections can be asymptomatic or associated with arthritis, pericarditis, polyserositis, pneumonia, conjunctivitis, enteritis, diarrhea and reproductive failure [4], [6], [7], [8], [9], [10]. Experimental infections in specific pathogen free and gnotobiotic pigs using *C. suis* field strains have resulted in conjunctivitis, respiratory lesions and gastrointestinal lesions [11], [12], [13]. To date, transmission of *C. suis* to humans has not been reported [14]. However, *C. suis* and *C. trachomatis* are closely related [1] and, therefore, the zoonotic potential of *C. suis* is likely.

Chlamydia suis can be difficult to grow in cell culture or embryonated eggs [14]. A molecular diagnostic test, able to specifically detect *C. suis* in both animal and human samples, is crucial to gain insights into the prevalence, the economic impact and the zoonotic 'threat' of *C. suis* infections. In 2006, Sachse et al. [15], developed a 23S rDNA gene-based array tube (AT) micro array for detection and differentiation of *Chlamydia* spp. The assay detects a single copy of the cloned target DNA [16]. Borel *et al.* [17] calculated the sensitivity of the AT micro array over an entire panel of human and animal clinical samples, including five human pharyngeal swabs and four tissues of naturally infected diseased pigs. The sensitivity of the AT micro array was slightly lower than for real-time PCR. When excluding long-time stored swab samples from the calculation, the sensitivity was equivalent to that of real-time PCR. However, in our hands, *C. suis* species identification by use of the AT microarray was often only possible on isolates of vaginal and rectal swabs and not directly on these clinical samples [18]. More recently, Pantchev et al., [19] developed a 23S rDNA gene-based real-time PCR for *C. suis*. However, the PCR was strictly developed for the examination of veterinary samples as it cannot distinguish *C. suis* from the genetically closely related human pathogen *C. trachomatis* [19]. Therefore, it was the aim of the present study to develop a sensitive and specific *C. suis* real-time PCR for the purpose of examining clinical samples of both pigs and humans.

MATERIALS AND METHODS

Bacterial Cultures

For determination of sensitivity, specificity and detection limit of the *C. suis*-specific real-time PCR, the following 14 chlamydial strains were used: *C. suis* strains S45, R19, H7, R24, R27 and R33, *C. pneumoniae* (TW-183), *C. felis* (FP Baker), *C. caviae* (GIPC), *C. abortus* (S26/3), *C. psittaci* (92/1293), *C. pecorum* (1710S), *C. muridarum* (MoPn) and *C. trachomatis* (L2/434/BU). Bacteria were grown in cycloheximide treated McCoy cells (mouse fibroblast cells, obtained from Els De Coster, Labo Medische Analyse, CRI, Zwijnaarde, Belgium; original source: ATCC CRL-1696) using a standard methodology, as described previously [20]. Chlamydiae were released from infected monolayers (two passages) by freezing and thawing, followed by ultrasonication (Bransonic 12, BIOMEDevice, San Pablo, CA, USA). Cell culture harvest was centrifuged for 10 min (1000 × *g*, 4°C) and *Chlamydiae* were subsequently concentrated by ultracentrifugation for 45 minutes (50,000 × *g*, 4°C). Bacteria were resuspended in 2 mL sucrose phosphate glutamate buffer (SPG, 218 mM sucrose, 38 mM KH_2PO_4, 7 mM K_2HPO_4, 5 mM L-glutamic acid) and stored at –80°C until use.

DNA Extraction

Genomic DNA of bacterial cultures was prepared using the DNeasy Blood and Tissue Kit (Qiagen, Antwerp, Belgium). DNA extraction of clinical samples was performed using the G-spin Total DNA Extraction Mini Kit (Goffin Molecular Biotechnologies, Beek, The Netherlands). Both extraction methods were performed according to the instructions of the manufacturers.

Primers and Probes

Published 23S ribosomal RNA sequences of all known chlamydial species (Table 1) [1] were aligned using Clustal X software (default settings) [21]. Forward and reverse primers (Life technologies, Paisley, United Kingdom) and probes (Eurogentec, Seraing, Liège, Belgium) were designed using primer3 software (Table 2) [22]. Primer and probe specificity was checked by BLAST [21]. The optimal annealing temperature and specificity of the primers and probes within the genome was evaluated in 20 µl PCR reactions containing 3 µl of 2 µM of forward and reverse primers, 0.5 µl Super Taq 5U/µl (Sphaero Q, Gorinchem, The Netherlands), 2 µl Super Taq Reaction Buffer, 2 µl of each of 1.25 mM dNTP and 50 ng of genomic DNA of *C. suis* strain S45. The cycling conditions were as follows: 72°C for 5 minutes; 35 cycles of 95°C for 60s, 60°C for 60s and 72°C for 60s and a final step of 72°C for 5 min. CS23S-probe A and B were 5′ labeled with the reporter dye 6-carboxyfluorescein (FAM) and hexachloro-fluoresceine (HEX) respectively, and 3′ labeled with the quencher dye carboxytetramethylrhodamine (TAMRA).

Table 1: NCBI Nucleotide Accession numbers of the 23S rRNA gene sequences that were used for the alignment to design the primers and *C. suis* **specific probes**

Species	Strain	NCBI Nucleotide Accession Number
Chlamydia suis	R22	U68420
Chlamydia trachomatis	L2/434/BU	U68443
Chlamydia abortus	EBA	U76710
Chlamydia psittaci	NJ1	U68419
	6BC	U68447
Chlamydia pecorum	IPA	U68434
Chlamydia felis	FP Baker	U68457
Chlamydia muridarum	MoPn	U68436
	SFPD	U68437

Chlamydia caviae	GPIC	U68451
Chlamydia pneumoniae	TW-183	U76711

Table 2: Primers and probes of the *C. suis* **specific real**-time PCR

Oligonoculeotide	Sequence (5′- 3′)	Position[a]	Specificity
CS23S-F	*GCAGAGGAAAAGAAATC-GAAGA*	215–236	*C. suis and C. trachomatis*
CS23S-R	*CGGGACTATCACCCTGTATC*	359–378	*C. suis and C. trachomatis*
CS23S-ProbeA	FAM-CGAGCTGAAGAAGC-GAGGGGTTGTAG-TAMRA	280–305	*C. suis*
CS23S-ProbeB	HEX-CGAGCCGAAGAAGC-GAGGGGTTGTAG-TAMRA	280–305	*C. suis*

[a]Binding position from base 1 of the 23S rRNA gene of the chlamydial reference strain S45.

Inhibition Control Plasmid

The 23S rDNA amplicon of *C. suis* (S45) was amplified using 20 µl PCR reactions containing 3 µl of 2 µM of forward and reverse primers, 0.5 µl Super Taq 5U/µl (Sphaero Q, Gorinchem, The Netherlands), 2 µl Super Taq Reaction Buffer, 2 µl of each of 1.25 mM dNTP and 50 ng of genomic DNA. PCR products were purified using the Wizard SV Gel and PCR Clean-Up System (Promega, Madison, WI, USA) and cloned into pGem-T (Promega, Madison, WI, USA) according to the manufacturer's protocol. To verify the nucleotide sequence of the inserted fragment, sequence analyses were performed by the VIB Genetic Service Facility (University of Antwerp, Antwerp, Belgium) using vector associated T7 and SP6 priming sites. The resulting *C. suis*-species-specific internal inhibition control plasmid was designated pGemT::CSIC.

Species-specific Real-time PCR

Real-time PCR was performed with the Rotor-Gene Q Instrument (Qiagen Benelux, Venlo, The Netherlands) using 25 µl reaction mixture containing 2 µl of DNA template, 4 µl of primer mixture (300 nM forward and reverse primer), 2.5 µl of CS23S-probe A and B (200 nM), 12.5 µl of absolute qPCR mix (Thermoscientific, Acros Organics, Geel, Belgium) and 1.5 µl DNAse and RNAse free water. The cycling conditions were as follows: 95°C for 15 minutes, 50 cycles of 95°C for 15 seconds, 60°C for 60 seconds. All default program settings were used. Standard graphs of the Cycle threshold (Ct) values, obtained by testing tenfold serial dilutions (10^8 to 10^1) of the purified species-specific inhibition control plasmid, were used for quantification. DNA was always tested in the presence of control plasmid (50 copies/µl) to check for PCR inhibitors.

Analytical Sensitivity and Specificity

The analytical sensitivity of the assay was evaluated using decimal serial dilutions (10^8 to 10^1 copies/µl) of the inhibition control plasmid and of purified genomic DNA of the *C. suis* reference strain S45 from suspensions equivalent to 10^8 to 10^1 copies/µl. Three independent dilution series were analyzed and reactions were performed in triplicate. The specificity was evaluated using: i) genomic DNA from *C. peumoniae (TWAR)*, *C. felis (FP Baker)*, *C. caviae (GPIC)*, *C. abortus(S26/3)*, *C. psittaci (92/1293)*, *C. pecorum (1710S)*, *C. muridarum (MoPn)* and *C. trachomatis (*L2/434/BU*)* as well as ii) DNA from 23 and 38 bacterial species commonly found in swine and humans, respectively, and iii) genomic DNA from swine and human tissues (Table 3 and 4).

Table 3: Organisms of non-chlamydial origin found in swine and used for specificity testing

Organisms*		
Acinetobacter calcoaceticus	*Enterobacter sp.*	*Pseudomonas aeruginosa*
Bacteroides sp.	*Enterococcus sp*	*Proteus mirabilis*
Bordetella bronchiseptica	*Escherichia coli*	*Proteus vulgaris*
Bordetella pertussis	*Klebsiella sp.*	*Salmonella enterica subsp. enterica serovar Typhimurium*
Bordetella parapertussis	*Lactobacillus sp.*	*Staphylococcus aureus*
Brachyspira sp.		*Staphylococcus hyicus*
Brucella sp.	*Mycoplasma hyopneumoniae*	*Streptococcus suis*
Clostridium sp.	*Pasteurella multocida*	*Treponema sp.*

* Identification of micro-organisms was performed by culture on selective media, biochemical identification and molecular characterization methods like PCR and DNA sequencing.

Table 4: Organisms of non-chlamydial origin found in humans and used for specificity testing

Organisms*		
Acinetobacter baumannii	*Klebsiella oxytoca*	*Ruminococcus*
Bacteroides sp.	*Klebsiella pneumoniae*	*Salmonella sp.*
Bifidobacterium sp.	*Lactobacillus sp.*	*Serratia marcescens*
Bordetella bronchiseptica	*Stenotrophomonas sp*	*Shigella sp.*
Citrobacter braakii	*Moraxella catarrhalis*	*Staphylococcus aureus*
Citrobacter freundii	*Morganella morganii*	*Staphylococcus epidermidis*
Clostridium sp	*Mycoplasma pneumoniae*	*Moraxella catarrhalis*
Corynebacterium sp.	*Peptostreptococcus sp.*	*Streptococcus agalactiae*
Enterobacter sp.	*Streptococcus pneumoniae sp.*	*Streptococcus mitis*
Enterococcus sp.	*Propionibacterium sp.*	*Streptococcus nilleri*
Escherichia coli	*Pseudomonas aeruginosa*	*Streptococcus pyogenes*
Eubacterium sp.	*Proteus mirabilis*	*Veillonella sp.*
Faecalibacterium sp.	*Proteus vulgaris*	

* Identification of micro-organisms was performed by culture on selective media, biochemical identification and molecular characterization methods like PCR and DNA sequencing.

Case Studies

Pig farms

To validate the performance of the *C. suis* real-time PCR, vaginal swabs originating of 10 to 15 sows from three Dutch pig (A, B, C) farms were tested. Two of the three farms were dealing with cases of reproductive failure (decrease of the conception rate from 90 to <65%, 1–3% abortions, sows that delivered only two to five piglets of a non-uniform weight, white to yellow non-smelling vulval liquid, irregular return to estrus). Clinical disease did not improve following medication with doxycycline [400 g (ton feed) $^{-1}$] plus a combination of trimethoprim [120 g (ton feed) $^{-1}$] and sulfamethoxazole [600 g (ton feed) $^{-1}$] for 10–14 days during gestation and half of these drug concentrations during lactation. Antibody detection or PCR for *Leptospira* species, *Mycoplasma* species, *Brucella suis*, porcine reproductive and respiratory syndrome virus, Aujeszky's disease, swine influenza, porcine circovirus 2 and porcine parvovirus, were negative. Tests were performed at the Dutch Animal Health Service (Deventer, The Netherlands) and by the Flemish Animal Health Service (Drongen, Belgium). The third farm reported no clinical problems and was included as a negative control.

Animals were sampled for *C. suis* culture and PCR. For culture, rayon-tipped aluminium-shafted swabs (Copan; Fiers, Kuurne, Belgium) were immersed in chlamydia transport medium [20] immediately after vaginal sampling and stored at −80°C for culture. For PCR, vaginal swabs immersed in DNA/RNA stabilization buffer (Roche) were stored at −80°C. All DNA extractions were performed using the G-spin Total DNA Extraction Mini Kit (Goffin Molecular Biotechnologies, Beek, The Netherlands) according to the manufacturer,s protocol.

Purified DNA of vaginal swabs was analyzed by the newly developed *C. suis*-specific real-time PCR. Presence of viable *Chlamydiae* was determined by inoculation on McCoy cells (in duplicate) followed by direct immunofluorescence (anti-LPS/FITC) staining on the first culture drum at six days post inoculation (Imagen, Oxoid, Drongen, Belgium). If positive, the second culture drum was used for *C. suis* identification by use of the newly developed real-time PCR. Positively identified *C. suis* isolates were also examined by the *tet*(C) PCR, designed for demonstrating the tetracycline resistance gene *tet*(C) [23], For this purpose, we used genomic DNA of the Tc^R *C. suis* strain R19 [24] as positive control.

All animal samples were received from veterinary practitioners and were delivered for diagnostic purpose. Therefore, approval of the veterinary ethical committee was not required.

Slaughterhouse employees

Validation of the *C. suis* PCR for human diagnosis was performed by sampling abattoir workers. During a yearly medical check up, employees of a Belgian pig slaughterhouse were asked (informed consent) to voluntarily provide a pharyngeal and conjunctival swab, as well as a swab of fresh stool. Rayon-tipped aluminium-shafted swabs were immersed in chlamydia transport medium [20] and stored at −80°C until culture. Swabs for PCR were immersed in DNA/RNA stabilization buffer and stored at −80°C. Purified DNA was analyzed by the newly developed *C. suis*-specific real-time PCR. Presence of viable *Chlamydiae*, identification of viable *C. suis* and the presence of the *tet*(C) gene in *C. suis* strains was examined as mentioned for the pig farms.

The case study was approved by the medical ethical committee (approval EC UZG 2011/459). Participants provided their written informed consent and the consent procedure was approved by the medical ethical committee of Ghent University.

RESULTS

Primers and Probes

It was impossible to design *C.°suis*-specific primers in the 23S rDNA target region, such that the forward and reverse primers would also not anneal to *C.°trachomatis* DNA. However, we were able to design *C. suis*-specific probes for real-time PCR. The differentially labeled probes CS23S-A and CS23S-B were *C. suis*-specific and they were both required for coverage of the selected target region of all known *C. suis* strains (Table 2). Forward (CS23S-F) and reverse (CS23S-R) primers generated a PCR product of 159 bp.

Internal Inhibition Control

Standard curves made using 10^6 to 10^1 copies of the pGemT::CSIC control plasmid (Figure 1) showed slopes around −3.103, which correlates to an efficiency of 109%, with correlation coefficients > 98%.

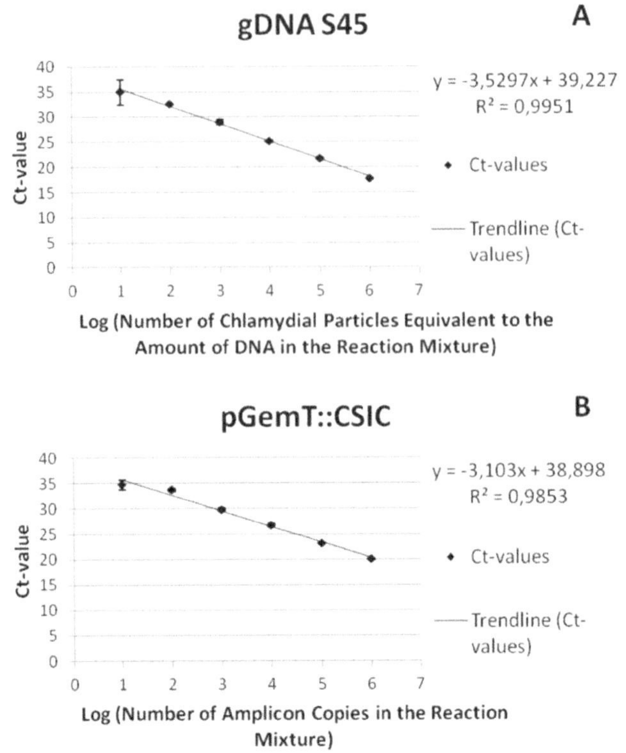

Figure 1: Real-time PCR quantitation of chlamydial S45 genomic DNA (A) and pGemT::CSIC control plasmid (B).

The standard curve data points are the average of 3 replications, standard deviations are shown as error bars. The equations and R^2 linearity values resulting from the linear regression analysis, are shown on the graphs.

Species-specific Real-time PCR

The species-specific real-time PCR detected all known *C. suis* strains. Amplification of genomic DNA extracts of all *C. suis* strains resulted in the expected PCR products with amplification curves exceeding the 0.02 threshold before cycle 25. The species-specific PCR was able to detect 10 rDNA copies (or two copies/μl) of both genomic and control plasmid DNA. Standard curves resulting from the sensitivity analysis are shown in Figure 1. The slopes were around −3.5, which correlates to an efficiency of 91%, with correlation coefficients > 99%. The species-specificity test revealed a Ct-value of 19 for DNA of 10^6 $TCID_{50}$ *C. suis.* Real-time PCR on genomic DNA of *C. pneumoniae, C. felis, C. caviae, C. pecorum, C. abortus, C.*

muridarum and *C. trachomatis* generated no signal. The real-time PCR gave no results for genomic DNA of other pathogens commonly found in the eyes, respiratory or urogenital tract of swine and humans, neither with DNA from swine or human tissues.

Case studies

Pig farms

Chlamydia suis was not detected on the clinically healthy farm C. The pathogen was present on farms A and B, both dealing with reproductive failure. *Chlamydia suis* DNA was found in all (100%) vaginal swabs of farms A and B. Viable *C. suis* organisms were present in 7 of 10 (70%) and 11 of 15 (73.3%) vaginal swabs of farms A and B, respectively. Chlamydial isolates of farm A and B were examined for the presence of the *tet*(C) gene. The *tet*(C) gene was discovered in two of seven (28.5%) and 6 of 11 (54.5%) *C. suis* isolates of farms A and B, respectively. Clinical symptoms disappeared after treatment with enrofloxacin (fluoroquinolone, Baytril 5%; Bayer Healthcare), as previously described [25]. Enrofloxacin was added to the sperm diluter [2 ml (l diluter) $^{-1}$] and it was also used to rinse the sow's reproductive tract (Baytril 5%, 5 ml+95 ml distilled water) immediately before artificial insemination. Infected sows were re-examined two weeks after treatment and were negative in both PCR and culture.

Slaughterhouse employees

Only 12 of 84 (14.3%) employees voluntarily participated. They provided 12 conjunctival, 12 pharyngeal and 12 stool swabs. Pharyngeal swabs and stool were all *C. suis* negative. Two of 12 (16.6%) conjunctival swabs were positive by the newly developed *C. suis* real-time PCR, revealing a Ct-value of 26 and 28, respectively. Real-time PCR results on all swabs could be confirmed by culture. Culture harvest of both positively identified *C. suis* isolates was negative by the *tet(C)* PCR. Positive employees worked daily for three years in the pig gut washing area and for eight years in the animal reception area, respectively. Clinical signs were absent.

DISCUSSION

Chlamydial infections are treated with tetracyclines. In 1998, tetracyline resistant (Tc^R) *C. suis* strains emerged in the U.S. pig industry (Iowa and

Nebraska), and are currently also known to be prevalent in the Belgian, Cypriote, German, Israeli, Italian and Swiss pig industry [2], [3], [9], [26], [27]. The present study adds an additional country to the list, as the current study is the first to discover Tc^R *C. suis* strains in the Dutch pig industry. The *C. suis* Tc^R phenotype is manifested through the tetracycline resistance gene *tet*(C). *Tet*(C) is integrated into the chlamydial chromosome [28], but it is a transposable genetic element, which is also present in other bacteria such as *E. coli*.

The international economic consequences of *C. suis* infections are not yet fully established, but the financial loss due to severe reproductive failure and the need for antibiotic treatment currently worries pig producers all over the world. Moreover, pig farmers are aware of the existence of Tc^R *C. suis* strains and the risk of importing Tc^R *C. suis* contaminated sperm for artificial insemination [18] (and P. Delava, personal communication 2014). Furthermore, emergence of Tc^R chlamydial strains might pose a potential threat for public health if this species turns out to be a zoonotic pathogen. Suchland *et al.* demonstrated *in vitro* horizontal transfer of the tetracycline resistance gene among chlamydial species [29]. Therefore, contact between Tc^R and Tc^S *Chlamydia* spp. could lead to transfer of the *tet(C)* gene and the subsequent phenotype, which could then be propagated and selected for in patients that are treated with tetracyclines. However, the lack of evidence of antibiotic resistance leading to treatment failure in humans, seems to indicate that horizontal transfer of the *tet(C)* gene is rather rare. Sandoz and Rockey [30] state that unsufficiency of the current diagnostic methods, could be the cause of this lack of evidence.

In this respect, studying the presence of *C. suis* in humans, dealing with pigs on a regular basis, is meaningful. Thus, there is an urgent need for a highly sensitive and specific molecular diagnostic test to study the epidemiology of *C. suis* in pigs and to study a possible transfer of *C. suis* to humans.

Real-time PCR technology offers the possibility to automatically combine amplification, specific hybridization and detection in one single test. It allows specific and sensitive gene quantification with a minimal contamination risk, as the AmpErase UNG system can be incorporated in order to prevent post-PCR carry-over of amplified DNA. The technique has been used successfully to detect C. *pneumoniae* [31], [32], [33], [34], C. *trachomatis* [35], C. *felis* [36], [37], [38], C. *pecorum* [39], C. *psittaci* [40], and *C. abortus* [19].

Although, the high genomic sequence similarity between *C. suis* and *C. trachomatis* made it difficult to design a sensitive *C. suis*-specific PCR, the analytical sensitivity of the assay was 10 rDNA copies/reaction (or two copies/μl). Besides, the *C. suis* real-time PCR contains two differentially labeled probes. Albeit differentially labeling is not required for the detection of all *C. suis* strains, their use creates the ability to differentiate the two 'groups' of *C. suis* strains which are distinguished by one single SNP in the target region.

The *C. suis*-specific real-time PCR is suitable for diagnosis in swine as successfully demonstrated in the present case report. Importantly, the *C. suis*-specific real-time PCR is also suitable for human samples, as it distinguishes *C. suis* from i) the human pathogens *C. trachomatis* and *C. pneumoniae* and ii) the currently known zoonotic species: *C. felis*, *C. abortus* and *C. psittaci*.

The detection of *C. suis* DNA as well as viable *C. suis* in the eyes of two slaughterhouse employees illustrates the value of this assay in future investigations on the zoonotic potential of *C. suis*. Becker et al. [41] showed intensively raised pigs to be pre-disposed to chlamydial associated conjunctivitis. Transmission from infected pig eyes to humans is perhaps not unusual in highly irritant environments like a slaughterhouse and intensive pig farms. *C. suis* positive persons had no clinical complaints and they showed no disease symptoms. We detected only low amounts of *C. suis* in the eyes, and perhaps *C. suis* only ended up in the eyes by touching face/eyes with contaminated hands. Possibly, it was just an eye "contaminant" and not a real infection. Perhaps serology would have clarified this issue. However, employees refused to give blood during our experiment. Thus, future studies in humans should include the detection of *C. suis*-specific antibodies in serum or at least in conjunctival swabs. None of the participating employees was infected with *C. trachomatis* and both *C. suis* isolates were tetracycline sensitive. Therefore, further research is needed, sampling larger human risk populations, to estimate the prevalence of *C. trachomatis* and Tc^R *C. suis* strains in people working with pigs. Besides, the potential of *C. suis* to cause pathology in the human eye needs to be examined in more detail, especially since *C. suis* is phylogenetically highly related to *C. trachomatis*, the etiology of human trachoma, and, as a recent study by Deborah Dean also found *C. suis* in the human eye [42].

In conclusion, we designed and validated a *C. suis*-specific real-time PCR for use in swine and humans. The assay can be used for detection and monitoring of *C. suis* strains in pigs and for examining their zoonotic potential more extensively. Additionally, the real-time PCR could be useful for preventing the spread of *C. suis* strains, for instance through international trade of boars and sperm.

ACKNOWLEDGMENTS

F. Boyen (Department of Pathology, Bacteriology and Poultry Diseases, UGent, Belgium), M. Uyttendaele (Department of Food Safety and Food Quality, UGent, Belgium), G. Claeys (Department of Clinical Biology, Microbiology and Immunology, UGent, Belgium) and P. Butaye (Department of Pathology, Bacteriology and Poultry Diseases, UGent, Belgium) are acknowledged for providing human or swine pathogens. A. Dumont, L. De Vogelaere and L. Devlieger are acknowledged for excellent assistance during the experiments.

REFERENCES

1. Everett KDE, Bush RM, Andersen AA (1999) Emended description of the order Chlamydiales, proposal of *Parachlamydiaceae* fam. nov. and *Simkaniaceae* fam. nov., each containing one monotypic genus, revised taxonomy of the family *Chlamydiaceae*, including a new genus and five new species, and standards for the identification of organisms. International Journal of Systematic Bacteriology 49: 415–440.
2. Camenisch U, Lu ZH, Vaughan L, Corboz L, Zimmermann DR, et al. (2004) Diagnostic investigation into the role of *Chlamydiae* in cases of increased rates of return to oestrus in pigs. The Veterinary record 155: 593–596.
3. Di Francesco A, Baldelli R, Cevenini R, Magnino S, Pignanelli S, et al. (2006) Seroprevalence to *chlamydiae* in pigs in Italy. Veterinary Record 159: 849–850.
4. Eggemann G, Wendt M, Hoelzle LE, Jager C, Weiss R, et al. (2000) Prevalence of *Chlamydia* infections in breeding sows and their importance in reproductive failure. DTW Deutsche tierarztliche Wochenschrift 107: 3–10.
5. Vanrompay D, Geens T, Desplanques A, Hoang TQT, De Vos L, et al. (2004) Immunoblotting, ELISA and culture evidence for Chlamydiaceae in sows on 258 Belgian farms. Veterinary Microbiology 99: 59–66.
6. Sarma DK, Tamuli MK, Rahman T, Boro BR, Deka BC, et al. (1983) Isolation of *Chlamydia* from a Pig with Lesions in the Urethra and Prostate-Gland. Veterinary Record 112: 525–525.
7. Woollen N, Daniels EK, Yeary T, Leipold HW, Phillips RM (1990) Chlamydial Infection and Perinatal-Mortality in a Swine Herd. Journal of the American Veterinary Medical Association 197: 600–601.
8. Willigan DA, Beamer PD (1955) Isolation of a transmissible agent from pericarditis of swine. Journal of the American Veterinary Medical Association 126: 118–122.
9. Andersen AA, Rogers D (1998) Resistance to tetracycline and sulfadiazine in swine *C. trachomatis* isolates. Proceedings of the Ninth International Symposium on Human Chlamydial Infection. Edited by: Stephens.International Chlamydia Symposium, San Francisco, Calif: p. 313–316.
10. Zahn I, Szeredi L, Schiller I, Kunz US, Burgi E, et al. (1995) Immunohistological Determination of *Chlamydia psittaci/Chlamydia*

pecorum and *C. trachomatis* in the Piglet Gut. Journal of Veterinary Medicine Series B-Zentralblatt Fur Veterinarmedizin Reihe B-Infectious Diseases and Veterinary Public Health 42: 266–276.

11. Reinhold P, Kirschvink N, Theegarten D, Berndt A (2008) An experimentally induced *Chlamydia suis* infection in pigs results in severe lung function disorders and pulmonary inflammation. Veterinary Research 39: 35.
12. Pospischil A, Borel N, Chowdhury EH, Guscetti F (2009) Aberrant chlamydial developmental forms in the gastrointestinal tract of pigs spontaneously and experimentally infected with *Chlamydia suis* . Veterinary Microbiology 135: 147–156.
13. Guscetti F, Schiller I, Sydler T, Heinen E, Pospischil A (2009) Experimental enteric infection of gnotobiotic piglets with Chlamydia suis strain S45. Veterinary Microbiology 135: 157–168.
14. Schautteet K, Vanrompay D (2011) *Chlamydiaceae* infections in pig. Veterinary research 42: 29. [PMC free article]
15. Sachse K, Hotzel H, Slickers P, Ellinger T, Ehricht R (2005) DNA microarray-based detection and identification of *Chlamydia* and *Chlamydophila* spp. Molecular and Cellular Probes 19: 41–50.
16. Ehricht R, Slickers P, Goellner S, Hotzel H, Sachse K (2006) Optimized DNA microarray assay allows detection and genotyping of single PCR-amplifiable target copies. Molecular and Cellular Probes 20: 60–63.
17. Borel N, Kempf E, Hotzel H, Schubert E, Torgerson P, et al. (2008) Direct identification of *chlamydiae* from clinical samples using a DNA microarray assay-A validation study. Molecular and Cellular Probes 22: 55–64.
18. Schautteet K, De Clercq E, Miry C, Van Groenweghe F, Delava P, et al. (2013) Tetracycline-resistant *Chlamydia suis* in cases of reproductive failure on Belgian, Cypriote and Israeli pig production farms. Journal of medical microbiology 62: 331–334.
19. Pantchev A, Sting R, Bauerfeind R, Tyczka J, Sachse K (2010) Detection of all *Chlamydophila* and *Chlamydia* spp. of veterinary interest using species-specific real-time PCR assays. Comparative Immunology Microbiology and Infectious Diseases 33: 473–484.
20. Vanrompay D, Ducatelle R, Haesebrouck F (1992) Diagnosis of Avian Chlamydiosis - Specificity of the Modified Gimenez Staining on Smears and Comparison of the Sensitivity of Isolation in Eggs

and 3 Different Cell-Cultures. Journal of Veterinary Medicine Series B-Zentralblatt Fur Veterinarmedizin Reihe B-Infectious Diseases and Veterinary Public Health 39: 105–112.

21. Thompson JD, Gibson TJ, Plewniak F, Jeanmougin F, Higgins DG (1997) The CLUSTAL_X windows interface: flexible strategies for multiple sequence alignment aided by quality analysis tools. Nucleic Acids Research 25: 4876–4882.
22. Skaletsky SRaHJ (2000) Primer3 on the WWW for general users and for biologist programmers. Krawetz S, Misener S, (eds) Bioinformatics Methods and Protocols: Methods in Molecular Biology Humana Press, Totowa, NJ, pp 365–386.
23. Dugan J, Rockey DD, Jones L, Andersen AA (2004) Tetracycline resistance in *Chlamydia suis* mediated by genomic islands inserted into the chlamydial inv-like gene. Antimicrobial Agents and Chemotherapy 48: 3989–3995.
24. Rogers DG, Andersen AA (1996) Intestinal lesions caused by two swine chlamydial isolates in gnotobiotic pigs. Journal of Veterinary Diagnostic Investigation 8: 433–440.
25. Schautteet K, De Clercq E, Miry C, Van Groenweghe F, Delava P, et al.. (2012) Tetracycline-resistant *Chlamydia suis* in cases of reproductive failure on Belgian, Cypriote and Israeli pig production farms. Journal of medical microbiology.
26. Lenart J, Andersen AA, Rockey DD (2001) Growth and development of tetracycline-resistant *Chlamydia suis* . Antimicrobial Agents and Chemotherapy 45: 2198–2203.
27. Borel N, Regenscheit N, Di Francesco A, Donati M, Markov J, et al. (2012) Selection for tetracycline-resistant Chlamydia suis in treated pigs. Veterinary Microbiology 156: 143–146.
28. Dugan J, Rockey DD, Jones L, Andersen AA (2004) Tetracycline resistance in Chlamydia suis mediated by genomic islands inserted into the chlaravdial inv-like gene. Antimicrobial Agents and Chemotherapy 48: 3989–3995.
29. Suchland RJ, Sandoz KM, Jeffrey BM, Stamm WE, Rockey DD (2009) Horizontal Transfer of Tetracycline Resistance among *Chlamydia* spp. In Vitro. Antimicrobial Agents and Chemotherapy 53: 4604–4611.
30. Sandoz KM, Rockey DD (2010) Antibiotic resistance in Chlamydiae. Future Microbiology 5: 1427–1442.

31. Welti M, Jaton K, Altwegg M, Sahli R, Wenger A, et al. (2003) Development of a multiplex real-time quantitative PCR assay to detect *Chlamydia pneumoniae*, *Legionella pneumophila* and *Mycoplasma pneumoniae* in respiratory tract secretions. Diagnostic Microbiology and Infectious Disease 45: 85–95.
32. Kuoppa Y, Boman J, Scott L, Kumlin U, Eriksson I, et al. (2002) Quantitative detection of respiratory *Chlamydia pneumoniae* infection by real-time PCR. Journal of Clinical Microbiology 40: 2273–2274.
33. Hardick J, Maldeis N, Theodore M, Wood BJ, Yang S, et al. (2004) Real-time PCR for *Chlamydia pneumoniae* utilizing the Roche lightcycler and a 16S rRNA gene target. Journal of Molecular Diagnostics 6: 132–136.
34. Tondella MLC, Talkington DF, Holloway BP, Dowell SF, Cowley K, et al. (2002) Development and evaluation of real-time PCR-based fluorescence assays for detection of *Chlamydia pneumoniae* . Journal of Clinical Microbiology 40: 575–583.
35. Solomon AW, Holland MJ, Burton MJ, West SK, Alexander NDE, et al. (2003) Strategies for control of trachoma: observational study with quantitative PCR. Lancet 362: 198–204.
36. Helps C, Reeves N, Egan K, Howard P, Harbour D (2003) Detection of *Chlamydophila felis* and feline herpesvirus by multiplex real-time PCR analysis. Journal of Clinical Microbiology 41: 2734–2736.
37. Helps C, Reeves N, Tasker S, Harbour D (2001) Use of real-time quantitative PCR to detect *Chlamydophila felis* infection. Journal of Clinical Microbiology 39: 2675–2676.
38. Sykes JE, Allen JL, Studdert VP, Browning GF (2001) Detection of feline calicivirus, feline herpesvirus 1 and *Chlamydia psittaci* mucosal swabs by multiplex RT-PCR/PCR. Veterinary Microbiology 81: 95–108.
39. DeGraves FJ, Gao DY, Hehnen HR, Schlapp T, Kaltenboeck B (2003) Quantitative detection of Chlamydia psittaci and C-pecorum by high-sensitivity real-time PCR reveals high prevalence of vaginal infection in cattle. Journal of Clinical Microbiology 41: 1726–1729.
40. Geens T, Dewitte A, Boon N, Vanrompay D (2005) Development of a Chlamydophila psittaci species-specific and genotype-specific real-time PCR. Veterinary Research 36: 787–797.

41. Becker A, Lutz-Wohlgroth L, Brugnera E, Lu ZH, Zimmermann DR, et al. (2007) Intensively kept pigs pre-disposed to chlamydial associated conjunctivitis. Journal of Veterinary Medicine Series a-Physiology Pathology Clinical Medicine 54: 307–313.
42. Dean D, Rothschild J, Ruettger A, Kandel RP, Sachse K (2013) Zoonotic Chlamydiaceae species associated with trachoma, Nepal. Emerging Infectious Diseases 19: 1948–1955.

CHAPTER 12

First Report of Corynebacterium Pseudotuberculosis from Caseous Lymphadenitis Lesions in Black Alentejano Pig (Sus scrofa domesticus)

Manuela Oliveira[1], Cynthia Barroco[1], Carla Mottola[1], Raquel Santos[1], Abdelhak Lemsaddek[2], Luis Tavares[1] and Teresa Semedo-Lemsaddek[1]

[1] CIISA/Faculdade de Medicina Veterinária, Universidade de Lisboa, Avenida da Universidade Técnica, 1300-477 Lisboa, Portugal

[2] BioFIG, Centro para a Biodiversidade, Genómica Integrativa e Funcional, Faculdade de Ciências, Universidade de Lisboa, 1749-016 Lisboa, Portugal

Citation: Oliveira, M., Barroco, C., Mottola, C. et al. First report of Corynebacterium pseudotuberculosis from caseous lymphadenitis lesions in Black Alentejano pig (Sus scrofa domesticus). BMC Vet Res 10, 218 (2014). https://doi.org/10.1186/s12917-014-0218-3.

ABSTRACT

Background

Corynebacterium pseudotuberculosis is the etiologic agent of caseous lymphadenitis, a common disease in small ruminant populations throughout the world and responsible for a significant economic impact for producers.

Case presentation

To our knowledge, this is the first characterization of *C. pseudotuberculosis* from caseous lymphadenitis lesions in Black Alentejano pig (*Sus scrofa domesticus*). In this study, phenotypic and genotypic identification methods allocated the swine isolates in *C. pseudotuberculosis* biovar ovis. The vast majority of the isolates were able to produce phospholipase D and were susceptible to most of the antimicrobial compounds tested. Macrorestriction patterns obtained by Pulsed Field Gel Electrophoresis (PFGE) grouped the *C. pseudotuberculosis* in two clusters with a high similarity index, which reveals their clonal relatedness. Furthermore, swine isolates were compared with *C. pseudotuberculosis* from caprines and PFGE patterns also showed high similarity, suggesting the prevalence of dominant clones and a potential cross-dissemination between these two animal hosts.

Conclusions

This work represents the first report of *Corynebacterium pseudotuberculosis* from caseous lymphadenitis lesions in Black Alentejano pig and alerts for the importance of the establishment of suitable control and sanitary management practices to control the infection and avoid further dissemination of this important pathogen to other animal hosts.

BACKGROUND

Corynebacterium pseudotuberculosis is an important animal pathogen, being the etiological agent of caseous lymphadenitis (CLA) or cheesy gland disease in small ruminants [1]. CLA is frequently detected in major sheep and goat production areas around the world. It is characterized by abscess formation in the skin, internal and external lymph nodes and internal organs. This disease causes significant economic impact on the small ruminant industry through decreased meat yield, damaged wool and leather, decreased reproductive efficiency, culling of affected animals and increased morbidity and mortality rates [2]. CLA can become endemic in a herd or flock and once established it is difficult to eradicate due to its poor response to therapeutics, ability to persist in the environment and difficulties in detecting subclinical infected animals [1]-[3]. It is also easily spread amongst animals due to direct contact with superficial wounds or draining abscesses [3]. Nowadays, the most common treatment for CLA is abscess drainage followed by disinfection with an iodine solution and antibiotic therapy.

C. pseudotuberculosis is a relatively homogenous taxonomic group that can be distinguished from most *Corynebacterium* species by the production of Phospholipase D (PLD) and urease and the inability to ferment starch. The importance of *C. pseudotuberculosis* as an animal pathogen has prompted characterization studies on its toxins, particularly the haemolytic toxin PLD [4]. The differentiation between biovars is also important for infection epidemiology, as they are host specific. Evaluation of the ability to reduce nitrates allows discriminating between the negative isolates from biovar ovis, usually related with CLA in sheeps and goats, and the positive members of biovar equi, found in horses and bovines [3],[4].

Other species belonging to the *Corynebacterium* genus have already been related to CLA. *Corynebacterium ulcerans* was isolated from abscessed lymph nodes in wild boar (*Sus scrofa*) [5],[6]. However, to our understanding, there is only one report available regarding *C. pseudotuberculosis* isolation from asymptomatic swine [7]. In this work we describe the first phenotypic and genotypic characterization of *C. paratuberculosis* clinical isolates from caseous lymphadenitis lesions in the Black Alentejano pig (*Sus scrofa domesticus*), Alentejo region, South Portugal.

CASE PRESENTATION

Data presented refers to two "Alentejana" breed swine farms, A and B, located in "Odemira" district, "Alentejo" region, South Portugal, where purulent lymphadenitis cases were detected (Figure 1). In farm A and B, respectively with 700-800 and 400-500 animals, piglets are maintained in camping with the sows until weaning at 5 weeks, and afterwards are reared in extensive system, in lots with around 200 animals each. Farms only reared swine, but they were not isolated, as animals' movement could be observed. There was no commercial relationship between farms A and B. In both farms, CLA was detected in piglets from weaning until 10 month of age, in all rearing lots, affecting up to 10% of the animals. Lesions were observed mainly in the mandibular and retropharyngeal lymphnodes, but were also found in other locations. About 1% of the animals died as a consequence of polyarthritis and/or infection dissemination to other organs. In the remaining cases, infection was controlled and was not responsible for rejections at slaughter.

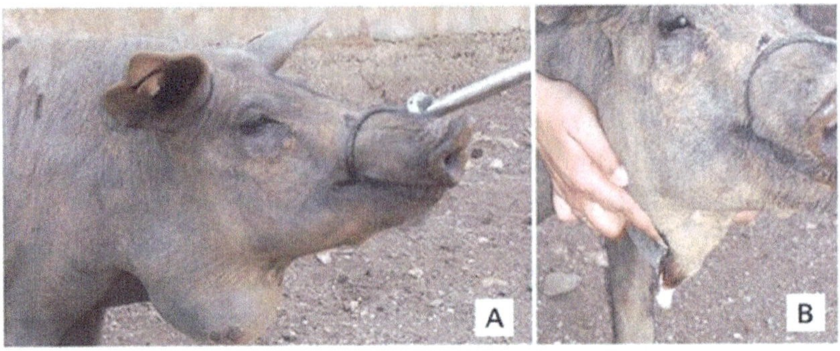

Figure 1: Caseous lymphadenitis lesions in Black Alentejano pig (*Sus scrofa domesticus*) from a farm located in the Odemira district, "Alentejo" region, south Portugal; A) Mandibular abscess; B) Abscess purulent content.

Ten bacterial isolates were obtained from ten swabs collected from abscesses in ten Black Alentejano pigs (*Sus scrofa domesticus*). Samples were collected as part of routine diagnostics, from live animals using sterile materials and after performing the asepsis of the abscess skin, and kept refrigerated at 4°C until transport to the laboratory. Samples were plated on Columbia Agar (COS) (43041, bioMérieux, Basingstoke, UK) and incubated aerobically at 37°C for 24 h. Isolates were initially checked for Gram staining, morphology and production of cytochrome c oxidase. From the swine samples it was possible to obtain pure cultures, formed by Gram-positive, oxidase negative, pleomorphic rods. Identification was performed using biochemical identification galleries (12136A, API Coryne®, bioMérieux, Basingstoke, UK) according to the manufacturer's instructions. Subsequently, in order to distinguish between *C. pseudotuberculosis* biovar equi and biovar ovis, the ability to reduce nitrates was also evaluated [3]. Isolates were identified as *C. pseudotuberculosis* by Api Coryne®, all of which were nitrate reductase negative, hence identified as *C. pseudotuberculosis* biovar ovis. Although other microorganisms have already been related to lymphadenitis in several animals, such as *Actinomyces hyovaginalis* in goats, sheeps and pigs [8],[9], *Francisella tularensis* in humans [10], *Staphylococcus aureus* ssp. *anaerobius* in lambs [11] and *C. ulcerans* in wild boars and roe deer [5],[6], *C. pseudotuberculosis* is the main bacterial species responsible for CLA [1]. Results from this study confirm that this bacterial species is also the main responsible for CLA development in the Black Alentejano pig (*Sus scrofa domesticus*) at the sampled farms, as it was possible to isolate this agent from all collected samples, in pure cultures.

As *C. pseudotuberculosis* is frequently related to CLA in small ruminants but not in pigs, we included in this study nine *C. pseudotuberculosis* caprine isolates belonging to a collection of clinical isolates from the Bacteriology Laboratory from the Faculty of Veterinary Medicine from the University of Lisbon, Portugal, for comparison purposes.

PLD production by the 19 isolates was determined by observation of haemolytic antagonism with β haemolysin from *Staphylococcus aureus* and haemolytic synergism with *Streptococcus agalactiae*, according to Literák et al. [12]. cAMP phenotypic assay demonstrated that all, except one swine isolate, were PLD positive.

For assessment of the antimicrobial resistance profile, 13 antimicrobial compounds (Oxoid, Basingstoke, Hampshire, UK) commonly used in veterinary medicine were selected, as follows: Amoxycillin/Clavulanic acid (AMC, CT0223B, 30 μg), Ampicillin (AMP, CT0003B, 10 μg), Chloramphenicol (C, CT0013B, 30 μg), Cephalexin (CL, CT0007B, 30 μg), Gentamicin (CN, CT0024B, 10 μg), Cefotaxime (CTX, CT0166B, 30 μg), Enrofloxacin (ENR, CT0639B, 5 μg), Nalidixic acid (NA, CT0031B, 30 μg), Penicillin G (P, CT0043B, 10 units), Streptomycin (S, CT0047B, 10 μg), Sulfamethoxazole/Trimethoprim (SXT, CT0052B, 25 μg), Tetracycline (TE, CT0054B, 30 μg) and Vancomycin (VA, CT0058B, 30 μg). The compounds were tested by the Disk Diffusion (DD) Method, according to the Clinical and Laboratory Standards Institute [13] guidelines. Isolates were susceptible to the majority of the antimicrobial compounds tested. All *C. pseudotuberculosis*, except one, were resistant to nalidix acid and streptomycin. Although isolates were susceptible to most antimicrobial compounds tested, CLA infections are difficult to eradicate, as the bacterial agent remains enclosed in caseous abscesses, where the penetration of therapeutic drugs is extremely difficult.

Isolate molecular characterization was performed by Multiplex PCR and Pulsed Field Gel Electrophoresis (PFGE). For DNA isolation, 4-5 bacterial colonies of 48 hour cultures were resuspended in 100 μL TE [10 mM Tris-Cl (pH 8.0), 1 mM EDTA (pH 8.0)]. Afterwards, samples were incubated at 96°C for 5-7 minutes and centrifuged at 15000 g during 5 minutes. Supernatants were stored at -20°C.

Primers targeting the 16S rRNA, *rpoB* and *pld* genes of *C. pseudotuberculosis* previously described by other authors were used [14]-[16]. Reference strain *C. pseudotuberculosis* CECT 808 was included as positive control.

Multiplex PCR was performed in a final reaction volume of 25 μL containing 2.5 U NZYTaq 2x Green Master Mix (NZYTech, Portugal), and 10 μM of each of the primers 16S-F/16S-R, *rpoB*-F/*rpoB*-R and *pld*-F/*pld*-R. Reactions were carried out in a thermal cycler (MyCycler; Bio-Rad laboratories, Hemel Hempstead, UK), using the conditions described by Pacheco et al. 2007 [16].

All *Corynebacterium* isolates under analysis generated the amplicons of ~446 bp and ~816 bp, corresponding respectively to the *rpoB* and 16S RNA genes. Regarding the *pld* gene product (200 bp) it was detected in only 14 isolates. This discrepancy between PLD phenotypic and genotypic assays (18 vs. 14 positive isolates) led us to perform another PCR amplification using the primers directed to the *pld* gene described by Pacheco et al. but in a single reaction, i.e., 16S-F/16S-R, *rpoB*-F/*rpoB*-R were excluded. In this PCR amplification the 19 isolates under analysis showed a positive result. Thus, the false negative results obtained in the previous multiplex PCR were probably due to reagent consumption, which prevented *pld* amplification. Since the other multiplex targets were 16S rRNA, *rpoB* (two housekeeping genes present in more than one copy in the bacterial genome) this is not totally surprising. These findings highlight the need of performing both phenotypic and genotypic PLD analysis and, in case of discrepancy, converting the multiplex in a PCR directed to a single target to further confirm the negative results.

For PFGE analysis, isolates were grown in Brain-Heart Infusion broth with 0.1% Tween 20 for 48 hour at 37°C and DNA plugs were prepared as previously described by Connor et al. [17]. Macrorestrition was carried overnight with 20 U of the endonuclease *Sfi*I (Takara BIO INC, Saint-Germain-en-Laye, France) at 50°C. PFGE was performed using a Chef DRII system (Bio-Rad laboratories, Hemel Hempstead, UK), and gels consisted of 1% agarose (Sigma-Aldrich Química, S.L) in 0.5X Tris-borate-EDTA buffer (Bio-Rad laboratories, Hemel Hempstead, UK). The Lambda Ladder PFG Marker (New England Biolabs, Ipswich, USA) was used as a molecular weight marker. Electrophoresis was performed at 6 V/cm at 14°C for 23 hour with an initial switch time of 5 s and final switch time of 20 s. After staining with ethidium bromide, gel images were acquired with the ImageMaster (PharmaciaBiotech, GE Healthcare, UK).

BioNumerics 6.5 software (Applied Maths, Belgium) was used to register PFGE macrorestriction patterns and clustering analysis performed using the Dice similarity coefficient and the unweighted-pair group method with arithmetic mean (UPGMA).

As *C. pseudotuberculosis* is often related with CLA in small ruminants, goat isolates (n = 9) from caseous lymphadenitis lesions, belonging to a large collection of clinical isolates from the Laboratory of Microbiology and Immunology of the Faculty of Veterinary Medicine of the University of Lisbon, Portugal, were used in this study for comparison and genotypic characterization, as previously mentioned.

From the PFGE macrorestriction profiles obtained for the nineteen isolates under analysis (ten *C. pseudotuberculosis* from pigs and nine from goats) a similarity dendogram was built using the Bionumerics software (Figure 2). Above 55%, isolates grouped in two clusters: cluster I, including eight swine isolates (four from each farm) and one caprine isolate; and cluster II, with eight caprine and two swine isolates (one from each farm). The higher similarity index observed within each cluster reveals the clonal relationship between the *C. pseudotuberculosis* under analysis, despite animal host species and farm of origin. Similar results were reported by other authors who also found high genomic similarities between *C. pseudotuberculosis* isolated from distinct animal species [3],[17],[18].

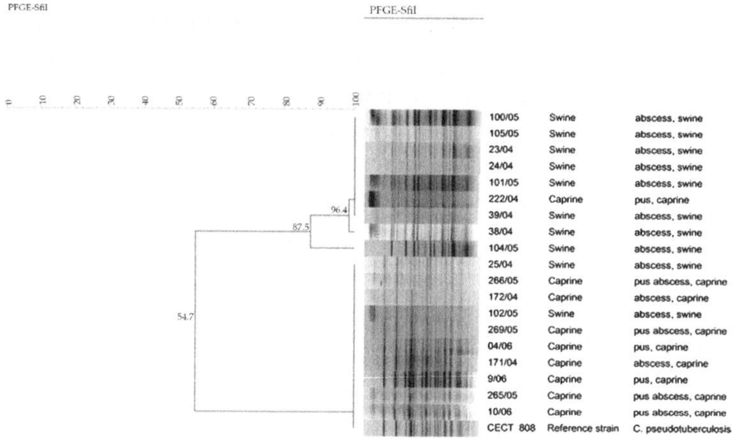

Figure 2: Dendrogram based on PFGE patterns from *C. pseudotuberculosis* strains analyzed using the endonuclease *Sfi* I. Similarity was calculated with Dice correlation coefficient -r- and clustering was performed with UPGMA.

Overall, our study suggests the prevalence of dominant clones and a putative cross-dissemination between swine and goats. As the farms under study only rear this swine breed, animals could have been infected by pigs coming from other farms that also have sheep and goats, as the production system was not closed.

CONCLUSIONS

This work represents the first report of *Corynebacterium pseudotuberculosis* from caseous lymphadenitis lesions in Black Alentejano pig (*Sus scrofa domesticus*). It also alerts for the importance of the establishment of suitable control and sanitary management practices to control the infection and avoid further dissemination of this important pathogen to other animal hosts. The presence of this microorganism in the two farms studied confirms its spreading ability, as well as their clonal relationships established by PFGE analysis. Further studies including a larger number of isolates should be performed in order to fully characterize these agents and identify possible routes of transmission.

AUTHORS' CONTRIBUTIONS

MO conceived the study and participated in its coordination, helped to draft the manuscript, having also participated in bacterial isolation and in the phenotypic assays; CB carried out the phenotypic assays and the molecular genetic studies; CM participated in the molecular genetic studies; RS participated in results analysis and helped to draft the manuscript; AL participated in the sequence alignment; LT participated in the study design and coordination; TSL participated in the design, coordination and execution of the molecular genetic studies and helped to draft the manuscript. All authors read and approved the final manuscript.

ACKNOWLEDGEMENTS

Authors would like to acknowledge Lina Cavaco, DVM, PhD and Filipe Nunes, DVM, PhD, for all the help in isolates collection and identification and Diogo Costa, DVM, for allowing access to the farms. This work was supported by CIISA ("Centro de Investigação Interdisciplinar em Sanidade Animal") from the Faculty of Veterinary Medicine, University of Lisbon, Project PEst-OE/AGR/U10276/2014, funded by "Fundação para a Ciência e Tecnologia" (FCT), Lisbon, Portugal.

REFERENCES

1. Dorella FA, Pacheco LC, Oliveira SC, Miyoshi A, Azevedo V: *Corynebacterium pseudotuberculosis*: microbiology, biochemical properties, pathogenesis and molecular studies of virulence. Vet Res. 2006, 37: 1-18. 10.1051/vetres:2005056.
2. Williamson LH: Caseous lymphadenitis in small ruminants. Vet Clin North Am Food Anim Pract. 2001, 17: 359-371.
3. Soares SC, Silva A, Trost E, Blom J, Ramos R, Carneiro A, Ali A, Santos AR, Pinto AC, Diniz C, Barbosa EG, Dorella FA, Aburjaile F, Rocha FS, Nascimento KK, Guimarães LC, Almeida S, Hassan SS, Bakhtiar SM, Pereira UP, Abreu VA, Schneider MP, Miyoshi A, Tauch A, Azevedo V: The pan-genome of the animal pathogen *Corynebacterium pseudotuberculosis* reveals differences in genome plasticity between the biovar ovis and equi strains. PLoS ONE. 2013, 8: e53818-10.1371/journal.pone.0053818.
4. Coyle MB, Lipsky BA: Coryneform bacteria in infectious diseases: Clinical and laboratory aspects. Clin Microbiol Rev. 1990, 3: 227-246.
5. Contzen M, Sting R, Blazey B, Rau J: *Corynebacterium ulcerans* from Diseased Wild Boars. Zoonoses Public Health. 2011, 58: 479-488. 10.1111/j.1863-2378.2011.01396.x.
6. Eisenberg T, Kutzer P, Peters M, Sing A, Contzen M, Rau J: Nontoxigenic tox-bearing *Corynebacterium ulcerans* infection among game animals, Germany. Emerg Infect Dis. 2014, 20: 448-452.
7. Zhao HK, Yonekawa K, Takahashi T, Kikuchi N, Hiramune T, Yanagawa R: Isolation of *Corynebacterium pseudotuberculosis* from the cervical canal of clinically normal sows. Res Vet Sci. 1993, 55: 356-359. 10.1016/0034-5288(93)90107-Q.
8. Schumacher VL, Hinckley L, Gilbert K, Risatti GR, Londoño AS, Smyth JA: *Actinomyces hyovaginalis*-associated lymphadenitis in a Nubian goat. J Vet Diagn Invest. 2009, 21: 380-384. 10.1177/104063870902100315.
9. Foster G, Wragg P, Koylass MS, Whatmore AM, Hoyles L: Isolation of *Actinomyces hyovaginalis* from sheep and comparison with isolates obtained from pigs. Vet Microbiol. 2012, 157: 471-475. 10.1016/j.vetmic.2012.01.003.
10. Markoc F, Koseoglu RD, Koc S, Gurbuzler L: Tularemia in differential diagnosis of cervical lymphadenopathy: cytologic features of tularemia

lymphadenitis. Acta Cytol. 2014, 58: 23-28. 10.1159/000355869.

11. Møller K, Agerholm JS, Ahrens P, Jensen NE, Nielsen TK: Abscess disease, caseous lymphadenitis, and pulmonary adenomatosis in imported sheep. J Vet Med B Infect Dis Vet Public Health. 2000, 47: 55-62. 10.1046/j.1439-0450.2000.00310.x.
12. Literák I, Horváthová A, Jahnová M, Rychlík I, Skalka B: Phenotype and genotype characteristics of the Slovak and Czech *Corynebacterium pseudotuberculosis* strains isolated from sheep and goats. Small Rumin Res. 1999, 32: 107-111. 10.1016/S0921-4488(98)00174-6.
13. Performance Standards for antimicrobial Susceptibility Testing; Seventeenth Informational Supplement. CLSI document M100-S17. 2007
14. Çetinkaya B, Karahana M, Atila E, Kalina R, De Baereb T, Vaneechoutteb M: Identification of *Corynebacterium pseudotuberculosis* isolates from sheep and goats by PCR. Vet Microbiol. 2002, 88: 75-83. 10.1016/S0378-1135(02)00089-5.
15. Khamis A, Raoult D, La Scola B: *rpoB* gene sequencing for identification of *Corynebacterium* species. J Clin Microbiol. 2004, 42: 3925-3931. 10.1128/JCM.42.9.3925-3931.2004.
16. Pacheco LG, Pena RR, Castro TL, Dorella FA, Bahia RC, Carminati R, Frota MN, Oliveira SC, Meyer R, Alves FS, Miyoshi A, Azevedo V: Multiplex PCR assay for identification of *Corynebacterium pseudotuberculosis* from pure cultures and for rapid detection of this pathogen in clinical samples. J Med Microbiol. 2007, 56: 480-486. 10.1099/jmm.0.46997-0.
17. Connor KM, Fontaine MC, Rudge K, Baird GJ, Donachie W: Molecular genotyping of multinational ovine and caprine *Corynebacterium pseudotuberculosis* isolates using pulsed-field gel electrophoresis. Vet Res. 2007, 38: 613-623. 10.1051/vetres:2007013.
18. Dorneles EM, Santana JA, Ribeiro D, Dorella FA, Guimarães AS, Moawad MS, Selim SA, Garaldi AL, Miyoshi A, Ribeiro MG, Gouveia AM, Azevedo V, Heinemann MB, Lage AP: Evaluation of ERIC-PCR as genotyping method for *Corynebacterium pseudotuberculosis* isolates. PLoS ONE. 2014, 9: e98758-10.1371/journal.pone.0098758.

CHAPTER 13

Longitudinal Study of Staphylococcus Aureus Colonization and Infection in a Cohort of Swine Veterinarians in the United States

Jisun Sun[1], My Yang[1], Srinand Sreevatsan[1], Jeffrey B. Bender[1], Randall S. Singer[2], Todd P. Knutson[1], Douglas G. Marthaler[1] and Peter R. Davies[1]

[1] Department of Veterinary Population Medicine, University of Minnesota, 385 ASVM, 1988 Fitch Ave, St. Paul, MN 55108, USA
[2] Department of Veterinary and Biomedical Sciences, University of Minnesota, St. Paul, MN 55108, USA

ABSTRACT

Background

People working with pigs are at elevated risk of harboring methicillin resistant *S. aureus* (MRSA) in their nose, which is attributable to occupational

Citation: Sun, J., Yang, M., Sreevatsan, S. et al. Longitudinal study of Staphylococcus aureus colonization and infection in a cohort of swine veterinarians in the United States. BMC Infect Dis 17, 690 (2017). https://doi.org/10.1186/s12879-017-2802-1

exposure to animals harboring livestock adapted *S. aureus*. To obtain insight into the biological nature of occupationally related nasal culture positivity, we conducted a longitudinal study of 66 swine veterinarians in the USA.

Methods

The study cohort resided in 15 US states and worked predominantly with swine. Monthly for 18 months, participants self-collected nasal swabs and completed a survey to report recent exposure to pigs and other animals; the occurrence of work related injuries; and any relevant health events such as skin and soft tissue infections or confirmed staphylococcal infections. Nasal swabs were cultured using selective methods to determine the presence of MRSA and methicillin susceptible *S. aureus* (MSSA), and isolates were characterized by *spa* typing and MLST.

Results

Prevalences of *S. aureus* (64%, monthly range from 58 to 82%) and MRSA (9.5%; monthly range from 6 to15%) were higher than reported for the US population (30% and 1.5% respectively). Predominant *spa* types were t034 (ST398, 37%), t002 (ST5, 17%) and t337 (ST9/ST398 13%), a distribution similar to that found in a concurrent study in pigs in the USA. Veterinarians were classified into three groups: Persistent carriers (PC, 52%), Intermittent carriers (IC, 47%) and Non-carriers (NC, 1%). Persistent carriage of a single *spa* type was observed in 14 (21%) of participants, and paired (first and last) isolates from PC subjects had minor genetic differences. Swabs from PC veterinarians carried higher numbers of *S. aureus*. Among IC veterinarians, culture positivity was significantly associated with recent contact with pigs.

Conclusions

Exposure to pigs did not lead to prolonged colonization in most subjects, and the higher numbers of *S. aureus* in PC subjects suggests that unknown host factors may determine the likelihood of prolonged colonization by *S. aureus* of livestock origin. Exposure to *S. aureus* and persistent colonization of swine veterinarians was common but rarely associated with *S. aureus* disease.

BACKGROUND

Working and living in close contact with domestic animals facilitates bidirectional interspecies transmission of microbiota. Concerns about the importance of animals as sources of antibiotic resistant pathogens have been heightened by the unveiling of healthy livestock as reservoirs of methicillin-resistant *Staphylococcus aureus* (MRSA) in many parts of the world [1,2,3,4,5,6,7]. While most research has focused on the ST398 lineage of livestock associated MRSA that predominates in Europe, several genotypes of *S. aureus* are adapted to livestock, and their relative prevalence varies geographically and among livestock species [2, 8,9,10].

In developed countries, approximately 20 to 30% of healthy people harbor *S. aureus* in the nasal cavity, and nasal colonization is associated with elevated risk of clinical infections [11, 12]. The most recent (2003–2004) national data for the USA estimated 28.6% and 1.5% of the population to harbor *S. aureus* and MRSA, respectively, in their nasal cavities [13]. *S. aureus* colonization is heterogeneously distributed across subsets of the population classified as 'persistent' (or permanent), 'intermittent' or 'non' carriers, although the criteria for defining persistent carriage vary [14, 15]. Bacterial, host, microbiome, and other environmental factors may influence the likelihood and duration of nasal colonization of humans with *S. aureus* [16,17,18], but detailed longitudinal studies of the dynamics of nasal carriage and bacterial genetic diversity are necessary to better understand this phenomenon [14]. Because some lineages of *S. aureus* are known to be host-adapted to particular avian or mammalian species [19, 20], and subtle genomic changes can alter host tropism [21], regular exposure of people to *S. aureus* of animal origin further complicates the poorly understood biology of nasal staphylococcal colonization.

S. aureus is considered part of the normal bacterial flora of pigs [22], and people working with live pigs are at elevated risk of being culture positive for *S. aureus* and MRSA. Notably, the predominant genotypes detected in humans with animal contact are typically those present in the animal populations with which they have contact [1, 23,24,25,26,27,28,29]. Because *S. aureus* are among the most numerous bacteria in bioaerosols of swine barns [30, 31], discriminating between transient contamination of superficial anatomical sites of people (e.g., upper airways or skin) and sustained colonization is problematic, particularly for workers with regular animal contact. Transient contamination may be the most common outcome in people after short term exposure to MRSA positive swine herds

[28, 32, 33]. To date there have only been 2 substantial longitudinal studies of MRSA and *S. aureus* colonization in occupationally exposed swine workers, both in the Netherlands. A study of 110 farm workers sampled 6 times over a year reported that 38% were 'persistent nasal carriers' of MRSA, but the possibility of repeated exposure and recontamination could not be eliminated [34]. A study of 137 swine veterinarians sampled 5 times over a 2 year period classified 13% of subjects to be persistently colonized with MRSA based on consistent molecular typing of isolates [24]. The rather different estimates (38% vs 13%) of persistent carriage reported in these 2 studiesmay be an artifact of the different sampling protocols and/or experimental subjects (farmers vs. veterinarians).

Fundamental questions remain about the capacity for *S. aureus* lineages disseminated from animals to colonize and cause disease in humans. Veterinarians are likely more informative subjects than farmers for elucidating long term colonization patterns following interspecies exposure as they typically are exposed to multiple herds rather than a single animal population. The goal of this study was to analyze long term patterns of *S. aureus* (including MRSA) colonization in an intensively sampled cohort of US swine veterinarians.

METHODS

The specific aims of the study were to describe the frequency and duration of positive *S. aureus* and MRSA nasal cultures in a cohort of veterinarians having regular contact with varied populations of commercial swine in the USA, and to characterize the genotypes of the isolates detected. The intensity of sampling (monthly for 18 months) was designed to enable more detailed understanding of *S. aureus* colonization patterns than in previous studies of swine workers.

Recruitment of Study Participants

Participants for the study were recruited at the annual meeting of the American Association of Swine Veterinarians (AASV) in Denver, CO in 2012. Eligible veterinarians were members of the AASV who were US residents and typically had regular (i.e., > twice per week) professional contact with pigs. A total of 71 veterinarians provided written consent to be research subjects, of which 68 subsequently participated in sample collection. Two participants withdrew during the course of the study (one due to emigration, one due to leaving swine practice), yielding a final cohort of 66 veterinarians who completed the longitudinal sampling protocol. Participants resided in 15 US states (IA, IL, IN, MI, MN, NE, SD, TX, OK, AL, MO, PA, NC, MD, OH), predominantly in the major swine producing regions of the Midwest and Southeast.

Sample Submission and Survey Data

Collection materials were mailed to the participants who were given written instructions for self-collection of nasal swabs, as well as an instructional video via YouTube. Starting in July 2012, participants were contacted by monthly email and requested to collect and submit a nasal swab via mail. The email message included a link to a survey using an online tool (http://www.surveymonkey.com) for veterinarians to provide information related to recent pig contact (e.g., time since last pig contact, hours worked in the previous week, number of farms visited in the previous week), and events of physical injury and selected health events (occurrence of skin or soft tissue infections, or confirmed staphylococcal infections) occurring in the month preceding sampling. To encourage compliance, sample collection was conducted at the convenience of the participants, and follow-up emails were sent to non-responders to encourage response rates. To determine quantitative bacteriology of *S. aureus*, a one-time cross-sectional sampling was performed on 41 available subjects who attended the 2014 AASV meeting in Dallas, TX.

Bacteriology

Samples were refrigerated on arrival at the University of Minnesota, and processed in 3 to 4 batches each month as samples typically were received over a 10–14 day period. For the quantitative bacteriology, all samples were collected and processed as one batch within 24 h of collection.

Isolation of *S. aureus* was performed using the methods described previously [35]. Nasal swabs were double enriched in Mueller-Hinton broth (BBL™, MD, USA) supplemented with NaCl (6.5%) and in Phenol-Red Mannitol broth (BBL™, MD, USA) supplemented with 4µg/ml Oxacillin (Sigma-Aldrich, MO, USA). Broths with a color change were inoculated onto chromogenic agar plate (BBL CHROM agar MRSA, MD, USA) and Factor plate (Veterinary Diagnostic Laboratory, University of Minnesota, MN, USA) to culture MRSA and *S. aureus*, respectively. Two colonies per sample were collected for further characterization. DNA was extracted from colonies with 19.5 μl 10 mM Tris-HCl and 0.5 μl Lysostaphin (both Sigma-Aldrich, MO, USA) at 37 °C for 30 min. PCR was used to detect *mecA* and perform *spa* typing. The primers for *mecA* were [F: 5′ GTA GAA ATG ACT GAA CGT CCG ATA A 3′, R: 5′ CCA ATT CCA CAT TGT TCG GTC TAA 3′], and for *spa* [F: 5′ AGA CGA TCC TTC GGT GAG C 3′, R: 5′ GCT TTT GCA ATG TCA TTT ACT G 3′]. PCR master mix (USB HotStart-IT FideliTaq, affymetrix, CA, USA) was used to amplify the DNA under the following conditions: 95 °C for 2 min, 94 °C for 30s, 55 °C for 30s, 72 °C for 1 min with 30 cycles and 72 °C for 10 min. All PCR products were visualized in 1% agarose gel with SYBR Safe dye in 1X TAE buffer (Tris-Acetate-EDTA, Thermo Fisher Scientific Inc., MA USA) for 40 min at 200 V.

Quantitative Bacteriology

Nasal swabs collected from the 41 subjects were transported in one batch on ice to the laboratory and placed in 1 ml Mueller-Hinton broth tubes within 24 h. Ten-fold dilutions were prepared from 100ul broth (up to 10^{-4}) and 100ul from each dilution was spread on a Factor plate and incubated at 37 °C for 22 h. before counting by observers blinded to the carrier status of subjects to determine CFU/swab.

Molecular Typing and Analysis

All selected *S. aureus* isolates were subtyped using *spa* typing [36]. After amplification of *spa*, PCR products were cleaned up with Illustra Exoprostar, (GE Healthcare Bio-sciences, PA, USA) and sequenced at the University of Minnesota Genomics Center. Sequences aligned using Sequencher 5.1 software (Gene Codes Corporation, MI, USA) were submitted to the Ridom *spa* typing database (http://spa.ridom.de/index.shtml).

Multi-locus sequence typing (MLST) of *S. aureus* was performed following methods reported previously via the MLST database of *S. aureus* (http://saureus.mlst.net) [37]. MLST typing was performed purposively so that at least one isolate from each *spa* type detected was also evaluated by MLST.

Definition of Carrier Status

Consistent with previous studies [14, 15, 24], we classified subjects by carrier status to be non-carriers (NC), intermittent carriers (IC), or persistent carriers (PC). A carrier index (range 0 to 1) was defined as the proportion of sampling events that yielded a *S. aureus* (including MRSA) isolate. Non-carriers were defined as subjects that were never positive for *S. aureus* (including MRSA), and intermittent carriers were culture positive at least once occasion with a carrier index of <0.8. Persistent carriers had a carrier index of 0.8 or greater. The cut-off of 0.8 was based on *post hoc* evaluation of the frequency distribution of the carrier index and was considered conservative (i.e., biased against false positive misclassification of persistent carriage). PC subjects were further classified as true persistent carriers (TPC) if a single *spa* type of *S. aureus* was recovered at all positive sampling events.

PCR Testing for *scn, chp, sak* for the Immune Evasion Cluster (IEC)

A purposively selected subset of 116 isolates was tested for presence of *scn, chp,* and *sak* of the IEC, considered as markers of adaptation to humans. The selection protocol utilized 4 categories of veterinarians 1) TPC subjects colonized with 1 *spa* type; 2) PC subjects colonized with more than 1 *spa* type; 3) IC subjects; 4) IC subjects colonized at least once with MRSA. For TPC subjects (category 1), the first and last isolates obtained were selected for each veterinarian. For category 2 subjects in whom the detected genotype changed over time, 2 isolates from each *spa* type were chosen up to 4 isolates per veterinarian (i.e., 2 predominant *spa* types). For category 3 subjects, only the predominant *spa* type was selected. Both MRSA and MSSA were detected from 4 IC subjects during the study period (category 4), and one MRSA isolate and an isolate of the predominant MSSA *spa* type were selected. Thus, a total of 24 MRSA and 92 MSSA were tested for *scn, chp, sak* following the methods previously described [38]. Annealing temperatures for the *scn, chp* and *sak* were 63 °C, 51.5 °C and 53 °C, respectively. All PCR products were electrophoresed in 1% agarose gel stained with SYBR Safe dye in 1X TAE buffer (Tris-Acetate-EDTA, Thermo Fisher Scientific Inc., MA USA) for 40 min at 200 V and visualized on a UV transilluminator. ATCC 700698 (Mu3), ATCC 700699 (Mu50) and ATCC 25904 (Newman) were used as positive controls for *scn, chp* and *sak*, respectively.

Whole Genome Sequencing

Eighteen isolates were selected from 9 persistent carriers (three subjects per sequence type) to evaluate genomic variation across the study period. Isolates from the first (month 1) and last (mostly month 18) samples collected for each subject were included. One veterinarian was colonized with ST5-t062 for 11 months, then ST398-t011 *S. aureus* for the remaining 7 months. For this subject, ST5-t062 isolates from month1 and month11 were selected. Another veterinarian was colonized with ST398-t034 MRSA followed by a closely related *spa* type (ST398-t011 MRSA) for the final 4 months. Genomic DNA was extracted from overnight cultures in LB (Lysogeny broth, BD Difco™, NJ, USA) using the Qiagen Blood and Tissue Kit (Valencia, CA, USA) following the manufacturer's instructions. Approximately 10 ng of extracted DNA per sample was sent to University of Minnesota Genomics Center (Minneapolis, MN, USA). Independent next generation sequence (NGS) libraries (Nextera DNA Library Preparation Kit, CA, USA) were created for each sample, pooled onto a single lane HiSeq 2500 rapid-run, and 250 bp paired-end reads were generated. Runs yielded an average of 2.3 million reads per sample, and 82% of reads had a quality score (Phred + 33) greater than 30 (details described in Additional file 1). To estimate genetic distances, the isolates were aligned using Mauve (ver.2) and single nucleotide polymorphisms (SNPs) from aligned genomes were extracted and imported into MEGA 7.0 software to generate a maximum likelihood phylogenetic tree employing 100 bootstrap iterations.

Comparison with *S. aureus* Isolates from Veterinarians and from Pigs in the USA

The prevalence and genotypic characterization of a geographically diverse sample of *S. aureus* collected from pigs in the USA was recently published [8]. The majority (36 of 38) of farms included in that study were served by 36 veterinary participants of the current study, providing a congruent time-space window to underpin the comparison of genotypes detected in US pigs and swine veterinarians. *Spa* types from the current study and from the study of pigs were categorized to be 'shared' (if detected in both species), 'swine only' or 'human only'. A minimal spanning tree (MST) for clustering of *spa* types was constructed using the Bionumerics 7.1 (Applied Maths, SintMartens-Latem, Belgium).

Statistical Analysis

Univariate analysis was performed to evaluate associations between culture positivity of *S. aureus* and working activities related with occupational exposure using R studio (Version 0.99.892). The exposures related to animal contact were: last contact with pigs (categorical variable, 0: same day; 1: previous day, 2: 2 days previously, 3: 3 days, 4: more than 3 days); hours of pig contact (continuous); and the number of farms visited in the previous week (continuous). As these self-reported exposure variables were correlated, multivariable analysis was limited to last contact with pigs, which was considered the most reliable and relevant variable with respect to transient contamination. To account for repeated observations on the same subjects over time, a two-way nested mixed-effects model was used, with samples nested within veterinarians, and the fixed effect of time since last contact with pigs (Days: same day vs. more than 1 day). The model was performed in R software environment 3.0, via lme4 package as follows:

Logit (pi)=Intercept+Daysi+Veti+SampleijLogit (pi)=Intercept+Daysi +Veti+Sampleij

In addition, wearing a facemask at last pig contact, and occurrence of injuries by livestock or soft tissue infections were also evaluated. The numbers $\log_{10}$CFU (colony forming unit) of putative *S. aureus* colonies per swab was compared between intermittent carriers and persistent carriers using Mann-Whitney U test.

RESULTS

Two veterinarians withdrew from the study due to emigration or altered work circumstances, leaving a cohort of 66 swine veterinarians who provided monthly nasal swabs. One veterinarian stopped working with swine after 7 months, but completed the sampling protocol and was retained in the study. Compliance with swab submission and survey completion was over 99% for both nasal swab submission (1179/1188) and survey submission (1177/1188). The median interval between sample collection and sample processing was 4 days (IQ range: 3–5 days; range: 1-35 days) and was not associated with the likelihood of culture positivity ($p = 0.99$).

Overall, *S. aureus* was detected in 63.7% (757/1188) of monthly nasal swab samples (yielding 1356 *S. aureus* isolates characterized) and MRSA in 9.5% (113/1188) samples (yielding 213 MRSA isolates characterized). The monthly apparent prevalence of *S. aureus* ranged from 58% to 82%, while apparent prevalence of MRSA ranged from 6 to 15%, and there was no indication of seasonal or longer term trends in prevalence over the course of the study (Fig. 1). MRSA was detected at least once during the study in 18 (27%) of subjects.

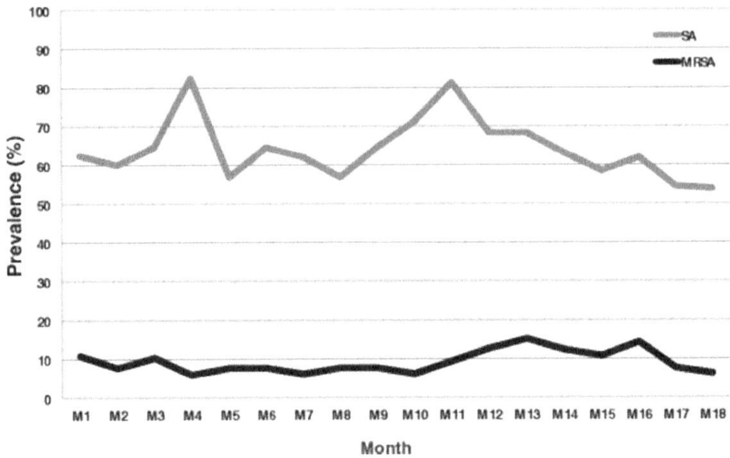

Figure 1: Proportion of *S. aureus* and MRSA positive nasal swabs from a cohort of swine veterinarians sampled monthly from July 2012 to December 2013.

At the individual level, the proportion of positive sampling events ranged from 0% (one veterinarian) to 100% (18 veterinarians) and was clearly bimodally distributed (Fig. 2). In univariate analyses, the likelihood of a culture positive result was negatively associated with the interval between the last contact with pigs and collection of the sample ($P = 0.001$), and positively associated with the hours of pig contact per week ($P = 0.02$), but not associated with the number of farms visited ($P = 0.09$) in the previous week. Multivariable analysis to adjust for repeated observations on the same veterinarians showed the odds or a positive culture were reduced by 33% for samples collected one day or longer after pig contact (OR = 0.67; 95% CI: 0.46–0.96) compared with same day collection. Three subjects reported having a staphylococcal infection during the study (2 MSSA, 1 MRSA), all of which were described as localized and did not lead to hospitalization or time off work. This corresponds with an incidence of 2.5 cases per 1000 person months.

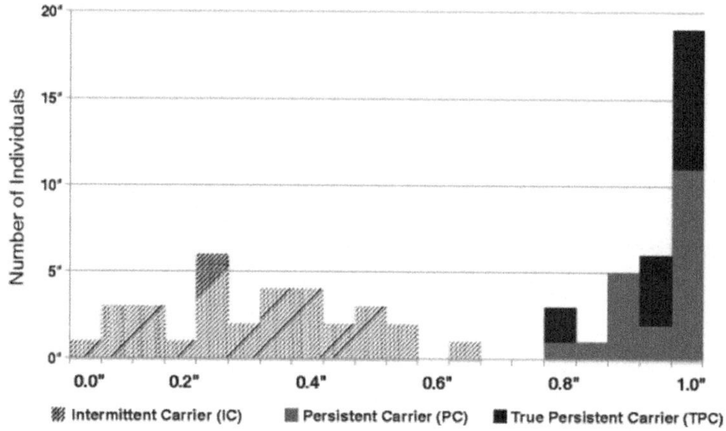

Figure 2: Histogram of the proportion of sampling events yielding a *S. aureus* isolate ("Carrier index").

Of the 66 subjects, 31 (47%) were classified as intermittent carriers and 34 (52%) were classified as persistent carriers of *S. aureus*. Based on consistent detection of a specific *spa* type at all culture positive samplings, 14 (21%) veterinarians were classified as true persistent carriers (TPC). The TPC group included the veterinarian who ceased working with pigs after 7 months but remained positive with ST398/t034 (methicillin susceptible) for the remaining 10 months. The majority (60%) of MRSA isolations were from 4 PC subjects who were positive for MRSA (all ST398) on at least 15 occasions. One of the PC MRSA subjects, who consistently harbored ST398/t034 MRSA, worked exclusively in one production system where this *spa* type occurs at high prevalence, and which was used as a MRSA positive control farm in the related swine study [8].

Whole Genome Sequencing

Comparison of the isolates from 9PC subjects (3 each of sequence types ST398, ST9 and ST5) generally indicated close genetic similarity between the first and last isolates collected. All paired isolates within subjects had 0.01 and 0.2 nucleotide differences per site within ST5 and ST398 respectively (Fig. 3). Pairs of isolates of the same genotype (ST9-t337) showed greater similarity within subjects than between subjects (Fig. 3b).

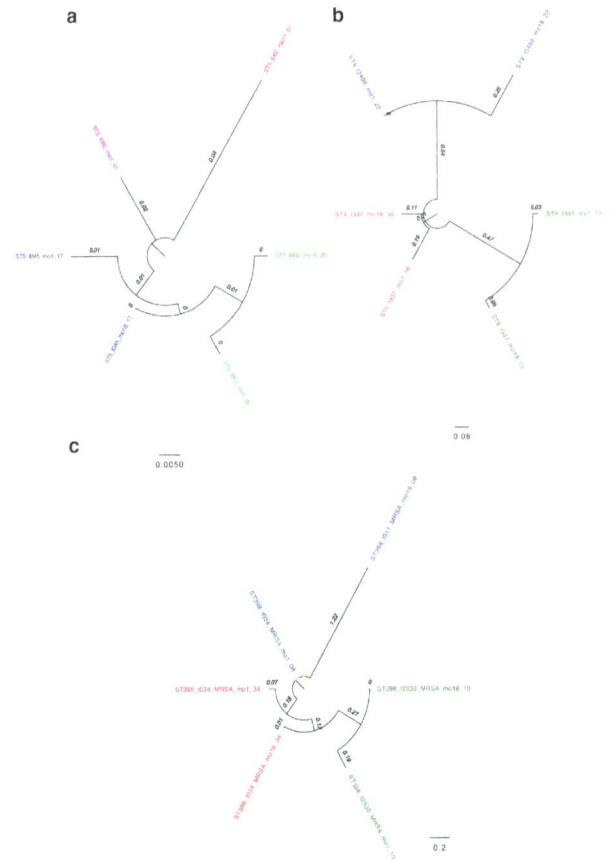

Figure 3: Genomic distance between isolates from beginning of study and the end of study in persistent carriers displayed by Sequence Type (a) ST5, b) ST9, c) ST398). Within sequence type, the isolates from same person were colored with same color. Names of each isolate were described by ST_spa type_(if MRSA)_sampling month_vetID. The bar scales on each ST indicate number of nucleotide difference per site.

Overall, the *S. aureus* isolates were distributed among 27 *spa* types within 8 MLST sequence types (Table 1). Three sequence types (ST398, ST5, ST9) constituted over 94% of the *S. aureus* isolates. Over the course of the study, ST398 isolates were detected at least once in 63 (83%) veterinarians; ST5 isolates in 43 (65%) veterinarians, ST9 in 29 (56%) veterinarians; and other MLST types in 8 (12%) veterinarians. The 3 predominant MLST types (ST398, ST5, ST9) were all isolated at least once from 23 (35%) subjects. Within each of these 3 MLST types, a single *spa* type (t034, t002, and t337, respectively) constituted approximately 70% of isolates. Two of these sequence types, ST398 (80.8%) and ST5 (14.1%), accounted for almost 95% of all MRSA isolates, with remainder being ST8/t008, a common human MRSA variant considered unlikely to be of swine origin.

Table 1: Numbers (%) of *spa* **types of** *S. aureus* **and MRSA isolated from swine veterinarians**, by MLST type

Sequence type	***Spa* type**	***S. aureus* (*n* = 1356)**	**MRSA (*n* = 213)**
ST398	t034	462 (34.1)	116 (54.5)
	t571	71 (5.2)	0
	t011	63 (4.6)	12 (5.6)
	t337[a]	20 (1.5)	0
	t3446[a]	19 (1.4)	0
	t1250	8 (0.6)	0
	t2330	6 (0.4)	44 (20.7)
	t2876	24 (1.8)	0
	t7160	13 (1.0)	0
	t1255	3 (0.2)	0
		650 (51.8)	172 (80.8)
ST5	t002	238 (17.6)	27 (12.7)
	t045	69 (5.1)	0
	t062	19 (1.4)	0
	t242	0	3 (1.4)
	t570	12 (0.9)	0
	t856	3 (0.2)	0
		341 (25.1)	30 (14.1)
ST9	t337[a]	178 (13.1)	0
	t2498	47 (3.5)	0
	t10494	11 (0.8)	0
	t3446[a]	9 (0.7)	0
	t1334	3 (0.2)	0
	t1430	2 (0.1)	0
		289 (18.4)	0
ST8	t008	0	11 (5.1)
	t2196	18 (1.3)	0
ST30	t338	4 (0.3)	0
	t363	1 (0.1)	0

ST72	t126	30 (2.2)	0
ST278	t330	22 (1.6)	0
ST2007	t8314	1 (0.1)	0

[a]Two sequence types were identified within t337 and t3446 isolates

The detection patterns of specific *spa* types over time varied enormously from highly consistent presence of individual *spa* types in TPC subjects, to very inconsistent patterns with multiple *spa* types detected in individual veterinarians over time (Fig. 4a, b). More than one *spa* type was detected in 71 (6%) of the monthly nasal swab samples. It is also notable, that in some of the PC and IC subjects a single *spa* type was detected over multiple consecutive months, but thereafter other *spa* types were detected.

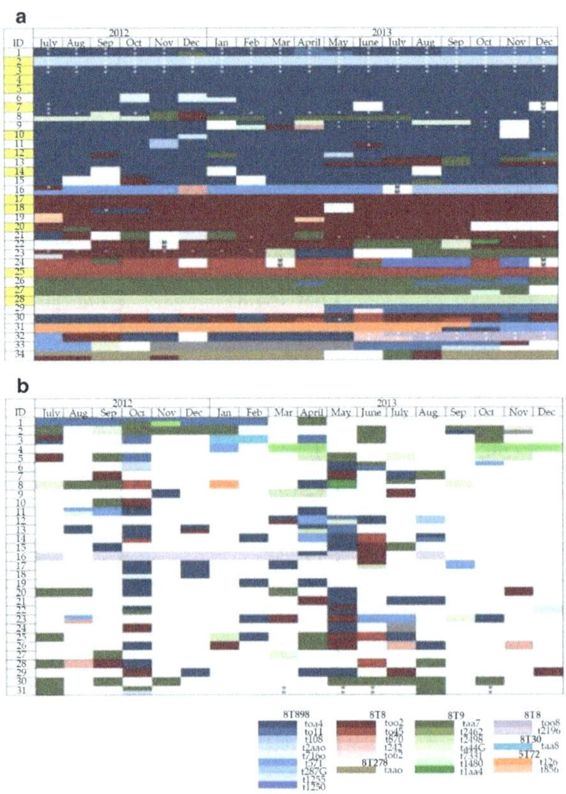

Figure 4: Patterns of detection of *S. aureus spa* types in veterinarians categorized as permanent carriers (a) and intermittent carriers (b). Missing samples are indicated as 'M', white spaces indicate culture negative events; white dots signify methicillin resistant isolates (typically 2 isolates typed per month). Yellow boxes in (a) indicate true persistent carriers. Colors also reflect the MSLT

type of the major sequence types being ST398 (blue shades), ST5 (red shades), and ST9 (green shades).

In the quantitative comparison of PC and IC veterinarians among 41 AASV members sampled in 2014, PC veterinarians (18 of 22; 82%) were more likely to be nasal culture positive than IC (4 of 19, 21%). Culture positive swabs from PC subjects harbored approximately 2 logs more *S. aureus* per swab than positive swabs from IC subjects ($P = 0.05$). *Spa* typing of isolates from this non-selective procedure yielded the same spa types found at other months for each TPC subject, and the same spa type detected in the most recent positive culture for 2 PC subjects.

IEC Genes Testing

Approximately 10% (one MRSA and 11 MSSA) of the 116 tested isolates were positive for two or three of the IEC genes (Table 2). A single isolate (t2196, ST8) was only positive for the *scn* and *sak* genes. The majority of isolates positive for IEC genes were *spa* types likely to be of human origin apart from the t011 (ST398), t5883 (ST398) and t002 (ST5) isolates.

Table 2: *S. aureus* **isolates testing positive for IEC**

	Month[a]	ID	MRSA	*spa* type (ST)	*scn*	*sak*	*chp*
1	04	24	MSSA	t126 (ST72)	+	+	+
2	04	57	MSSA	t330 (ST278)	+	+	+
3	04	61	MSSA	t062 (ST5)	+	+	+
4	06	19	MSSA	t5883 (ST398)	+	+	+
5	07	05	MSSA	t338 (ST30)	+	+	+
6	09	44	MSSA	t2196 (ST8)	+	+	–
7	12	61	MSSA	t062 (ST5)	+	+	+
8	12	61	MSSA	t011 (ST398)	+	+	+
9	16	57	MSSA	t330 (ST278)	+	+	+
10	17	41	MRSA	t008 (ST8)	+	+	+
11	17	66	MSSA	t002 (ST5)	+	+	+
12	18	66	MSSA	t002 (ST5)	+	+	+

[a]Month of sampling from month 1 to month 18

Comparison with *S. aureus* Isolates from Veterinarians and from Pigs in the USA

Spa types from the current study and a previous pig study [8] were categorized as 'shared' if detected in both species, 'swine only' or 'human only'. Thirteen *spa* types were shared and accounted for 83% and 92% of all isolates from veterinarians and pigs, respectively. Twenty *spa* types were found only among swine isolates while 14 *spa* types were identified only among veterinary isolates (Table 3). A minimal spanning tree (MST) analysis for clustering of *spa* types from the swine and veterinary studies was constructed (Fig. 5). As expected, isolates were clustered together by sequence type. ST9 isolates were more likely to be found among pig isolates while *spa* types belonging to ST398 and ST5 were relatively more frequent among isolates from veterinarians.

Table 3: ***Spa*** **type comparison between swine and veterinary isolates**

Shared	Swine only	Human only
t002	t899	t008
t011	t5883	t045
t034	t5838	t062
t10494	t5462	t1250
t1255	t3232	t126
t1334	t306	t1430
t242	t2582	t2196
t2498	t2462	t2330
t337	t2315	t2876
t3446	t1793	t330
t570	t14851	t338
t571	t1419	t363
t8314	t11744	t856
	t11374	t922
	t11241	
	unknown1	
	unknown2	
	unknown3	
	unknown4	
	unknown5	

*Repeat succession of unknown types: Unknown1 (r07r16r23r23r02r12r17 r23r02r34), Unknown2 (r07r16r16r16r23r23r02r12r23r02r34), Unknown3 (r07r16r16r23r02r12r23r02r34), Unknown4 (r08r475r2r25r2r25r34r34r25), Unknown5 (r07r16r23r23r02r23r02r34)

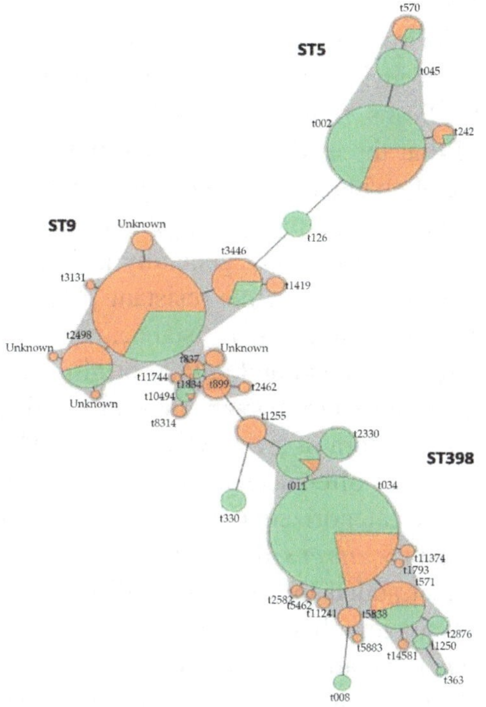

Figure 5: Genetic relatedness of *S. aureus* isolates from swine (n = 1193) and veterinarians (n = 1659). Each node in this minimum spanning tree depicts one of 38 *spa* types identified from swine and swine veterinarians. The size of circles denotes the number of isolates. Swine isolates and vet isolates are color coded with orange and green, respectively.

DISCUSSION

The high prevalence of *S. aureus* recovered from nasal cultures reflects the increased exposure to *S. aureus* occurring in livestock environments [4, 26, 39]. Although some influence of methodological differences cannot be eliminated, the overall *S. aureus* prevalence (64%) in swine veterinarians is approximately double that estimated in studies of the overall US population [13] and in other developed countries [11, 12]. However, it is very similar to a 72% prevalence in a study of Dutch swine veterinarians [24]. The prevalence of MRSA (9.5%) was also higher than reported in the

US population (1.5%), but was substantially lower than in the Dutch swine veterinarians (44%), which likely reflects the lower prevalence of MRSA in the US swine industry relative to the Netherlands [4, 8, 40].

A substantial majority (>84%) of *S. aureus* isolates in this study were deemed likely to be of swine origin based on several criteria. A parallel study of 38 US swine farms, sampled by a subset of the veterinarians in the current study, found that three MLST sequence types (ST9, ST398, ST5) constituted over 99% of swine isolates, with *spa* types t337, t034, and t002 predominating [8]. The same sequence types and *spa* types were similarly predominant among the veterinarians, generally lacked the IEC genes, and were predominantly (85%) tetracycline resistant [41]. The absence of the IEC genes, together with tetracycline resistance, has been used previously to differentiate isolates of human and animal origin [42, 43] and our data suggest that *S. aureus* acquired from swine may largely displace *S. aureus* of human origin in the nasal flora of swine veterinarians. However, given that more than one *spa* type was often detected in individual samples, and only 2 isolates per sample were categorized, it is possible that carriage of relatively low numbers of human *S. aureus* by veterinarians went undetected [44], and that animal contact added to, rather than displaced human *S. aureus*. Regardless, the data suggest that animal exposure alters the composition of the nasal *S. aureus* populations of swine workers. Any resultant impact on risk of clinical infection will depend on the relative persistence, transmissibility, and pathogenicity of *S. aureus* of swine origin compared with human adapted variants.

The primary goal of this study was to understand the persistence of *S. aureus* of swine origin in occupational groups that are in close contact with pigs. Several previous studies of livestock workers have examined this question, with varying outcomes [3,4,5, 26, 29, 32, 39]. Most studies focused on MRSA alone, and the frequency and duration of sampling has varied widely. A common obstacle to inference has been the inability to differentiate repeated contamination of the nasal mucosa from true persistent colonization [45], particularly for farmers who are repeatedly exposed to the same herd (and *S. aureus* populations) [39]. We specifically studied veterinarians because they generally visit multiple farms and therefore should be exposed to more diverse *S. aureus* populations.

Longitudinal studies of nasal carriage of *S. aureus* in humans typically classify subjects as persistent carriers, intermittent carriers, and non-carriers [14, 46]. It is believed that a subset (usually of the order of 20%) of healthy people are persistently colonized with *S. aureus* [15, 47], and this is associated with higher risk of *S. aureus* clinical infections [48, 49]. Both host genetic factors and microbial factors may play a role in determining duration of carriage [16], and it was recently reported that presence of *S. lugdunensis* in the nose may suppress *S. aureus* populations [18]. Currently, there is no accepted consensus for defining persistent carriers, and categorization of individuals will be influenced by study design (particularly the frequency and duration of sampling), and the cut-off (carrier index) used to define carriage status [46]. We employed a *post-hoc* epidemiological approach to establish a cut-off (carriage index >0.8) to define persistent carriage. The same criterion has been employed in previous studies [14, 46]. The bimodal distribution of the frequency of culture positive samples among veterinarians (Fig. 2) suggests that even in environments with high exposure to *S. aureus* of animal origin, individual host characteristics are likely important determinants of the persistence of colonization. This inference is further supported by the substantially higher numbers of *S. aureus* recovered from nasal swabs from PC compared with IC veterinarians, which is consistent with quantitative studies performed in humans both with [39] and without [50] known livestock association.

The proportion of persistent carriers (52%, 34/66) was considerably higher, and the proportion of non-carriers (<2%, 1/66) considerably lower, than reported in people without livestock contact [14, 46, 51]. This may reflect unusually frequent exposure to *S. aureus* that can occur in intensive swine facilities [52]. Very similar rates of apparent persistent carriage of *S. aureus* were also observed in recent longitudinal studies of swine farmers (52%, [39]) and swine veterinarians (47%, [24]) in the Netherlands, although the sampling protocols and criteria for defining persistence in those studies differed.

The associations observed between culture positivity and time since last pig contact and hours of pig contact in the prior week are consistent with previous studies indicating that occupational exposures to animals often result in transient contamination [28, 32, 53]. Collection of swabs on the same day following pig contact substantially increases the odds of a positive culture, therefore sampling schedules will influence estimates of prevalence. The variability of *spa* types detected over time in the IC group (Fig. 4b) also attests to multiple exposures of swine veterinarians to diverse *S. aureus* populations. Although early research of persistent carriage in humans suggested that people were colonized over the long term by a single *S. aureus* variant, more recent studies indicate that *S. aureus* variants are often replaced over time [14, 46, 51]. Ritchie (2015) sampled 122 healthy young adults weekly over 13 weeks and described 3 patterns of persistent carriage, being continuous carriage of a single *spa* type; an abrupt change from one *spa* type to another; and periods of co-carriage with two *spa* types. Although 'strain turnover' was observed in both intermittent and persistent carriers in that study, the majority (63%) of PC subjects were colonized by a single *spa* type. We made similar observations in the veterinary cohort where all 3 patterns of PC were observed, and a substantial proportion (41%; 14 of 34 PC) of the PC veterinarians were classified as 'true persistent colonization' due to the repeated presence of a single *spa* type over 18 months. This slightly lower proportion (41% vs. 63%) of TPC could be an artifact of the longer sampling period (18 months vs. 13 weeks) providing more opportunity for strain turnover, or the additional exposure to *S. aureus* variants that occurs in the livestock environment. Our observations with WGS provided further evidence that true persistent colonization with animal origin of *S. aureus* can occur in some subjects. The paired isolates from each persistent carrier (18 months apart) had shorter genomic distances than seen between isolates of same *spa* type but from different individuals. A similar inference was reached in a previous study using whole genome mapping of isolates from 16 Dutch veterinarians [54]. Although it is not possible to eliminate the possibility of repeated reacquisition of the same variant from pigs by veterinarians, this is considered unlikely over an extended period unless they were exposed to homogeneous swine populations harboring few *S. aureus* variants. However, repeated reacquisition from pigs can be eliminated for one subject in our study who remained culture positive for ST398-t034 MSSA for 9 months after leaving swine practice. Current evidence suggests that a substantial proportion of swine veterinarians become colonized with *S. aureus* of swine origin for at least 18 months to 2 years [24], and can harbor substantial numbers of these organisms. Furthermore, although several studies indicate that ST398 MRSA are less transmissible among people than MRSA of human origin, transmission from veterinarians to their families, who also may become persistently infected, has been clearly demonstrated [54, 55]. Also, persistent colonization of humans with swine *S. aureus* isolates may be a mechanism for transmission among herds. This is arguably inconsequential given that *S. aureus* is part of the normal commensal flora of swine, but could be problematic for efforts to prevent MRSA transmission among herds [56].

The fact that some swine workers harbor substantial numbers of *S. aureus* with genotypes consistent with swine origin has implications regarding occupational health. However, the human health consequences of livestock associated *S. aureus* are not well defined, and available information is largely limited to ST398 MRSA. In the USA, to date *S. aureus* of animal origin appear to have had negligible impact on human health. The reported incidence of *S. aureus* infections in this study (2.5 per 1000 person months, or 3% per veterinary-year) is similar to that reported in a study of Iowa residents (2.7 per 1000 person months). Furthermore, *spa* types linked to livestock represented only 1% of *S. aureus* isolates, and 0.24% of MRSA isolates, from human clinical infections in the pig dense state of Iowa [57]. In contrast, a substantial study in Denmark showed increased likelihood of infection with specific livestock associated MRSA, but not increased MRSA infection risk overall, in people living in pig dense areas [58]. Transmission from swine workers into the broader community is a plausible explanation. However, despite the substantial exposure to *S. aureus* of swine origin that occurs in intensive production environments, there are remarkably few reports of medically significant infections occurring in swine workers [45, 59]. To date, we are unaware of any studies demonstrating increased risk of clinical *S. aureus* infections in livestock workers. A recent prospective study of swine farm workers in Holland found that carriage of ST398 MRSA was not associated with elevated risk of infections, healthcare contact, or measures of reduced quality of life [60]. Somewhat surprisingly, the small numbers of fatal human cases of ST398 infection have occurred in people without known livestock contact, but who were generally medically compromised [61]. Even though there are a small number of fatal human cases of ST398 infections, it is important to better understand the interchange of *S. aureus* between humans and animals and the implications for the transfer of resistance elements.

CONCLUSIONS

Swine veterinarians are continually exposed to *S. aureus* of swine origin when working with pigs. This longitudinal study confirmed that the outcomes of this occupational exposure range from short term contamination of the nasal cavities (reflected in higher prevalence of detection in samples collected soon after animal exposure) to apparent long term colonization (at least 18 months) in some individuals in whom the same spa type was detected repeatedly throughout the study. Exposure to pigs did not lead to prolonged colonization in most subjects, and the higher numbers of *S. aureus* in PC subjects suggests that unknown host factors may determine the likelihood of prolonged colonization by *S. aureus* of livestock origin. Although exposure to *S. aureus* and persistent colonization of swine veterinarians was common, the few reports of clinical *S. aureus* infections reported were of minimal medical significance.

ACKNOWLEDGEMENTS

The dedicated collaboration of the study participants is greatly appreciated.

REFERENCES

1. Voss A, Loeffen F, Bakker J, Klaassen C, Wulf M. Methicillin-resistant Staphylococcus Aureus in pig farming. Emerg Infect Dis. 2005;11(12):1965–6.
2. Chuang YY, Huang YC. Livestock-associated meticillin-resistant Staphylococcus Aureus in Asia: an emerging issue? Int J Antimicrob Agents. 2015;45(4):334–40.
3. Khanna T, Friendship R, Dewey C, Weese JS. Methicillin resistant Staphylococcus Aureus colonization in pigs and pig farmers. Vet Microbiol. 2008;128(3–4):298–303.
4. Smith TC, Gebreyes WA, Abley MJ, Harper AL, Forshey BM, Male MJ, Martin HW, Molla BZ, Sreevatsan S, Thakur S, Thiruvengadam M, Davies PR. Methicillin-resistant Staphylococcus Aureus in pigs and farm workers on conventional and antibiotic-free swine farms in the USA. PLoS One. 2013;8(5):e63704.
5. Graveland H, Wagenaar JA, Heesterbeek H, Mevius D, van Duijkeren E, Heederik D. Methicillin resistant Staphylococcus Aureus ST398 in veal calf farming: human MRSA carriage related with animal antimicrobial usage and farm hygiene. PLoS One. 2010;5(6):e10990.
6. Van den Eede A, Martens A, Lipinska U, Struelens M, Deplano A, Denis O, Haesebrouck F, Gasthuys F, Hermans K. High occurrence of methicillin-resistant Staphylococcus Aureus ST398 in equine nasal samples. Vet Microbiol. 2009;133(1–2):138–44.
7. Nemati M, Hermans K, Lipinska U, Denis O, Deplano A, Struelens M, Devriese LA, Pasmans F, Haesebrouck F. Antimicrobial resistance of old and recent Staphylococcus Aureus isolates from poultry: first detection of livestock-associated methicillin-resistant strain ST398. Antimicrob Agents Chemother. 2008;52(10):3817–9.
8. Sun J, Yang M, Sreevatsan S, Davies PR. Prevalence and characterization of Staphylococcus Aureus in growing pigs in the USA. PLoS One. 2015;10(11):e0143670.
9. Alba P, Feltrin F, Cordaro G, Porrero MC, Kraushaar B, Argudin MA, Nykasenoja S, Monaco M, Stegger M, Aarestrup FM, Butaye P, Franco A, Battisti A. Livestock-associated methicillin resistant and methicillin susceptible Staphylococcus Aureus sequence type (CC)1 in European farmed animals: high genetic relatedness of isolates from Italian cattle herds and humans. PLoS One. 2015;10(8):e0137143.

10. Feltrin F, Alba P, Kraushaar B, Ianzano A, Argudin MA, Di Matteo P, Porrero MC, Aarestrup FM, Butaye P, Franco A, Battisti A. A livestock-associated, multidrug-resistant, methicillin-resistant Staphylococcus Aureus clonal complex 97 lineage spreading in dairy cattle and pigs in Italy. Appl Environ Microbiol. 2015;82(3):816–21.
11. Sivaraman K, Venkataraman N, Cole AM. Staphylococcus Aureus nasal carriage and its contributing factors. Future Microbiol. 2009;4(8):999–1008.
12. den Heijer CD, van Bijnen EM, Paget WJ, Pringle M, Goossens H, Bruggeman CA, Schellevis FG, Stobberingh EE, Study Team APRES. Prevalence and resistance of commensal Staphylococcus Aureus, including meticillin-resistant S aureus, in nine European countries: a cross-sectional study. Lancet Infect Dis. 2013;13(5):409–15.
13. Gorwitz RJ, Kruszon-Moran D, McAllister SK, McQuillan G, McDougal LK, Fosheim GE, Jensen BJ, Killgore G, Tenover FC, Kuehnert MJ. Changes in the prevalence of nasal colonization with Staphylococcus Aureus in the United States, 2001-2004. J Infect Dis. 2008;197(9):1226–34.
14. Muthukrishnan G, Lamers RP, Ellis A, Paramanandam V, Persaud AB, Tafur S, Parkinson CL, Cole AM: Longitudinal genetic analyses of Staphylococcus aureus nasal carriage dynamics in a diverse population. BMC Infect Dis 2013, 13:221–2334–13-221.
15. Wertheim HF, Melles DC, Vos MC, van Leeuwen W, van Belkum A, Verbrugh HA, Nouwen JL. The role of nasal carriage in Staphylococcus Aureus infections. Lancet Infect Dis. 2005;5(12):751–62.
16. Sollid JU, Furberg AS, Hanssen AM, Johannessen M. Staphylococcus Aureus: determinants of human carriage. Infect Genet Evol. 2014;21:531–41.
17. Frank DN, Feazel LM, Bessesen MT, Price CS, Janoff EN, Pace NR. The human nasal microbiota and Staphylococcus Aureus carriage. PLoS One. 2010;5(5):e10598.
18. Zipperer A, Konnerth MC, Laux C, Berscheid A, Janek D, Weidenmaier C, Burian M, Schilling NA, Slavetinsky C, Marschal M, Willmann M, Kalbacher H, Schittek B, Brotz-Oesterhelt H, Grond S, Peschel A, Krismer B. Human commensals producing a novel antibiotic impair pathogen colonization. Nature. 2016;535(7613):511–6.
19. Guinane CM, Ben Zakour NL, Tormo-Mas MA, Weinert LA, Lowder

BV, Cartwright RA, Smyth DS, Smyth CJ, Lindsay JA, Gould KA, Witney A, Hinds J, Bollback JP, Rambaut A, Penades JR, Fitzgerald JR. Evolutionary genomics of Staphylococcus Aureus reveals insights into the origin and molecular basis of ruminant host adaptation. Genome Biol Evol. 2010;2:454–66.

20. Lowder BV, Fitzgerald JR. Human origin for avian pathogenic Staphylococcus Aureus. Virulence. 2010;1(4):283–4.
21. Viana D, Comos M, McAdam PR, Ward MJ, Selva L, Guinane CM, Gonzalez-Munoz BM, Tristan A, Foster SJ, Fitzgerald JR, Penades JR. A single natural nucleotide mutation alters bacterial pathogen host tropism. Nat Genet. 2015;47(4):361–6.
22. Linhares L. A pilot study of the epidemiology of Staphylococcus Aureus in multiple site swine production. Retrieved from the University of Minnesota Digital Conservancy. https://conservancy.umn.edu/handle/11299/162367.
23. van Cleef BA, Verkade EJ, Wulf MW, Buiting AG, Voss A, Huijsdens XW, van Pelt W, Mulders MN, Kluytmans JA. Prevalence of livestock-associated MRSA in communities with high pig-densities in The Netherlands. PLoS One. 2010;5(2):e9385.
24. Verkade E, van Benthem B, den Bergh MK, van Cleef B, van Rijen M, Bosch T, Kluytmans J. Dynamics and determinants of Staphylococcus Aureus carriage in livestock veterinarians: a prospective cohort study. Clin Infect Dis. 2013;57(2):e11–7.
25. Cuny C, Nathaus R, Layer F, Strommenger B, Altmann D, Witte W. Nasal colonization of humans with methicillin-resistant Staphylococcus Aureus (MRSA) CC398 with and without exposure to pigs. PLoS One. 2009;4(8):e6800.
26. Bisdorff B, Scholholter JL, Claussen K, Pulz M, Nowak D, Radon K. MRSA-ST398 in livestock farmers and neighbouring residents in a rural area in Germany. Epidemiol Infect. 2012;140(10):1800–8.
27. Fang HW, Chiang PH, Huang YC. Livestock-associated methicillin-resistant Staphylococcus Aureus ST9 in pigs and related personnel in Taiwan. PLoS One. 2014;9(2):e88826.
28. Frana TS, Beahm AR, Hanson BM, Kinyon JM, Layman LL, Karriker LA, Ramirez A, Smith TC. Isolation and characterization of methicillin-resistant Staphylococcus Aureus from pork farms and visiting veterinary students. PLoS One. 2013;8(1):e53738.

29. Armand-Lefevre L, Ruimy R, Andremont A. Clonal comparison of Staphylococcus Aureus isolates from healthy pig farmers, human controls, and pigs. Emerg Infect Dis. 2005;11(5):711–4.
30. Gibbs SG, Green CF, Tarwater PM, Mota LC, Mena KD, Scarpino PV. Isolation of antibiotic-resistant bacteria from the air plume downwind of a swine confined or concentrated animal feeding operation. Environ Health Perspect. 2006;114(7):1032–7.
31. Friese A, Schulz J, Hoehle L, Fetsch A, Tenhagen BA, Hartung J, Roesler U. Occurrence of MRSA in air and housing environment of pig barns. Vet Microbiol. 2012;158(1–2):129–35.
32. van Cleef BA, Graveland H, Haenen AP, van de Giessen AW, Heederik D, Wagenaar JA, Kluytmans JA. Persistence of livestock-associated methicillin-resistant Staphylococcus Aureus in field workers after short-term occupational exposure to pigs and veal calves. J Clin Microbiol. 2011;49(3):1030–3.
33. van den Broek IV, van Cleef BA, Haenen A, Broens EM, van der Wolf PJ, van den Broek MJ, Huijsdens XW, Kluytmans JA, van de Giessen AW, Tiemersma EW. Methicillin-resistant Staphylococcus Aureus in people living and working in pig farms. Epidemiol Infect. 2009;137(5):700–8.
34. van Cleef BA, van Benthem BH, Verkade EJ, van Rijen MM, Kluytmans-van den Bergh MF, Graveland H, Bosch T, Verstappen KM, Wagenaar JA, Bos ME, Heederik D, Kluytmans JA. Livestock-associated MRSA in household members of pig farmers: transmission and dynamics of carriage, a prospective cohort study. PLoS One. 2015;10(5):e0127190.
35. Linhares LL, Yang M, SreevatsanS, Munoz-Zanzi CA, Torremorell M, Davies PR. The effect of anatomic site and age on detection of Staphylococcus Aureus in pigs. J Vet Diagn Invest 2015;27(1):55-60.
36. Harmsen D, Claus H, Witte W, Rothganger J, Claus H, Turnwald D, Vogel U. Typing of methicillin-resistant Staphylococcus Aureus in a university hospital setting by using novel software for spa repeat determination and database management. J Clin Microbiol. 2003;41(12):5442–8.
37. Enright MC, Day NP, Davies CE, Peacock SJ, Spratt BG. Multilocus sequence typing for characterization of methicillin-resistant and methicillin-susceptible clones of Staphylococcus Aureus. J Clin Microbiol. 2000;38(3):1008–15.

38. Mehrotra M, Wang G, Johnson WM. Multiplex PCR for detection of genes for Staphylococcus Aureus enterotoxins, exfoliative toxins, toxic shock syndrome toxin 1, and methicillin resistance. J Clin Microbiol. 2000;38(3):1032–5.
39. van Cleef BA, van Benthem BH, Verkade EJ, van Rijen M, Kluytmans-van den Bergh MF, Schouls LM, Duim B, Wagenaar JA, Graveland H, Bos ME, Heederik D, Kluytmans JA. Dynamics of methicillin-resistant Staphylococcus Aureus and methicillin-susceptible Staphylococcus Aureus carriage in pig farmers: a prospective cohort study. Clin Microbiol Infect. 2014; 20(10):O764-71.
40. Dierikx CM, Hengeveld PD, Veldman KT, de Haan A, van der Voorde S, Dop PY, Bosch T, van Duijkeren E. Ten years later: still a high prevalence of MRSA in slaughter pigs despite a significant reduction in antimicrobial usage in pigs the Netherlands. J Antimicrob Chemother. 2016;71(9):2414–8.
41. Sun J, Singer RS, Hau SJ, Nicholson TL, Davies PR. Antimicrobial resistance of *Staphylococcus Aureus* isolated from swine and swine veterinarians in the USA. 2016.
42. Cuny C, Abdelbary M, Layer F, Werner G, Witte W. Prevalence of the immune evasion gene cluster in Staphylococcus Aureus CC398. Vet Microbiol. 2015;177(1–2):219–23.
43. Stegger M, Liu CM, Larsen J, Soldanova K, Aziz M, Contente-Cuomo T, Petersen A, Vandendriessche S, Jimenez JN, Mammina C, van Belkum A, Salmenlinna S, Laurent F, Skov RL, Larsen AR, Andersen PS, Price LB. Rapid differentiation between livestock-associated and livestock-independent Staphylococcus Aureus CC398 clades. PLoS One. 2013;8(11):e79645.
44. Votintseva AA, Miller RR, Fung R, Knox K, Godwin H, Peto TE, Crook DW, Bowden R, Walker AS. Multiple-strain colonization in nasal carriers of Staphylococcus Aureus. J Clin Microbiol. 2014;52(4):1192–200.
45. Goerge T, Lorenz MB, van Alen S, Hubner NO, Becker K, Kock R. MRSA colonization and infection among persons with occupational livestock exposure in Europe: prevalence, preventive options and evidence. Vet Microbiol. 2017;200:6–12.
46. Ritchie SR, Isdale E, Priest P, Rainey PB, Thomas MG. The turnover of strains in intermittent and persistent nasal carriers of Staphylococcus Aureus. J Inf Secur. 2016;72(3):295–301.

47. Williams REO. Healthy carriage of Staphylococcus Aureus: its prevalence and importance. Bacteriol Rev. 1963;27(1):56–71.
48. Verhoeven PO, Gagnaire J, Botelho-Nevers E, Grattard F, Carricajo A, Lucht F, Pozzetto B, Berthelot P. Detection and clinical relevance of Staphylococcus Aureus nasal carriage: an update. Expert Rev Anti-Infect Ther. 2014;12(1):75–89.
49. von Eiff C, Becker K, Machka K, Stammer H, Peters G. Nasal carriage as a source of Staphylococcus Aureus bacteremia. Study group. N Engl J Med. 2001;344(1):11–6.
50. Verhoeven PO, Grattard F, Carricajo A, Lucht F, Cazorla C, Garraud O, Pozzetto B, Berthelot P. An algorithm based on one or two nasal samples is accurate to identify persistent nasal carriers of Staphylococcus Aureus. Clin Microbiol Infect. 2012;18(6):551–7.
51. Ghasemzadeh-Moghaddam H, Neela V, van Wamel W, Hamat RA, Shamsudin MN, Hussin NS, Aziz MN, Haspani MS, Johar A, Thevarajah S, Vos M, van Belkum A. Nasal carriers are more likely to acquire exogenous Staphylococcus aureus strains than non-carriers. Clin Microbiol Infect. 2015;21(11):998.e1–7.
52. Bos ME, Verstappen KM, van Cleef BA, Dohmen W, Dorado-Garcia A, Graveland H, Duim B, Wagenaar JA, Kluytmans JA, Heederik DJ. Transmission through air as a possible route of exposure for MRSA. J Expo Sci Environ Epidemiol. 2016;26(3):263–9.
53. Graveland H, Wagenaar JA, Bergs K, Heesterbeek H, Heederik D. Persistence of livestock associated MRSA CC398 in humans is dependent on intensity of animal contact. PLoS One. 2011;6(2):e16830.
54. Bosch T, Verkade E, van Luit M, Landman F, Kluytmans J, Schouls LM. Transmission and persistence of livestock-associated methicillin-resistant Staphylococcus Aureus among veterinarians and their household members. Appl Environ Microbiol. 2015;81(1):124–9.
55. Verkade E, Kluytmans-van den Bergh M, van Benthem B, van Cleef B, van Rijen M, Bosch T, Schouls L, Kluytmans J. Transmission of methicillin-resistant Staphylococcus Aureus CC398 from livestock veterinarians to their household members. PLoS One. 2014;9(7):e100823.
56. Grontvedt CA, Elstrom P, Stegger M, Skov RL, Skytt Andersen P, Larssen KW, Urdahl AM, Angen O, Larsen J, Amdal S, Lotvedt SM, Sunde M, Bjornholt JV. Methicillin-resistant Staphylococcus Aureus

CC398 in humans and pigs in Norway: a "one health" perspective on introduction and transmission. Clin Infect Dis. 2016;63(11):1431–8.

57. Nair R, Wu J, Carrel M, O'Brien A, Quick M, Farina S, Wardyn S, Thapaliya D, Grenier D, Smith TC. Prospective multicenter surveillance identifies Staphylococcus Aureus infections caused by livestock-associated strains in an agricultural state. Diagn Microbiol Infect Dis. 2016;85(3):360-366.

58. Larsen J, Petersen A, Sorum M, Stegger M, van Alphen L, Valentiner-Branth P, Knudsen LK, Larsen LS, Feingold B, Price LB, Andersen PS, Larsen AR, Skov RL: Meticillin-resistant Staphylococcus aureus CC398 is an increasing cause of disease in people with no livestock contact in Denmark, 1999 to 2011. Euro Surveill 2015, 20(37). doi:10.2807/1560–7917.ES.2015.20.37.30021.

59. Omland O, Hoffmann L. Occupational acquisition of methicillin-resistant Staphylococcus Aureus in humans--a description of MRSA carrier and infected cases from the region of North Jutland in Denmark. Ann Agric Environ Med. 2012;19(4):637–40.

60. VAN Cleef BA, VAN Benthem BH, Verkade EJ, VAN Rijen MM, Kluytmans-VAN DEN Bergh MF, Graveland H, Bosch T, Verstappen KM, Wagenaar JA, Heederik D, Kluytmans JA: Health and health-related quality of life in pig farmers carrying livestock-associated methicillin-resistant Staphylococcus Aureus. Epidemiol Infect 2016, 144(8):1774–1783.

61. Nielsen RT, Kemp M, Holm A, Skov MN, Detlefsen M, Hasman H, Aarestrup FM, Kaas RS, Nielsen JB, Westh H, Kolmos HJ. Fatal septicemia linked to transmission of MRSA clonal complex 398 in hospital and nursing home, Denmark. Emerg Infect Dis. 2016;22(5):900–2.

SECTION 4: CANINES

CHAPTER 14

The Canine Oral Microbiome

Floyd E. Dewhirst[1,2], Erin A. Klein[1] , Emily C. Thompson[1] , Jessica M. Blanton[1] , Tsute Chen[1] , Lisa Milella[3] , Catherine M. F. Buckley[4] , Ian J. Davis[4] , Marie-Lousie Bennett[5] , Zoe V. Marshall-Jones[4]

[1]Department of Molecular Genetics, The Forsyth Institute, Cambridge, Massachusetts, United States of America

[2]Department of Oral Medicine, Infection and Immunity, Harvard School of Dental Medicine, Boston, Massachusetts, United States of America

[3] The Veterinary Dental Surgery, Byfleet, United Kingdom

[4] WALTHAM Centre for Pet Nutrition, Melton Mowbray, United Kingdom

[5] Mars Pet Care Europe, Birstall, United Kingdom

ABSTRACT

Determining the bacterial composition of the canine oral microbiome is of interest for two primary reasons. First, while the human oral microbiome has been well studied using molecular techniques, the oral microbiomes of other mammals have not been studied in equal depth using culture independent methods. This study allows a comparison of the number of bacterial taxa,

---

Citation: Dewhirst FE, Klein EA, Thompson EC, Blanton JM, Chen T, Milella L, et al. (2012) The Canine Oral Microbiome. PLoS ONE 7(4): e36067. https://doi.org/10.1371/journal.pone.0036067.

based on 16S rRNA-gene sequence comparison, shared between humans and dogs, two divergent mammalian species. Second, canine oral bacteria are of interest to veterinary and human medical communities for understanding their roles in health and infectious diseases. The bacteria involved are mostly unnamed and not linked by 16S rRNA-gene sequence identity to a taxonomic scheme. This manuscript describes the analysis of 5,958 16S rRNA-gene sequences from 65 clone libraries. Full length 16S rRNA reference sequences have been obtained for 353 canine bacterial taxa, which were placed in 14 bacterial phyla, 23 classes, 37 orders, 66 families, and 148 genera. Eighty percent of the taxa are currently unnamed. The bacterial taxa identified in dogs are markedly different from those of humans with only 16.4% of oral taxa are shared between dogs and humans based on a 98.5% 16S rRNA sequence similarity cutoff. This indicates that there is a large divergence in the bacteria comprising the oral microbiomes of divergent mammalian species. The historic practice of identifying animal associated bacteria based on phenotypic similarities to human bacteria is generally invalid. This report describes the diversity of the canine oral microbiome and provides a provisional 16S rRNA based taxonomic scheme for naming and identifying unnamed canine bacterial taxa.

INTRODUCTION

Bacteria of the oral cavity have been studied with great interest since Anton van Leeuwenhoek first examined the plaque between his teeth with his crude microscope in 1683 [1]. Using cultivable methods, approximately 300 species from the human oral cavity have been isolated, characterized and formally named. Studies of the oral microbiota of other vertebrates have been less extensive. Unfortunately, bacteria from non-human sources were often misidentified and misclassified based on phenotypic similarity to human microorganisms. With the advent of molecular identification methods, primarily based on 16S rRNA sequence analysis, it has become apparent that bacteria from different vertebrate hosts are frequently unique, despite similar biochemical and other phenotypic traits. While molecular methods have been valuable in clarifying the identification and taxonomy of isolates, the greatest strength of these methods is in the identification of the majority of organisms which are currently uncultivated. Studies with molecular methods have demonstrated that the bacterial diversity in most environments is severely underestimated in surveys with cultivation-based methods [2], [3].

While the human oral microbiome has been surveyed using culture-independent methods [4], the canine oral microbiome has not. Previous canine studies were based primarily on culture-dependant methods and sometimes sought to identify species commonly found in human plaque [5], [6], [7], [8].

The primary purpose of this study was to identify major species of bacteria present in canine oral microbiome through an examination of subgingival plaque using culture-independent methods. This study reports on the analysis of 5,958 16S rRNA sequences from 65 clone libraries and provides 416 full 16S rRNA reference sequences (>1500 base) for the 353 taxa identified. As the vast majority of these taxa are not formally named, a provisional taxonomic scheme is presented based on assigning each taxon to the closest genus or higher taxa, and assigning it a unique Canine Oral Taxon number.

MATERIALS AND METHODS

Ethics Statement

Dogs were recruited in the UK from a kenneled population and from client owned dogs presented at a specialist veterinary clinic; informed client consent was obtained. Two studies were performed as follows: subgingival plaque was collected from 20 dogs in the first study (10 of which were from a kenneled population) and from 31 dogs in the second. The studies were approved by the WALTHAM Centre for Pet Nutrition ethical review committee, and run under licensed authority in accordance with the UK Animals (Scientific Procedures) Act 1986.

Plaque Collection and DNA Isolation

Animals were sampled under anesthesia. Each dog was given a premedication of 0.02 mg/kg acepromazine (ACP 2 mg/ml) and 0.02 mg/kg buprenorphine (Vetergesic 0.3 mg/ml) intramuscularly, then induced with 0.4 mg/kg propofol (Rapinovet 10 mg/ml) given intravenously, and maintained on 2% inhalational isoflurane. Initially supra-gingival and gingival margin plaque and calculus were removed using a Gracey curette to prevent contamination of the sub-gingival sample. A periodontal probe was then inserted under the gingival margin and swept along the tooth surface. Plaque from at least eight teeth was pooled. The resulting subgingival plaque pool from each dog was suspended in a 350 µl solution of 50 mM Tris (pH 7.6), 1 mM EDTA (pH 8.0) and 0.5% Tween 20 and was immediately stored at −20°C prior to DNA extraction. DNA extraction was performed using the DNeasy Tissue Kit (Qiagen, Valencia, California) following the manufacturer's instructions for the isolation of genomic DNA from Gram-positive bacteria (which also works well for Gram-negative bacteria). For the second study DNA extraction was performed using the Masterpure Gram Positive DNA Purification Kit (Epicentre, USA), according to the manufacturer's instructions with an additional overnight lysis as follows. Plaque samples were centrifuged at 5000× g for 10 minutes and the cell pellet resuspended in 150 µl of TE buffer (10 mM Tris-Cl and 0.5 mM EDTA, pH 9.0). Following vortexing, 1 µl Ready-Lyse Lysozyme (Epicentre, UK) was added and the lysis mix incubated overnight at 37°C for 18 hrs. Following the extraction, DNA was resuspended in TE buffer.

DNA Amplification

DNA samples purified from subgingival plaque of 20 dogs in study 1 were individually amplified with "universal" primers F24/Y36 (9-29F/1525-1541R) to construct 20 libraries. Purified DNA from the 10 of the 20 dogs was also combined into 4 pools (each pool from 2 or 3 dogs), and each pool was amplified individually with "Bacteroidetes-selective", F24/F01 or "Spirochaetes-selective", F24/M98, primers to give eight additional libraries. In study 2, DNA samples purified from subgingival plaque of 31 dogs were individually amplified with "universal" primers F24+AD35/C72 (9-27F [YM+B]/1492-1509R) to construct 31 libraries. The forward primer was a combination of 4 parts of the 4-fold degenerate 9–27 "YM" primer F24 and one part *Bifidobacteriales* primer AD35 (modified from Frank et al. [9], to give a 5-fold degenerate primer mix for enhanced phylogenetic coverage. Equal amount of DNA from 3 sets of ten to eleven dogs were pooled to give 3 DNA super-pools. The three super-pools were amplified individually with "Bacteroidetes-selective", F24/F01 and "Spirochaetes-selective", F24/M98, primers to give six additional libraries.

PCR was performed in thin-walled tubes with a Perkin-Elmer 9700 Thermocycler. One μl of the purified DNA template was added to a reaction mixture (50 μl final volume) containing 20 pmole of each primer, 40 nmole of dNTPs, 2.5 units of Platinum Taq polymerase (Invitrogen, Carlsbad, CA) in 10× PCR buffer (200 mM Tris-HCl pH 8.4, 500 mM KCl). In a hot start protocol, samples were preheated at 94°C for 4 min followed by amplification using the following conditions: denaturation at 94°C for 45 s, annealing at 60°C for 45 s, and elongation at 72°C for 1.5 min with an additional 1 s for each cycle. A total of 30 cycles were performed and then followed by a final elongation step at 72°C for 15 min. The size and amount of each amplicon was examined by electrophoresis in a 1% agarose gel. DNA was stained with SYBR Safe DNA gel stain (Invitrogen, Carlsbad, CA) and visualized under UV light. After checking that a strong amplicon of the correct size was produced, a second preparative gel was run and the full length amplicon band was cut out and purified using a Qiagen Gel Extraction kit (Qiagen, Valencia, CA).

Cloning and Library Screening Procedures

Size-purified PCR amplified DNA was cloned using a TOPO TA Cloning Kit as previously described [4]. Approximately 90 colonies were picked for each library. Clones were amplified using M13 forward and reverse primers and amplicon purified as previously described [4].

16S rRNA Sequencing

Purified DNA was sequenced using an ABI prism cycle-sequencing kit (BigDye® Terminator Cycle Sequencing kit) on an ABI 3100 Genetic Analyser (Applied Biosystems, Foster City, CA). The sequencing primers were used in a quarter-dye chemistry following the manufacturer's instructions.

16S rRNA Data Analysis

Approximately 500 bases of sequence were determined using primer Y31 (519–533R) to allow preliminary identification of clones. If the clone sequence appeared novel (differing by more than 7 bases from previously identified canine oral reference sequences), a full sequence of approximately 1,500 bases was obtained using 6 to 8 sequencing primers for full double strand coverage. The sequencing primers used over the course of the two studies evolved. Primers which failed to produce readable sequence for multiple taxa due to mismatches are labeled "limited" and were not used in subsequent studies. Primers which proved successful empirically and by alignment with human and canine oral reference sequences are labeled "general". Full 16S rRNA sequences were assembled from the ABI electropherogram files using Sequencher (Gene Codes Corporation, Ann Arbor, Michigan). Programs for data entry, editing, sequence alignment, secondary structure comparison, similarity matrix generation, and phylogenetic tree construction were written by F.E. Dewhirst [10]. Consensus neighbor-joining trees [11] were constructed from our aligned sequences using MEGA 4 [12]. The similarity matrices were corrected for multiple base changes at single positions by the method of Jukes and Cantor [13]. Comparisons with missing data were eliminated pairwise. The consensus trees were based on 1,000 bootstrap resamplings.

Sequences were checked for the possibility of being chimeric using a custom program [4] which checked the phylogenetic distance between the best BLAST match of the ends of each sequence with the canine reference set excluding self matches. Sequences whose ends diverged >5% were examined using Mallard [14] and heuristically for sequence consistency with phylogenetic neighbors in our overall sequence alignment sorted phylogenetically.

Nucleotide Sequences

The full 16S rRNA sequences for 416 clones representing 353 canine oral taxa were deposited in GenBank and received accession numbers JN713151-JN713566. The accession numbers are also included for each phylotype in Figs. 1,2, 3, and 4. The partial 16S RNA sequences (the 5′-end ~500 bases) of 5,959 clones were deposited in GenBank as JQ294075-JQ300033.

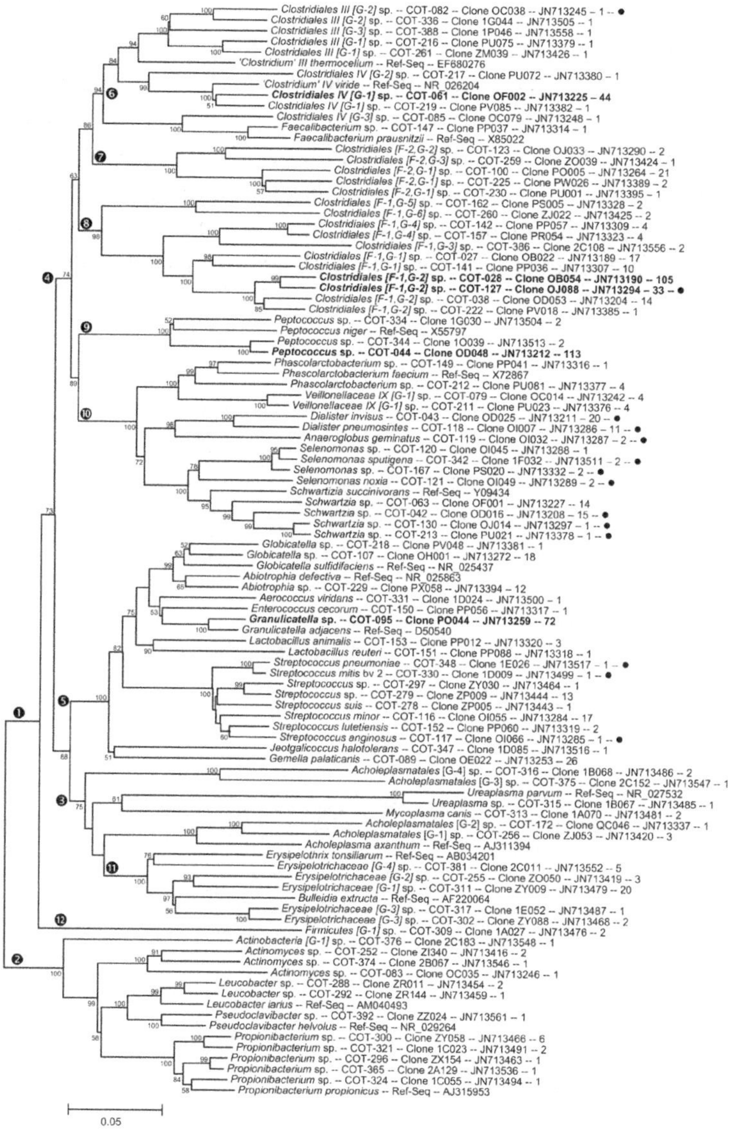

Figure 1: Consensus neighbor-joining tree for canine oral tax in phyla Actinobacteria, Firmicutes and Tenericutes.

The name of each taxon is followed by Canine Oral Taxon number, reference clone designation, GenBank accession number, and the number of clones out of 5958 that were identified as this taxon. Taxa marked with a filled circle are also found in the human oral cavity. Taxa for which there were 30 or more clones are shown in bold. The tree was constructed with MEGA 4 using the Jukes and Cantor correction neighbor-joining distance matrix. Comparisons with missing data were eliminated pairwise. The numbers to the left of the branches indicate the percent of time the clade was recovered out of 1,000 bootstrap resamplings. Only bootstrap percentages greater than 50 are shown. Roman numerals following a genus name indicate Collins' *Clostridia* cluster numbers [17] The scale bar shows 5% sequence divergence. The encircled numbers mark clades discussed in the text.

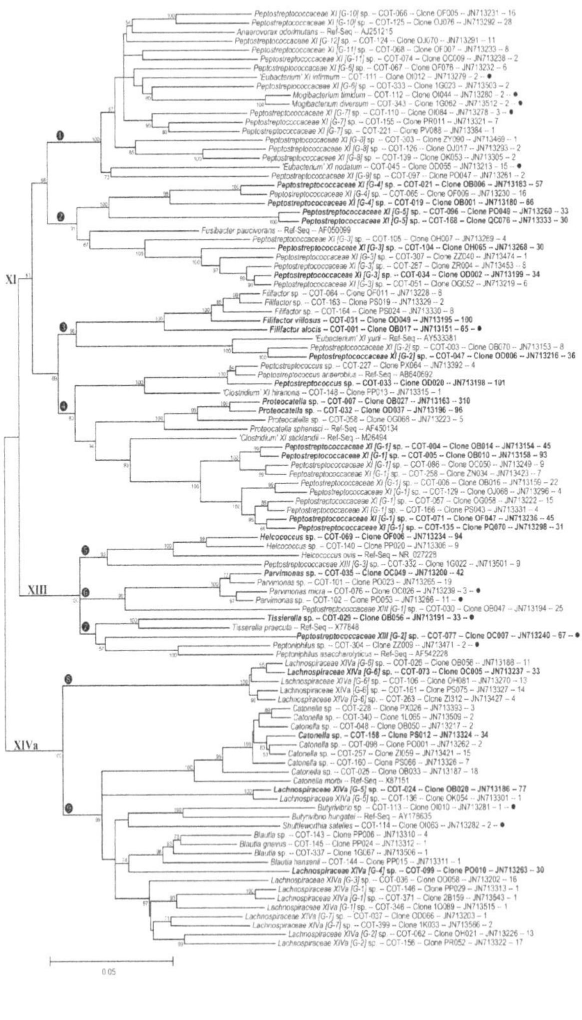

Figure 2: Consensus neighbor-joining tree for class *Clostridia*, families *Pepto-strepto-coccaceae* **and** *Lachnospiraceae*.

Labeling and methods used are as described in Fig. 1.

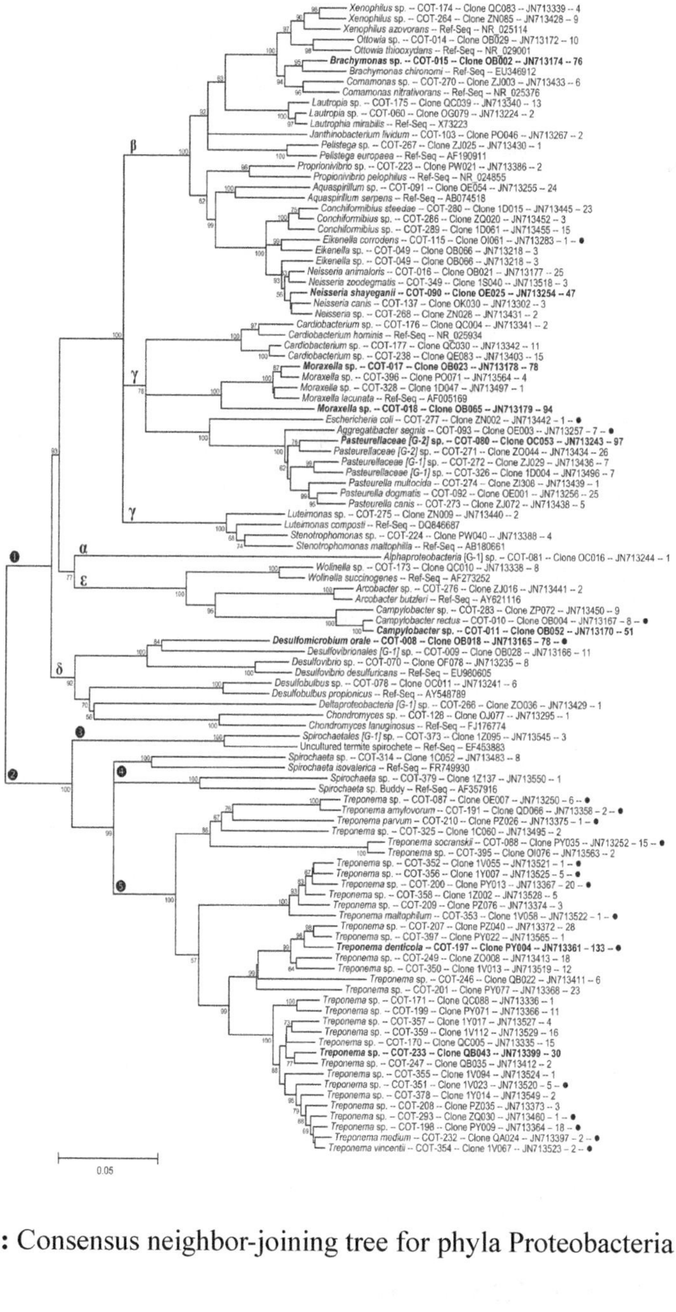

Figure 3: Consensus neighbor-joining tree for phyla Proteobacteria and Spirochaetes.

Labeling and methods used are as described in Fig. 1. The Greek letters mark the respective Proteobacteria classes.

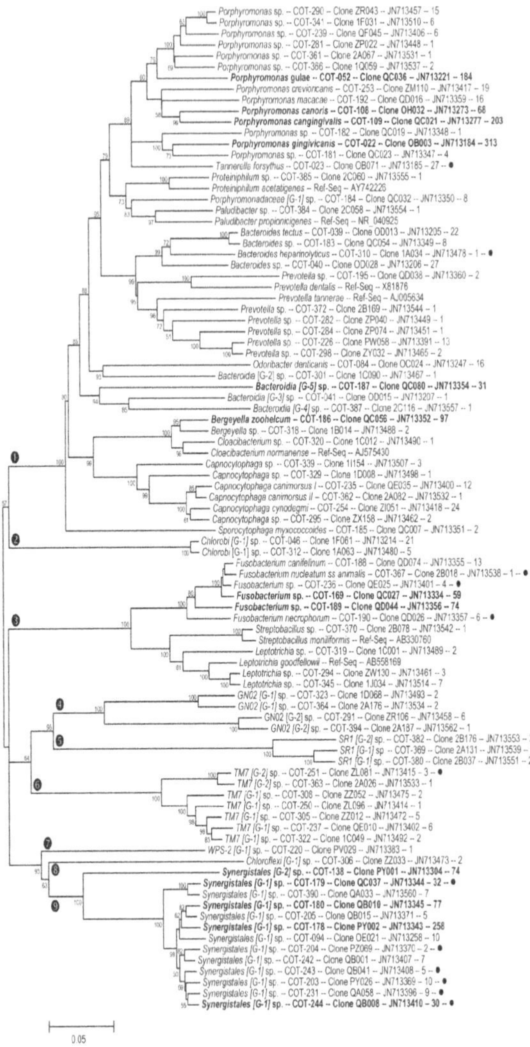

Figure 4: Consensus neighbor-joining tree for phyla Bacteroidetes, Fusobacteria, Chlorobi, Chloroflexi, Synergistetes and candidate divisions TM7, SR1, GN02 and WPS-2.

Labeling and methods used are as described in Fig. 1.

RESULTS AND DISCUSSION

Oral samples for 16S rDNA clone library construction came from a wide variety of dog breeds. The breed and age of each dog for each library is given.

The breeds examined include large (Saint Bernard) and small (Papillion) breeds, and those with long (Australian Collie) and short (Shih Tzu) snouts and ranged in age from 3 to 8 years old. While the breeds examined in this study are originally from geographically diverse locations, the dogs sampled are from a limited area of the United Kingdom. Thus, future studies employing samples from dogs living in different countries could well find additional canine microbial diversity. Because the 51 dogs examined came from 25 breeds, there was no attempt to compare microbiomes between breeds as the number of dogs/breed were too low.

Cloning Studies

A total of 6,025 clones were examined from 65 libraries of approximately 90 clones per library. Sixty-seven clones which had sequences shorter than 350 trimmed bases or which were found to be chimeric were excluded for a total of 5,958 validated clones used for analyses. The validated clones from the first cloning library were initially grouped into provisional phylotypes based on their 500 base partial sequences. A full sequence was then determined for a representative of each phylotype. The phylotypes were given arbitrary Canine Oral Taxon numbers (COT-001 through COT-399) in the order they were identified and the full length sequences used as a reference set against which subsequent clones were examined by BLASTN analysis. In this study, a phylotype or COT is defined as a set of one or more 16S rRNA sequences with greater than 98.5% full sequence similarity (23 or fewer base differences for a 1530 base sequence). This phylotype definition was chosen because the 16S rRNA sequence divergence for most strains of named oral species examined is less than 1.5% and inter-species divergence is usually greater than 1.5%. As subsequent clone libraries were screened, any clone with a partial 500-base sequence not matching a reference set sequence by at least 98% (7 base mismatches) was fully sequenced and added as a new reference sequence and given a COT number. Thus all 5,958 partial clone sequences match a reference sequence at a similarity of greater than 98%. Some taxa have two or more reference sequences because members of a taxon can differ by up to 23 base differences and appear <98% similar in their first 500 bases. A total of 353 phylotypes were identified. Seventy of these phylotypes (19.8%) were identified as named species based on greater than 98.5% sequence similarity to a type strains in BLASTN searches of GenBank [15] and Greengenes [16]. The remaining 284 phylotypes (80.2%) represent currently unnamed taxa. As this study made no attempt to cultivate members of the canine microbiome, we are not in a position to address what percent of the unnamed taxa are cultivable or as yet uncultivated as has been done for human taxa [4].

Taxonomy

Each canine taxon was placed in a phylum or candidate division based initially on BLASTN results against the Human Oral Microbiome Database (HOMD) [4], GenBank databases Reference RNA sequences (refseq_rna) and RNA and Nucleotide collection (nr/nt) [15], and using tools at Greengenes [16]. The Greengenes site was particularly useful for classifying and placing sequences from the rare phyla or candidate divisions Chlorobi, Chloroflexi, GN02 and WPS-2. The 16S rRNA sequences of all canine taxa were placed in an aligned database (hand-aligned based on secondary structure) and analyzed extensively by tree construction anchored to named reference sequences. As was previously done for the human oral microbiome [4], a provisional six level taxonomy was created consistent with the 16S rRNA tree structure. The 353 canine bacterial phylotypes were placed in 14 bacterial Phyla, 23 Classes, 37 Orders, 66 Families, and 148 Genera. The number of taxa and clones in each phylum or candidate division are shown in Table 1.

Table 1: Bacterial phyla identified in canine subgingival plaque

			Clones			
Phyla	Phylo-types	Universal 1525 R[a]	Universal 1492 R[b]	Selective Spiro[c]	Selective Bact[d]	Total
Firmicutes	162	1,148	1,379	0	213	2,740
Proteobacteria	52	224	569	0	68	861
Bacteroidetes	43	213	516	0	420	1,149
Spirochaetes	37	17	22	366	4	409
Synergistetes	13	1	5	511	9	526
Actinobacteria	12	1	11	0	8	20
Fusobacteria	10	0	112	0	58	170
TM7	7	0	7	0	13	20
Tenericutes	6	0	7	0	3	10
GN02	4	0	5	0	6	11
SR1	3	0	0	0	13	13
Chlorobi	2	1	12	0	13	26
Chloroflexi	1	0	2	0	0	2

WPS-2	1	1	0	0	0	1
Total	353	1,606	2,647	877	828	5,958

[a]Clones from libraries made using 9–27F (F24) and 1525–1541R (Y36) primers.

[b]Clones from libraries made using expanded coverage 9–27F (F24/AE35) and 1492–1509R (C72) primers.

[c]Clones from libraries made using "Spirochaetes-selective" F24/M98 primer pair.

[d]Clones from libraries made using "Bacteroidetes-selective" F24/F01 primer pair.

Shown in Figs 1, 2, 3, and 4 are consensus neighbor-joining trees based on the aligned full 16S rRNA sequences for the 353 canine taxa. Each taxon header includes name (genus and species), Canine Oral Taxon number (COT), clone designation, GenBank accession number, and number of clones identified for each taxon out of a total of 5,958. The 51 taxa with 30 or more clones are shown in bold as major taxa. Those 58 taxa marked with a filled circle are taxa shared with humans, as defined by the canine reference sequences sharing >98.5% similarity with reference sequences in the Human Oral Microbiome Database by BLASTN comparison (www.homd.org). Where a taxon is <90% similar to a named genus, it is designated using the family, or most specific higher taxa name, [G-1] sp. where "[G-1]" indicates it belongs to a novel genus. Family level grouping in the *Clostridia* (Figs 1 & 2) include the widely recognized classification of Collins *et al.* [17]. Thus, *Clostridium viride* is written *'Clostridium' IV viride* to indicate it is not in the genus *Clostridium sensu stricto* but rather is a member of Collins Cluster IV.

Firmicutes and Tenericutes

The majority of taxa in the Firmicutes are shown in Fig. 1 in the cluster marked by encircled "1". The Firmicutes families *Peptostreptococcaceae* and *Lachnospiraceae* are shown in Fig. 2. The phylum Tenericutes, previously the class *Mollicutes* within the Firmicutes [18], is marked with an encircled "3» in Fig. 1. The Firmicutes class *Erysipelotrichi*, marked with an encircled "11", branches within the "phylum" Tenericutes, demonstrating phylogenetic inconsistencies created by elevating class level branches within the Firmicutes to phylum level. One hundred sixty-two taxa were identified as members of the phylum Firmicutes.

Clostridia

The dominant class within the Firmicutes is *Clostridia*, containing 138 taxa. The *Clostridia* clade is shown in Fig. 1, marked encircled "4», and all taxa in Fig. 2. The cluster of 10 taxa, marked encircled "6» in Fig. 1, fall into unnamed genera in Collins Clusters III and IV, except for one taxa falling in the genus *Faecalibacterium*. Sixteen taxa fall into two family level Clusters with no named members, marked encircled "7" and "8", for novel families F-2 and F-1 respectively. Three taxa fall in the family *Peptococcaceae*, marked encircled "9", related to the human associated species *Peptococcus niger*. The family *Veillonellaceae*, previously *Acidaminococcaceae*, is marked encircled "10". We chose not to follow the suggestion of Marchandin *et. al.* [19], to elevate this family to a class as we believe it is taxonomically unjustified. The *Veillonellaceae* cluster contains members of the genera *Dialister, Anaeroglobus, Phascolarctobacterium, Schwartzia, Selenomonas*, and an unnamed genus. Nine of these taxa are also found in humans. Shown in Fig. 2 are those *Clostridia* taxa falling in Collins Clusters XI, XIII, and XIVa, with the first two clusters constituting the family *Peptostreptococceae* and the last cluster the family *Lachnospiraceae*. These two families contain the majority of the *Clostridia* taxa in both dogs and humans. In Collins Cluster XI, the cluster of taxa marked encircled "1» contains 18 taxa. Most are in 7 unnamed genera which may be unique to dogs. This cluster contains some named taxa shared with humans such as *'Eubacterium' XI infirmum, Mogibacterium timidum* and *M. diversum*, and *'Eubacterium' XI nodatum*. The cluster marked encircled "2" contains 11 taxa in 3 unnamed genera distantly related to *Fusibacter paucivorans*. The cluster marked encircled "3" contains five *Filifactor* species, including *F. alocis* and *F. villosus*, and two taxa related to human associated species *'Eubacterium' XI yurii*. The cluster marked encircled "4" contains 16 taxa in the genus *Peptostreptococcus sensu stricto, Proteocatella*, and an unnamed genus distantly related to *'Clostridium' XI sticklandii*. The validly named reference bacterium *Proteocatella sphensci* [20] was initially called *'Frigovirgula patagoniensis'* in GenBank (AF450134) and the name *'Frigovirgula'* unfortunately persists causing minor confusion. Within Collins Cluster XIII, clusters marked encircled "5», "6», & "7», are 11 taxa in the genera *Helcococcus, Parvimonas, Tissierella, Peptoniphilus,* and three unnamed genera. Five taxa, including *P. micra,* are shared with humans. Seventeen canine taxa fall in the *Lachnospiraceae* [21], Collins Cluster XIVa, with major subclusters marked encircled "8" and "9". The subclusters contain taxa in the genera *Blautia, Butyrivibrio, Catonella, Shuttleworthia,* as well as 7 unnamed genera. Two taxa, including *S. satelles*, are shared with humans.

Bacilli

The second most dominant class within the Firmicutes is the *Bacilli* with 18 taxa. The *Bacilli* clade is marked with an encircled "5» in Fig. 1. All taxa can be placed in the following genera: *Abiotrophia, Aerococcus, Enterococcus, Gemella, Globicatella, Granulicatella, Jeotgalicoccus, Lactobacillus* and *Streptococcus*. While three streptococcal species are shared with humans, streptococci appear to represent a minor genus in dog. This is not surprising as simple carbohydrates and sugars are not normally a major constituent of the canine diet and canine saliva has a pH of approximately 8.0 (WALTHAM, unpublished data 2011) which may be hostile to members of this aciduric genus.

Erysipelotrichi

Five taxa in this Firmicutes class, marked encircled "11" in Fig. 1, were identified. None were sufficiently close to reference species to place them in the genera *Erysipelothrix* or *Bulleidia*.

Novel Firmicutes Class

Firmicutes [G-1] sp. COT-309 appears to be a member of a novel deeply branching linage marked encircled "12" in Fig. 1. The closest named species had only 80% sequence similarity, however, a clone from the microbiome of fiber adherent species from rumen fluid was 93% similar (EU844484) supporting this canine taxa as a member of a mammal host associated lineage.

Tenericutes

Six members of this phylum were identified and are marked encircled "3" in Fig. 1, but excluding the Class *Erysipelotrichi* discussed above. In this tree, the "phylum» does not branch as a monophyletic entity. *Mycoplasma canis* and an *Ureaplasma parvum*-related taxon can be placed in named genera, but four additional taxa fall into unnamed genera.

Actinobacteria

Twelve Actinobacteria were identified and are marked encircled "2" in Fig. 1. Taxa in the genera *Actinomyces, Leucobacter, Pseudoclavibacter, Propionibacterium*, were identified as well as a deeply branching taxa Actinobacteria *[G-1]* sp. COT-376. None of these canine oral taxa are shared with humans. In study 1 using the standard 9–27F and 1525–1541R primers, only one Actinobacteria clone was recovered. Because the 1525–1541R primer has been reported to discriminate against Actinobacteria [22], we switched to the 1492–1505R primer in hopes of obtaining less biased coverage in our second study. Eleven clones were obtained with the revised "universal» primers and eight additional clones by fortuitous mispriming using the "Bacteroidetes-selective" primer set. It appears that no truly "universal" 16S rRNA primers exist and studies of diversity benefit from the use of multiple primer sets. *Actinomyces* sp. COT-083 fell in the genus *Actinomyces*, and is 97% similar to *Actinomyces coleocanis*, a species isolated from the vagina of a dog [23].

Proteobacteria

Fifty-two phylotypes were identified from the phylum *Proteobacteria*, and are marked with an encircled "1" in Fig. 3. The five classes are marked with Greek letters. The 22 *Betaproteobacteria* taxa include 11 from the mammalian host associated genera *Neisseria, Eikenella* and *Conchiformibius*. Whether the taxa associated with other genera in the *Betaproteobacteria* are truly part of the endogenous oral microbiome, or are transient common environmental bacteria remains to be determined. The 18 *Gammaproteobacteria* taxa include the host associated genera *Cardiobacterium*, *Moraxella* and species in the families *Pasteurellaceae* and *Enterbacteriaceae*. Taxa in the genera *Luteimonas* and *Stenotrophomonas* may be transient common environmental bacteria. One deeply branching *Alphaproteobacteria* taxon, distantly related to named species (81% similarity), was identified. The five *Epsilonproteobacteria* and six *Deltaproteobacteria* taxa are related to well-known mammalian host associated genera except for *Chondromyces*, which is generally associated with soil or decaying organic matter.

Spirochaetes

Thirty-seven phylotypes from the phylum Spirochaetes were identified and are marked by encircled "2" in Fig. 3. Thirty-four taxa are members of the genus *Treponema*, marked encircled "5", including the named species

T. amylovorum, T. denticola, T. maltophilum, T. medium, T. parvum, T. socranskii, and *T. vincentii* which are also found in the human oral cavity. A total of 14 canine *Treponema* spp. are shared with humans. Unlike previous studies of the human oral cavity [4], three taxa outside the genus *Treponema* were identified and marked encircled "3» and "4». *Spirochaeta* sp. COT-379 is most closely related to *Spirochaeta coccoides* (NR_042260; not shown) and *Spirochaeta* sp. Buddy. These two species are not helical cells, typical of spirochetes, but rather have a coccoid morphology. *Spirochaeta* sp. COT-314 is 92% similar to a strain isolated from the marine bristle worm *Alvinella pompejana* (AJ431240; not shown) and *Spirochaeta isovalerica. Spirochaetes [G1]* sp. COT-373 is a deeply-branching taxa with 93% similarity to a clone sequence from the termite gut, EF453883. Thus it appears that the diversity of spirochetes in the mammalian oral cavity may be broader than just the genus *Treponema*. The vast majority of the spirochete clones came from the 7 libraries produced using "spirochete-selective" primers (Table 1), which demonstrates the utility of using selective primers.

Bacteroidetes

Forty-three phylotypes were identified as members of the phylum Bacteroidetes, marked by encircled "1" in Fig. 4. Eleven named genera include: *Porphyromonas, Tannerella, Proteiniphilum, Paludibacter, Bacteroides, Prevotella, Odoribacter, Bergeyella, Cloacibacterium, Capnocytophaga* and *Sporocytophaga*, and 5 unnamed deeply branching genera what are not anchored to named taxa. The use of "Bacteroidetes-selective" primers with DNA from 7 super-pools produced 420 clones in the Bacteroidetes phylum and increased the depth and diversity of taxa identified over that produced from "universal" primers (Table 1).

There are naming issues for a number of species in the Bacteroidetes phylum. *Porphyromonas gingivicanis* and *Porphyromonas crevioricanis* were properly named and validly published by Hirasawa & Takada in 1994 [24]. Unfortunately, no 16S rRNA sequences for the type strains of these species were deposited by anyone for 12 years (see DQ677835 & DQ67736) and for 14 years by the authors (see AB430828 & AB430829). While these sequences were unavailable, Collins *et al*. named *Porphyromonas cansulci* [25] and deposited its 16S rRNA sequence in GenBank as entry X76260. "*Porphyromonas canis*" was invalidly named by Sakamoto & Benno in 1999 as GenBank entry AB034799. From the 16S rRNA sequences, we now know that *P. cansulci* is a synonym for *P. crevioricanis*, and that *"P. canis"* is an invalid synonym for *P. gingivicanis*. *Odoribacter denticanis* was named and validly published by Hardham *et al.* [26], but was challenged by Euzeby in comments in the List of Prokaryotic Names with Standing in Nomenclature (http://www.bacterio.cict.fr/) for not having a type strain available. This appears to be rectified as the type strain is now available from three national collections. This species was also previously referred to as "*Wernerella denticanis*" and "*Porphyromonas denticanis*". *Bacteroides* sp. COT-183 has been called "*Bacteroides denticanum*" by Elliott (see DQ156993) and "*Bacteroides denticanoris*" by Hardham *et al.* (see AY54431) in GenBank and patent filings, but never validly described in any publication.

Chlorobi

Two phylotypes from the phylum Chlorobi, marked with encircled "2" in Fig. 4, were identified. The original cultivable members of the phylum *Chlorobi*, previously called Green Sulfur Bacteria or *Chlorobia*, are phototropic organisms [27]. Cultivation independent molecular methods have identified members from diverse environments. Recently a non-photosynthetic member of the phylum, *Ignavibacterium album*, has been described [28]. Sequences in GenBank with greater than 84% similarity to canine *Chlorobi* phylotypes COT-046 & COT-312 have been recovered from manure drainage, penguin dropping sediment, hydrothermal worm mucus, and from an anaerobic digester. A sequence with 99% similarity to COT-046 has been recovered from the oral cavity of a cat (unpublished observation), supporting the association of this taxa with the oral cavity of mammals. Nine clones from skin swabs of the volar forearms of four human subjects (based on subject identification number in GenBank entries) have a sequence similarity of 99% to canine *Chlorobi* taxa (for example HM278300 and HM330153 to COT-046). These four human subjects appear to have had the volar surface of their arms licked by dogs prior to sampling as their clone libraries include 23 to 51 canine oral taxa.

Fusobacteria

Ten taxa from the phylum Fusobacteria, marked encircled "3" in Fig. 4, were identified, including the genera *Fusobacterium, Streptobacillus*, and *Leptotrichia*. The Fusobacteria spp. includes four taxa that overlap the human *F. nucleatum* cluster. *Streptobacillus* sp. COT-370 is closely related to the rat bite fever organism *S. moniliformis*. It was suggested previously that dogs may be colonized with *S. moniliformis* by eating rats [29], but the current study suggests that dogs may be naturally colonized with a distinct, but closely related species. It is notable that clones from this phylum were not present in the 10 libraries made by PCR with standard 9–27F and 1525–1541R primers, but were present (110 clones) in 21 libraries using an extend specificity 9–27F and 1492–1509R primers

GN02

our taxa from the as-yet-uncultured GN02 candidate division, marked encircled "4" in Fig. 4 were identified. GN02 is one of 15 candidate divisions proposed in a study of the Guerrero Negro hypersaline microbial mat [30]. The canine phylotypes were originally placed in this division using BLASTN searches of the Greengenes database. In the past year, related taxa from human mouth and skin have started to appear in GenBank as human microbiome data have been submitted (for example FJ976283 & HM249743).

SR1

Three taxa from the as-yet-uncultured SR1 candidate division, marked encircled "5" in Fig. 4 were identified. The SR1 division was named for clones identified in a study of sediment with microbial streamers from the Sulphur River in Parkers Cave, Kentucky [31]. The SR1 division was previously part of candidate division OP11, so older references to a closely related taxa from the human oral cavity referred the human taxon as OP11 clone X112 [32]. The human phylotype, now designated SR1 sp. HOT-345, has been identified in multiple clone libraries [4].

TM7

Seven canine phylotypes were identified as members of the candidate division TM7, which is marked with an encircled "6" in Fig. 4. The phylum TM7 is a major lineage of *Bacteria* with no known pure-culture representatives [33]. TM7 organisms have been recognized in 16S rRNA cloning studies of many habitats, including soils, fresh ground water, seawater, and mammalian clinical samples [33]. They have been recovered from the human oral cavity [4], [32], [34], the human distal esophagus [35], and mouse feces [36].

WPS-2

The candidate division WPS-2, marked with encircled "7" in Fig. 4, is known from only 39 environmental clones in Greengenes otu_4420, mainly from soils. The WPS-2 division was one of two named for clones identified in a study of Wittenberg polluted soil, Germany [37]. WPS-2 sp. COT-220 is closest to GenBank entry DQ520181, and is the 40th member of this rarely observed candidate division marked encircled "7» in Fig. 4. As this taxon was detected as a single clone, and no related clones have been identified from human or other mammalian sources, it remains to be determined if this taxon is part of the endogenous canine oral microflora, or an environmental transient.

Chloroflexi

A single phylotype of the Chloroflexi phylum was identified and is marked with encircled "8" in Fig. 4. The Chloroflexi phylum, previously called green non-sulfur bacteria, has many cultivated species [38], and several were named subsequent to the description in Bergey's Manual of Systematic Bacteriology [39]. The canine Chloroflexi sp. COT-306 is 96% similar to human oral taxon Chloroflexi sp. HOT-439 and 86% similarity to named species *Anaerolinea thermophila* [40] in the class *Anaerolineae* [41].

Synergistetes

The phylum Synergistetes is known mainly from clone sequences, but contains about a dozen cultivated species including *Synergistes jonesii*, a rumen bacterium that degrades toxic pyridinediols [42] and *Pyramidobacter piscolens*, a species from the human oral cavity [43]. Organisms from the Synergistetes phylum have previously been mistakenly included in the phylum *Firmicutes* or placed in the phylum Deferribacteres (a sister phylum of Synergistetes and Flexistipes) [32]. As marked by an encircled "9" in Fig. 4, 13 canine phylotypes were identified. Six canine phylotypes match previously identified human phylotypes at >98.5% similarity [4].

Primer Biases

The number of clones identified in each phylum for libraries generated with two different "universal" primer pairs, a "Spirochaetes-selective" pair, and a "Bacteroidetes-selective" pair are shown. A marked difference in the diversity recovered in clone libraries using different initial PCR primers is apparent. In study 1, the commonly used "universal" 9–27 YM forward (F24) and 1525–1541 reverse (Y36) primers produced more than one clone only for the four common phyla Firmicutes, Proteobacteria, Bacteroides, and Spirochaetes. In the second study, using expanded coverage 9–27 forward primers (F24/AD35) [9] and the "universal" 1492–1509 reverse primer (C72), clones from 12 phyla/candidate divisions were recovered. Of particular note is the recovery of Fusobacteria taxa only with the second set of "universal" primers and recovery of significantly more Actinobacteria clones with the second primer set. PCR with the "Spirochaetes-selective" reverse primer M98 (1483–1501) yielded expected results: organisms from the Spirochaetes and Synergistetes phyla. Bacteria in these two taxa have "GG" at position 1484-5 whereas most other bacteria have "CT". The "Bacteroidaetes-selective" reverse primer F01 (1487–1505) selects for organisms with a "CT" at position 1490-1 whereas most non Bacteroidaetes have other bases at these positions. While the F24/F01 primer set yielded mostly clones from the Bacteroidetes phylum, clones for 12 phyla/candidate divisions were recovered. The recovery of Chlorobi clones was expected based on perfect primer sequence match; the recovery of TM7 and SR1, which have a one base mismatch "TT", is also expected; but recovery of other taxa, such as Firmicutes, Proteobacteria and Fusobacteria, is somewhat unexpected as they have 2 base mismatches. While the "Spirochaetes–selective" primers are truly selective, the "Bacteroidetes-selective" primers produced clones from 12 of 14 phyla and appear to be useful in recovering a number of rare phyla/candidate divisions. The recovery of taxa from diverse phyla was clearly aided by using multiple primer sets for PCR of DNA prior to library construction. Because this study used taxa selective primers (as all studies ultimately do) to construct libraries, it is impossible to say anything valid about relative abundance of canine oral species from the abundance of clones recovered.

Taxa Abundance

The rank abundance of clones for each canine oral taxon is presented. Because a variety of primers with various biases were used for library construction, the clone abundance data reflect only clone numbers found in these libraries and cannot be used to validly infer the underlying population structure. With the caveat noted, the most prevalent taxa, *Porphyromonas gingivicanis* COT-022, constituted 5.3% of the clones. Clones from 28 taxa were recovered at level of greater than one percent. The 89 singleton clones were present as 0.017% of 5,958 clones identified. Of the 50 most common taxa, it is striking that 40, or 80%, are unnamed. The taxon rank abundance profile for this canine study is very similar to that previously found for the human oral microbiome [4]. In the human study of about 35,000 clones, it was estimated that the number of taxa necessary to identify

90%, 95% and 98% of the clones was 259, 423 and 655 taxa respectively. Assuming the canine and human oral cavities contain about equal microbial diversity and similar rank abundance profiles, 353 canine taxa should allow identification of about 93% of clones in a study of similar size. This estimate is approximate, but suggests that 353 taxa capture a significant portion of the microbiome. While the current study provides good initial coverage of the canine oral microbiome, the oral samples examined were limited to the subgingival sites. Further studies sampling other oral habitats such as teeth, tongue, cheek, hard and soft palates, and tonsils will no doubt expand the number of canine taxa to approach the more than 1,000 currently defined for the human oral microbiome [4]. One goal of the current study was to obtain essentially full length 16S rDNA reference sequences, which are required for recognition and placement of previously unrecognized rare taxa members such as those in candidate divisions GN02 and WPS-2. Future studies using next generation sequencing methods will no doubt sequence more deeply, producing tens to hundreds of thousands of short sequences. Studies require tradeoffs between sequence length (full length better for phylogenetic studies), and sequence number (higher numbers better for determining community composition).

Comparison of reference 16S rDNA sequences from the canine oral cavity with those of the human oral cavity reveals that only 16.4% of the taxa are shared by BLASTN analysis at a threshold of 98.5% sequence similarity (see taxa marked with filled circle in Figs. 1, 2, 3, and 4). This indicates that there is a large divergence in the oral microbiomes of divergent mammalian species. Of the 83.6% of taxa that differ, the differences are not only at the species level, but also at genus through phylum levels. It is apparent from the results presented here, however, that the majority of oral bacteria from divergent mammalian species are unique and the practice of naming mammalian (or even more distantly related animal) isolates after the most phenotypically similar species from humans is likely to be shown invalid by using molecular tools.

The Canine Oral Microbiome

The provisional taxonomic scheme presented, and the linked 16S rRNA reference sequences, provide the most comprehensive resource to date for identifying and referencing both the named and the 80% as yet unnamed canine oral taxa. This sequence based identification resource should facilitate future molecular studies of canine health and disease as well as the zoonotic potential of canine oral microbes in human and veterinary infectious diseases.

The taxonomic scheme presented here currently includes only those taxa for which clones were identified in this study. It is anticipated future efforts will expand this taxonomy and reference sequence set to include all named canine-associated species, and isolates of novel taxa, for which full length 16S rRNA sequences exist.

CONCLUSIONS

The results of this study provide the groundwork for describing the diversity of taxa present in the canine oral cavity. The provisional scheme of giving each taxon a canine oral taxon number and placing it in a phylogenetic context should facilitate future studies of the canine oral microbiome and its role in canine health and disease. The canine oral microbiome is widely divergent from that of human, hence these results will also help in the interpretation of human microbiome studies where canine oral bacteria appear to be present in large numbers in certain human skin samples and in veterinary and human medical studies where previously unnamed canine taxa are recovered from clinical samples.

ACKNOWLEDGMENTS

We thank Sara Barbuto for her assistance in DNA sequencing and Liang Yang for depositing sequences in GenBank.

REFERENCES

1. van Leeuwenhoek A. Letter of 17 September 1683 to the Royal Society, London. Royal Society, MS L. 1683;1. 69
2. Amann RI, Ludwig W, Schleifer KH. Phylogenetic identification and in situ detection of individual microbial cells without cultivation. Microbiological reviews. 1995;59:143–169.
3. Hugenholtz P, Goebel BM, Pace NR. Impact of culture-independent studies on the emerging phylogenetic view of bacterial diversity. Journal of Bacteriology. 1998;180:4765–4774.
4. Dewhirst FE, Chen T, Izard J, Paster BJ, Tanner AC, et al. The human oral microbiome. Journal of Bacteriology. 2010;192:5002–5017.
5. Allaker RP, de Rosayro R, Young KA, Hardie JM. Prevalence of *Porphyromonas* and *Prevotella* species in the dental plaque of dogs. The Veterinary record. 1997;140:147–148.
6. Fournier D, Mouton C, Lapierre P, Kato T, Okuda K, et al. *Porphyromonas gulae* sp. nov., an anaerobic, gram-negative coccobacillus from the gingival sulcus of various animal hosts. International journal of systematic and evolutionary microbiology. 2001;51:1179–1189.
7. Hardham J, Dreier K, Wong J, Sfintescu C, Evans RT. Pigmented-anaerobic bacteria associated with canine periodontitis. Veterinary microbiology. 2005;106:119–128.
8. Elliott DR, Wilson M, Buckley CM, Spratt DA. Cultivable oral microbiota of domestic dogs. Journal of clinical microbiology. 2005;43:5470–5476.
9. Frank JA, Reich CI, Sharma S, Weisbaum JS, Wilson BA, et al. Critical evaluation of two primers commonly used for amplification of bacterial 16S rRNA genes. Applied and Environmental Microbiology. 2008;74:2461–2470.
10. Paster BJ, Dewhirst FE. Phylogeny of campylobacters, wolinellas, *Bacteroides gracilis*, and *Bacteroides ureolyticus* by 16S ribosomal ribonucleic acid sequencing. Int J Syst Bacteriol. 1988;38:56–62.
11. Saitou N, Nei M. The neighbor-joining method: a new method for reconstructing phylogenetic trees. Mol Biol Evol. 1987;4:406–425.
12. Tamura K, Peterson D, Peterson N, Stecher G, Nei M, et al. MEGA5: Molecular Evolutionary Genetics Analysis Using Maximum Likelihood, Evolutionary Distance, and Maximum Parsimony Methods. Molecular biology and evolution. 2011;28:2731–2739.

13. Jukes T, Cantor C. Evolution of protein molecules; Munro H, editor. New York: Academic Press; 1969. pp. 21–132.

14. Ashelford KE, Chuzhanova NA, Fry JC, Jones AJ, Weightman AJ. New screening software shows that most recent large 16S rRNA gene clone libraries contain chimeras. Applied and Environmental Microbiology. 2006;72:5734–5741.

15. Johnson M, Zaretskaya I, Raytselis Y, Merezhuk Y, McGinnis S, et al. NCBI BLAST: a better web interface. Nucleic acids research. 2008;36:W5–9.

16. DeSantis TZ, Hugenholtz P, Larsen N, Rojas M, Brodie EL, et al. Greengenes, a chimera-checked 16S rRNA gene database and workbench compatible with ARB. Appl Environ Microbiol. 2006;72:5069–5072.

17. Collins MD, Lawson PA, Willems A, Cordoba JJ, Fernandez-Garayzabal J, et al. The phylogeny of the genus *Clostridium*: proposal of five new genera and eleven new species combinations. International journal of systematic bacteriology. 1994;44:812–826.

18. Ludwig W, Schleifer KH, Whitman WB. Revised road map to the phylum *Firmicutes*. In: De Vos P, Garrity G, Jones D, Kreig NR, Ludwig W, Rainey FA, editors. Bergey's manual of systematic bacteriology. Second ed. New York: Springer; 2009.

19. Marchandin H, Teyssier C, Campos J, Jean-Pierre H, Roger F, et al. *Negativicoccus succinicivorans* gen. nov., sp. nov., isolated from human clinical samples, emended description of the family *Veillonellaceae* and description of *Negativicutes* classis nov., *Selenomonadales* ord. nov. and *Acidaminococcaceae* fam. nov. in the bacterial phylum Firmicutes. International journal of systematic and evolutionary microbiology. 2010;60:1271–1279.

20. Pikuta EV, Hoover RB, Marsic D, Whitman WB, Lupa B, et al. *Proteocatella sphenisci* gen. nov., sp. nov., a psychrotolerant, spore-forming anaerobe isolated from penguin guano. International journal of systematic and evolutionary microbiology. 2009;59:2302–2307.

21. Janssen PH, Hugenholtz P. Fermentation of glycolate by a pure culture of a strictly anaerobic gram-positive bacterium belonging to the family *Lachnospiraceae*. Archives of microbiology. 2003;179:321–328.

22. de Lillo A, Ashley FP, Palmer RM, Munson MA, Kyriacou L, et al. Novel subgingival bacterial phylotypes detected using multiple

universal polymerase chain reaction primer sets. Oral microbiology and immunology. 2006;21:61–68.

23. Hoyles L, Falsen E, Foster G, Collins MD. *Actinomyces coleocanis sp. nov.*, from the vagina of a dog. International journal of systematic and evolutionary microbiology. 2002;52:1201–1203.
24. Hirasawa M, Takada K. *Porphyromonas gingivicanis* sp. nov. and *Porphyromonas crevioricanis* sp. nov., isolated from beagles. International journal of systematic bacteriology. 1994;44:637–640.
25. Collins MD, Love DN, Karjalainen J, Kanervo A, Forsblom B, et al. Phylogenetic analysis of members of the genus *Porphyromonas* and description of *Porphyromonas cangingivalis* sp. nov. and *Porphyromonas cansulci* sp. nov. International journal of systematic bacteriology. 1994;44:674–679.
26. Hardham JM, King KW, Dreier K, Wong J, Strietzel C, et al. Transfer of *Bacteroides splanchnicus* to *Odoribacter* gen. nov. as *Odoribacter splanchnicus* comb. nov., and description of *Odoribacter denticanis* sp. nov., isolated from the crevicular spaces of canine periodontitis patients. International journal of systematic and evolutionary microbiology. 2008;58:103–109.
27. Garrity GM, Holt GH. Phylum BXI. Chlorobi *phy. nov.* . In: Boone DR, Castenholz RW, editors. Bergey's Manual of Systematic Bacteriology. Second ed. New York: Springer; 2001. pp. 601–623.
28. Iino T, Mori K, Uchino Y, Nakagawa T, Harayama S, et al. *Ignavibacterium album* gen. nov., sp. nov., a moderately thermophilic anaerobic bacterium isolated from microbial mats at a terrestrial hot spring and proposal of *Ignavibacteria* classis nov., for a novel lineage at the periphery of green sulfur bacteria. International journal of systematic and evolutionary microbiology. 2010;60:1376–1382.
29. Wouters EG, Ho HT, Lipman LJ, Gaastra W. Dogs as vectors of *Streptobacillus moniliformis* infection? Veterinary microbiology. 2008;128:419–422.
30. Ley RE, Harris JK, Wilcox J, Spear JR, Miller SR, et al. Unexpected diversity and complexity of the Guerrero Negro hypersaline microbial mat. Applied and Environmental Microbiology. 2006;72:3685–3695.
31. Harris JK, Kelley ST, Pace NR. New perspective on uncultured bacterial phylogenetic division OP11. Applied and Environmental Microbiology. 2004;70:845–849.

32. Paster BJ, Boches SK, Galvin JL, Ericson RE, Lau CN, et al. Bacterial diversity in human subgingival plaque. J Bacteriol. 2001;183:3770–3783.

33. Hugenholtz P, Tyson GW, Webb RI, Wagner AM, Blackall LL. Investigation of candidate division TM7, a recently recognized major lineage of the domain Bacteria with no known pure-culture representatives. Applied and Environmental Microbiology. 2001;67:411–419.

34. Aas JA, Paster BJ, Stokes LN, Olsen I, Dewhirst FE. Defining the normal bacterial flora of the oral cavity. J Clin Microbiol. 2005;43:5721–5732.

35. Pei Z, Bini EJ, Yang L, Zhou M, Francois F, et al. Bacterial biota in the human distal esophagus. Proc Natl Acad Sci U S A. 2004;101:4250–4255.

36. Salzman NH, de Jong H, Paterson Y, Harmsen HJ, Welling GW, et al. Analysis of 16S libraries of mouse gastrointestinal microflora reveals a large new group of mouse intestinal bacteria. Microbiology. 2002;148:3651–3660.

37. Nogales B, Moore ER, Llobet-Brossa E, Rossello-Mora R, Amann R, et al. Combined use of 16S ribosomal DNA and 16S rRNA to study the bacterial community of polychlorinated biphenyl-polluted soil. Applied and Environmental Microbiology. 2001;67:1874–1884.

38. Yabe S, Aiba Y, Sakai Y, Hazaka M, Yokota A. *Thermosporothrix hazakensis* gen. nov., sp. nov., isolated from compost, description of *Thermosporotrichaceae* fam. nov. within the class *Ktedonobacteria* Cavaletti, et al. 2007 and emended description of the class *Ktedonobacteria*. International journal of systematic and evolutionary microbiology. 2010;60:1794–1801.

39. Garrity GM, Holt GH. Phylum BVI. Chloroflexi *phy. nov.* . In: Boone DR, Castenholz RW, editors. Bergey's Manual of Systematic Bacteriology. Second ed. New York: Springer; 2001. pp. 427–446.

40. Sekiguchi Y, Yamada T, Hanada S, Ohashi A, Harada H, et al. *Anaerolinea thermophila* gen. nov., sp. nov. and *Caldilinea aerophila* gen. nov., sp. nov., novel filamentous thermophiles that represent a previously uncultured lineage of the domain Bacteria at the subphylum level. International journal of systematic and evolutionary microbiology. 2003;53:1843–1851.

41. Yamada T, Sekiguchi Y, Hanada S, Imachi H, Ohashi A, et al.

Anaerolinea thermolimosa sp. nov., Levilinea saccharolytica gen. nov., sp. nov. and *Leptolinea tardivitalis* gen. nov., sp. nov., novel filamentous anaerobes, and description of the new classes *Anaerolineae* classis nov. and *Caldilineae* classis nov. in the bacterial phylum Chloroflexi. International journal of systematic and evolutionary microbiology. 2006;56:1331–1340.

42. Allison JJ, Mayberry WR, McSweeney CS, Stahl DA. *Synergistes jonesii*, gen. nov.: A rumen bacterium that degrades toxic pyridinediols. System Appl Microbiol. 1992;15:522–529.
43. Downes J, Vartoukian SR, Dewhirst FE, Izard J, Chen T, et al. *Pyramidobacter piscolens* gen. nov., sp. nov., a member of the phylum '*Synergistetes*' isolated from the human oral cavity. International journal of systematic and evolutionary microbiology. 2009;59:972–980.

CHAPTER 15

Correction: The Canine Oral Microbiome

Floyd E. Dewhirst[1,2], Erin A. Klein[1] , Emily C. Thompson[1] , Jessica M. Blanton[1] , Tsute Chen[1] , Lisa Milella[3] , Catherine M. F. Buckley[4] , Ian J. Davis[4] , Marie-Lousie Bennett[5] , Zoe V. Marshall-Jones[4]

[1] Department of Molecular Genetics, The Forsyth Institute, Cambridge, Massachusetts, United States of America

[2]Department of Oral Medicine, Infection and Immunity, Harvard School of Dental Medicine, Boston, Massachusetts, United States of America

[3] The Veterinary Dental Surgery, Byfleet, United Kingdom

[4] WALTHAM Centre for Pet Nutrition, Melton Mowbray, United Kingdom

[5] Mars Pet Care Europe, Birstall, United Kingdom

The published funding and competing interests statements were incorrect. The correct funding statement is: No current external funding exists for this study. The correct competing interests statement is: We have the following interests. CMFB, IJD, M-LB and ZVM-J were employees of WALTHAM Centre for pet nutrition during the course of this study and M-LB is currently at Mars Pet Care Europe. FED, EAK, ECT, JMB, TS were employees of the

Citation: Dewhirst FE, Klein EA, Thompson EC, Blanton JM, Chen T, Milella L, et al. (2012) Correction: The Canine Oral Microbiome. PLoS ONE 7(6): 10.1371/annotation/c2287fc7-c976-4d78-a28f-1d4e024d568f.https://doi.org/10.1371/annotation/c2287fc7-c976-4d78-a28f-1d4e024d568f

Forsyth Institute during the course of this study and LM is from the Veterinary Dental Surgery. FED has consulted for WALTHAM. FED CMFB, M-LB, and ZVM-J are inventors on patent WO2008137541, Dog Plaque Health, filed by WALTHAM. There are no other patents, products in development or other marketed products to declare. This does not alter our adherence to all the PLoS ONE policies on sharing data and materials, as detailed online in the guide for authors. The publisher apologizes for the error.

CHAPTER 16

Prevalence of Thermophilic Campylobacter Species in Swedish Dogs and Characterization of C. jejuni isolates

Mia Holmberg[1], Thomas Rosendal[1], Eva O Engvall[2], Anna Ohlson[3] and Ann Lindberg[1]

[1] Department of Epidemiology and Disease Control, National Veterinary Institute, SVA, SE-751 89 Uppsala, Sweden

[2] Department of Bacteriology, National Veterinary Institute, SVA, SE-751 89 Uppsala, Sweden

[3] Växa Sverige, SE-104 25 Stockholm, Sweden

ABSTRACT

Background

The aims of this study were to investigate the prevalence of *Campylobacter* species in Swedish dogs, to identify the species of the *Campylobacter* isolates and to genotype the *C. jejuni* isolates. Young and

---

Citation: Holmberg, M., Rosendal, T., Engvall, E.O. et al. Prevalence of thermophilic Campylobacter species in Swedish dogs and characterization of C. jejuni isolates. Acta Vet Scand 57, 19 (2015). https://doi.org/10.1186/s13028-015-0108-0

healthy dogs were targeted and the sampling was performed at 11 veterinary clinics throughout Sweden from October 2011 to October 2012. Faecal swab samples were collected and sent to the laboratory at the National Veterinary Institute (SVA) for isolation of *Campylobacter*, speciation and genotyping.

Results

Campylobacter spp. were isolated from 67 of the 180 sampled dogs which yields an overall prevalence of 37%. The most prevalent species of *Campylobacter* among the participating dogs was *C. upsaliensis* with 52 of the 67 identified isolates. A lower prevalence was observed for *C. jejuni* with seven identified isolates and one isolate was identified as *C. helveticus*. Multi-locus sequence typing (MLST) was carried out on the seven *C. jejuni* isolates and all sequence types that were found are also commonly found in humans. The dogs were divided into three age groups; 1) under 12 months, 2) 12 to 23 months and 3) 24 months and older. The highest prevalence was found in the two younger age groups. Dogs shedding *C. jejuni* were between 3–12 months of age while dogs shedding *C. upsaliensis* were found in all ages.

Conclusions

The present investigation finds that *Campylobacter* spp. known to cause campylobacteriosis in humans are present in Swedish dogs. The results suggest an age predisposition where dogs under 2 years of age are more likely to shed *Campylobacter* spp. than older dogs. The most commonly isolated species was *C. upsaliensis* followed by *C. jejuni,* which was only detected in dogs up to 12 months of age. All *C. jejuni* isolates identified in the present study were of the same MLST types that have previously been described both in humans and in animals. The awareness of the *Campylobacter* risk of healthy young dogs may be an important way to reduce the transmission from dogs to infants, young children and immunocompromised adults.

Keywords: Dog, *Campylobacter*, *C. upsaliensis*, *C. jejuni*, MLST, Prevalence

BACKGROUND

Campylobacteriosis is the most commonly reported zoonotic disease and the most common cause of bacterial enteritis in humans in many countries throughout the world [1]. In 2013, there were 8114 notified human cases of campylobacteriosis in Sweden. Of these, 41% were considered to be domestically acquired [2]. Humans can be infected with *Campylobacter* by several routes and the bacteria are commonly found in a wide range of animals including cats and dogs. Especially in young dogs (<1 year), *Campylobacter* is often found in faecal samples and the dogs usually shed the bacteria without showing any clinical signs [3,4]. Several studies have reported the presence of *Campylobacter* spp. in both healthy dogs and dogs with diarrhoea, but *Campylobacter* is generally not considered to be pathogenic to dogs [5-8]. No association was found between presence of *Campylobacter* and diarrhoea in a Norwegian study on dogs [8] whereas a study in Canada found significant higher infection rate in diarrhoeic dogs compared with healthy dogs [5]. In an investigation of dogs in Ireland, diarrhoeic dogs were more likely to be *Campylobacter* positive than dogs without diarrhoea, but the dogs with diarrhoea also had concurrent gastrointestinal parasites, inflammatory bowel disease or diabetes [9]. A study of dogs in Barbados found no difference in clinical disease in dogs with and without presence of *Campylobacter* in the faeces, but indicated that co-infection with parvovirus and *Campylobacter* was common [10]. An association between occurrence of diarrhoea and *Campylobacter* infection was described in a previous Swedish investigation of dogs from 1979, and 63% of the dogs with diarrhoea also had antibodies to canine parvovirus [11]. The true role of *Campylobacter* in canine gastroenteritis is uncertain as the bacteria may be found in clinically healthy dogs or often as a co-infection or intestinal carriage in diseased dogs.

In most studies, the predominant *Campylobacter* species isolated from dogs is *C. upsaliensis* and dogs are regarded as an important reservoir for this species [3-6,9,12]. The second most common *Campylobacter* species isolated from dogs, in many populations, is *C. jejuni* [3,4,6,9,12], which is also responsible for the majority of human infections [13,14]. The reported prevalence of *Campylobacter* in dogs varies widely between studies, ranging from 22% to 100% and is reported to depend on factors such as the age, diet and housing of the dogs. Previous prevalence investigations of *Campylobacter* in dogs have also varied by study design and method of bacterial isolation.

Human campylobacteriosis is frequently attributed to contact with contaminated food (especially poultry meat) or water [15,16]. In several studies, direct contact with pet animals has also been identified as a possible source of human *Campylobacter* infection [15,17-19]. Presence of a puppy in the household has been identified as risk factor for campylobacteriosis, especially in young children [20,21]. In a study by Wolfs *et al.* [19] evidence was presented for transmission of *C. jejuni* from a dog to a 3-week old infant. However, a study by Studahl and Andersson [16] did not find a significant association between human campylobacteriosis and contact with dogs.

The present survey is part of an ongoing more comprehensive *Campylobacter* source attribution study in Sweden. Strains from human cases of campylobacteriosis were collected during the same time period as samples were taken from dogs, cattle, pigs, sheep, poultry and wild birds. Other relevant sources of human campylobacteriosis such as retail poultry meat, raw water and bathing water were also sampled during this period.

The aim of this study was to update our knowledge on the prevalence of *Campylobacter* spp. in young dogs in Sweden by collecting samples from healthy, young dogs throughout a year. The aim was also to identify the species of the *Campylobacter* isolates and to genotype the *C. jejuni* isolates by multi-locus sequence typing (MLST) to enable comparison between *C. jejuni* isolates from dogs and humans.

METHODS

Study Population and Sampling

Selection of clinics to participate in the study

The sampling of dogs was performed at veterinary clinics throughout Sweden from October 2011 through October 2012. The geographical regions (counties) to be included in the survey were selected depending on the number of reported human cases of campylobacteriosis in recent years (based on data from the Public Health Agency of Sweden, www.folkhalsomyndigheten.se). The counties with the highest incidence of human campylobacteriosis were prioritised but the aim was also to cover a large part of the country. Requests for participation in the sampling process were sent to 53 veterinary clinics that had a previously established contact with the laboratory at Department of Bacteriology at the National Veterinary Institute (SVA, www.sva.se). A selection of 11 was made from the 18 veterinary clinics that had agreed upon participation in the study. One of them, located in the north of Sweden was included in the survey despite the lower incidence of human campylobacteriosis in favour of the better geographic coverage.

Sampling procedure

The requirements for dogs to be sampled were that they were under the age of two, weaned and healthy with no signs of diarrhoea. Young dogs were targeted because the prevalence of *Campylobacter* spp. is likely to be highest in young animals [4,6,22]. The aim was to collect a total of 200 samples in the study in order to obtain approximately 100 isolates. Based on a previous study in Sweden by Engvall *et al.* [4] an overall isolation rate of *Campylobacter* spp. of around 50% was expected. The clinics were instructed to sample one to two dogs per month. Each dog was sampled only once during the sampling period. Faecal samples were collected from freshly voided faeces and sent to SVA on swabs in Amies transport medium with charcoal (Amies agar gel swabs – with charcoal, Copan, Italy). The swab samples were sent to SVA by ordinary mail at the day of sampling for isolation of *Campylobacter*, species identification and genetic subtyping of *C. jejuni* isolates. Written consent of the animal owners was obtained for sampling along with details about the age of the dog and postal code of the owner.

Isolation and species identification of Campylobacter

Swab samples were cultured on modified charcoal, cefoperazone, desoxycholate agar (mCCDA), (Oxoid Inc, Basingstoke, Hampshire, UK) and incubated at 41. 5±1.0°C for up to 5 days in a microaerobic atmosphere created by use of Campygen (Oxoid Inc) or Anoxomat (Advanced Instruments, Inc., Norwood, Massachusetts, USA). Preliminary identification of *Campylobacter* spp. was based on colony and microscopic morphology and the following phenotypic and biochemical characteristics and tests; motility, oxidase-, catalase-, hippurate and indoxyl acetate reactions. Strains confirmed as *Campylobacter* spp. were stored at −80°C until further identification. Identification of *C. jejuni* was mainly based on a positive hippurate test. The species identification of all hippurate-negative isolates was done by mass spectrometry, time of flight, Maldi-Tof [23]. All isolates that were identified as *C. upsaliensis* by Maldi-Tof were further tested by polymerase chain reaction (PCR) for confirmation [24,25].

Genotyping of *C. jejuni*

MLST was carried out as previously described by Dingle *et al.* [26] on the isolates identified as *C. jejuni*. Alleles, sequence type (ST) and clonal complex were assigned using the pubMLST database (http://pubmlst.org/campylobacter). Sequence types that shared four or more alleles were considered to belong to the same clonal complex. Because the present study is part of a larger comparative study of *C. jejuni* between various sources in Sweden, genotyping was restricted to this *Campylobacter* species.

RESULTS

In total, 180 dogs were sampled from the 11 participating veterinary clinics. The sampling period was from October 2011 to October 2012; however, two samples received in November 2012 were also included in the study. For practical reasons, the number of samples received per month varied between the participating clinics. One clinic only sent in samples from one month. The location of the clinics and number of sampled and positive dogs per clinic is illustrated in Figure 1. One of the positive samples lacked information about clinic on the referral form. Most samples were received during the first seven months of the sampling period (October – April). The highest proportion of positive samples occurred in the winter months with a peak in March 2012 (57%). The number of samples per month and positive samples per month are shown in Figure 2.

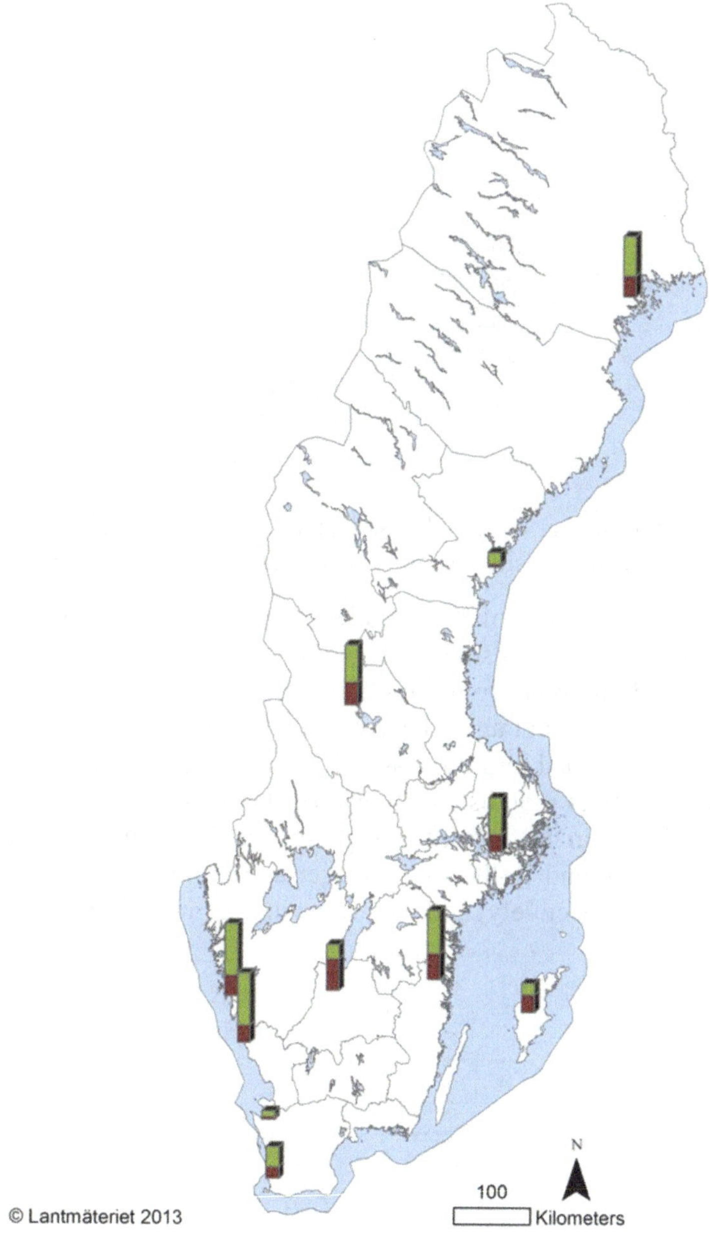

Figure 1: Spatial distribution of veterinary clinics where dogs were sampled. The height of the bars represents the number of sampled dogs (red = positive samples, green = negative samples).

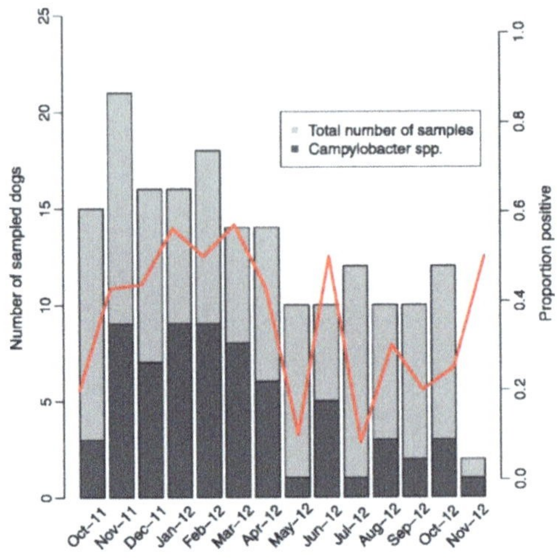

Figure 2: Total number of samples and samples with *Campylobacter* species per month. Proportion positive samples is indicated by the red line.

The age distribution of the sampled dogs ranged from one month to 11 years. The aim was to sample dogs under the age of two, although 17 samples turned out to be from dogs between 2 and 11 years and 9 samples lacked information about age. The average and median age was 12 months. The dogs were divided into three age groups; 1) under 12 months, 2) 12 to 23 months and 3) 24 months and older. Number of samples and prevalence of *Campylobacter* in the different age groups are shown in Figure 3. Highest prevalence was found in the two younger age groups (37% and 40% for group 1 and 2, respectively) and the lowest in age group 3 (12%).

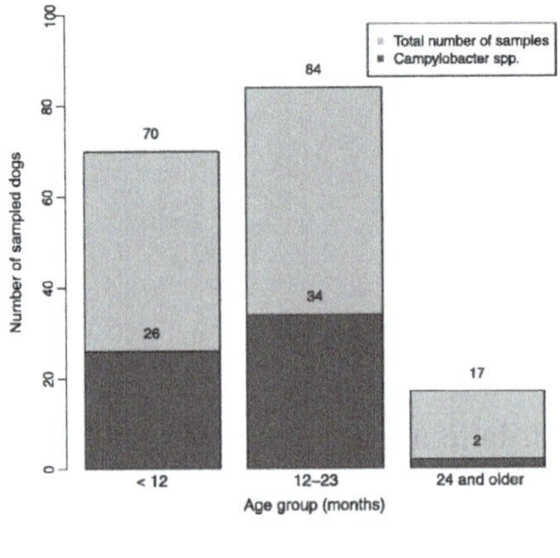

Figure 3: Number of dogs in different age groups that tested positive or negative for *Campylobacter* species.

Campylobacter spp. were isolated from 67 of the 180 sampled dogs which yields an overall prevalence of 37% in this material. The most common species of *Campylobacter* isolated from the dogs was *C. upsaliensis* with 52 of the 67 identified isolates (78%). A lower occurrence of *C. jejuni* was observed, with 7 identified isolates (9%) and 1 isolate (1%) was identified as *C. helveticus*. All *C. jejuni* samples were found in dogs up to 12 months of age. In addition there were 7 isolates of *Campylobacter* species that did not survive the storage process, prior to the speciation. These isolates were all hippurate negative which indicates that they were most likely not *C. jejuni*, but could not be identified to species level by Maldi-Tof or PCR.

Results from the MLST are shown in Table 1. The dogs shedding *C. jejuni* were sampled at different locations except for two that were sampled in the same clinic. All seven isolates had different ST types. However, two dogs, sampled at different locations, were shedding *C. jejuni* of the same clonal complex, ST Clonal complex 21.

Table 1: Results from multi-locus sequence typing (MLST) of *Campylobacter* isolates from dogs

Dog	Age (months)	Location of veterinary clinic	County of Sweden	MLST	MLST-complex
1	12	Västra Frölunda	Västra Götaland	50	ST-21 complex
2	4	Jönköping	Jönköping	883	ST-21 complex
3	12	Jönköping	Jönköping	52	ST-52 complex
4	4	Ängelholm	Skåne	45	ST-45 complex
5	3	Varberg	Halland	122	ST-206 complex
6	12	Gamleby	Kalmar	257	ST-257 complex
7	4	Gammelstad	Norrbotten	677	ST-677 complex

DISCUSSION

In this study, 37% of the tested dogs were found to be positive for *Campylobacter*. This proportion is similar to the previous prevalence estimates of other studies [6,9,27]. However, as we targeted young dogs and the mean age of the sampled dogs was 12 months this estimated prevalence may not be representative for the entire Swedish dog population. Moreover, the study design was such that each dog was only sampled once and only one isolate per dog was analysed. In longitudinal studies with another type of study design where dogs were sampled at several occasions considerably higher prevalences of 73-100% have been reported [3,28]. The most common *Campylobacter* species among the tested dogs (52/180 dogs) was by far *C. upsaliensis* which is in agreement with many previous studies on dogs where relatively high prevalences of *C. upsaliensis* have been found in many different populations and countries [5,7,9,22,28]. The lower prevalence (4%) that was observed for *C. jejuni* (7/180 dogs) is similar to some studies [12,27,29,30], but lower than in a previous Swedish study by Engvall *et al.* [4] where 11% of the dogs were positive for *C. jejuni*. However, in the previous study all samples were cultured on three selective plates (two CAT and one Preston agar plate). This probably contributed to the higher isolation rate compared to the present study. A recent study in Switzerland by Amar *et al.* [30] found a low prevalence of *C. jejuni* (6.3%) and an even lower prevalence (5.9%) of *C. upsaliensis* in healthy dogs. A higher prevalence of *C. jejuni* has been reported in stray dogs [10,28], which may be due to a different exposure to environmental sources of *Campylobacter* than what household dogs are normally exposed to. In a study from Spain, 19% (20/105) of dogs under 2 years of age living in a household, were found to have *C. jejuni* [22]. In the present study, one sample was identified as *C. helveticus* which is not so often identified in samples from dogs, but has previously been associated with cats [29].

Although *C. upsaliensis* is not isolated very frequently from humans in routine investigations it is possible that available data underestimate the prevalence due to the methods that are used at the public health laboratories which are primarily developed to detect *C. jejuni* and *C. coli*. In Sweden and many countries in Europe, human clinical *Campylobacter* isolates are not identified to species level which might contribute to a lack of data regarding the prevalence of *C. upsaliensis* in humans [31]. The role of *C. upsaliensis* in human disease is not very well established but it has been shown that *C. upsaliensis* can be a cause of gastroenteritis in both adults and children [32,33]. Labarca *et al.* [34] found that *C. upsaliensis* was the second most frequently isolated species in humans after *C. jejuni*. The authors also found that three dogs living in the households of two human patients infected with *C. upsaliensis* had the same *Campylobacter* species isolated in their stool specimens, but were not from the same clonal complex [34]. A study in Belgium reported that an outbreak in four day care centres was caused by *C. upsaliensis* [35]. Damborg *et al.* [36] found that a cluster of human *C. upsaliensis* strains was unrelated to dog strains of *C. upsaliensis* examined in the study by AFLP fingerprinting. However, the human and dog samples were not collected in the same countries which makes it difficult to interpret the results with regard to host specificity.

The age of each dog at sample collection was recorded and the dogs were divided into three age categories for comparison of prevalences. The results were in agreement with many other studies that have reported higher prevalence of *Campylobacter* in younger dogs or puppies compared with adult dogs [6,9,12,22,27,37]. These results suggest an age predisposition where young dogs are more susceptible to colonisation, possibly due to the development of immunity with age. Senior dogs have also been found to be at risk for *Campylobacter* colonisation [27,28]. Wieland *et al.* [29] found a significant association with age and *C. upsaliensis* but no association between age and presence of *C. jejuni*. Dogs shedding *C. jejuni* in the present study were all young, between 3–12 months while dogs shedding *C. upsaliensis* were found in all age categories, which is similar to what was described by Hald *et al.* [3].

Seasonality that is observed in human campylobacteriosis with peaks during the summer months [31] was not observed in this material as the highest incidence was observed in the winter months with a peak in March. Due to the low number of samples per month it is not possible to draw any conclusions regarding seasonality based on our findings. Carbonero *et al.* [22] found a significantly higher prevalence of *C. jejuni* in dogs during spring compared to winter. They also reported a higher prevalence of *C. upsaliensis* during the summer. Hald *et al.* [3] did not find seasonal variation in carrier rates among dogs in a longitudinal study.

Molecular typing techniques enable comparison of sequence types (STs) between humans and the potential source of *Campylobacter*. In our study, seven *C. jejuni* samples were isolated and further subtyped by MLST. The results show high heterogeneity as all isolates were of different STs and only two isolates (from dogs sampled in different areas) belonged to the same clonal complex, ST-21. Since only 7 *C. jejuni* isolates were identified and typed, the true extent of the *C. jejuni* population diversity in Swedish dogs cannot be estimated from this material. Mughini Gras *et al.* [21] showed a high degree of overlap between human and pet (dog and cat) *C. jejuni* STs. They identified two cases where identical *C. jejuni* STs (ST45 and ST658) were isolated from dogs and their owners. Four of the STs identified in the present study (ST50, ST45, ST257 and ST122) belong to those more frequently found STs in pet owners in the study by Mughini Gras *et al.* [21]. Manning *et al.* [38] found that a majority of the identified *C. jejuni* ST-complexes overlap between human and various animal sources. When comparing to the *Campylobacter* MLST database of *C. jejuni* isolates (http://pubmlst.org/campylobacter) all seven STs that were found are common human STs. In the study by Amar *et al.* [30] the most frequent STs identified in dogs include ST45 and ST21 that was found also in the present study. Parsons *et al.* [28] also reported that ST45 was the most common complex in dogs at a rescue kennel. Many studies have identified ownership or close contact with dogs as a potential risk factor for human campylobacteriosis. However, despite the use of molecular information it is often difficult to know whether the bacteria is transferred from dog to human or human to dog or was acquired from the same common source.

In general, it is likely that dogs and humans are exposed to common sources of *C. jejuni*. The relatively low prevalence of *C. jejuni* in dogs in this study suggests that the importance of dogs for human infections compared with other sources such as food products may be low. However, dogs and in particular puppies are likely to have close contact with their owner and children in a family. Considering this it is important to note that all *C. jejuni* positive dogs were puppies or young dogs up to 12 months old. As the prevalence of *C. upsaliensis* in dogs under two years of age is relatively high there is a risk for transmission of zoonotic *Campylobacter* from dogs to humans and especially young children that are more susceptible for infection.

CONCLUSION

The present investigation finds that *Campylobacter* spp. known to cause campylobacteriosis in humans are present in Swedish dogs. The results suggest an age predisposition where dogs under 2 years of age are more likely to shed *Campylobacter* spp. than older dogs. The most commonly isolated species was *C. upsaliensis* followed by *C. jejuni,* which was only detected in dogs up to 12 months of age. All *C. jejuni* isolates identified in the present study were the same MLST types as had been previously described both in humans and in animals. Further investigation is necessary to determine the similarity between these dog *C. jejuni* MLST types and those found in humans in Sweden during the same period. The awareness of the *Campylobacter* risk of healthy young dogs may be an important way to reduce the transmission from dogs to infants, young children and immunocompromised adults.

ACKNOWLEDGEMENTS

We are grateful to all dog owners who let their dogs be part of the study, we also gratefully acknowledge the veterinary clinics for their collaboration in the project. We would like to thank Mattias Myrenås, Boel Harbom and Ninni Pudas for their work in the lab with MLST, PCR and Maldi-tof. We would also like to thank the personnel at the Department of Bacteriology at the National Veterinary Institute for their work with analysing all the samples for presence of *Campylobacter*. This study was funded by the Swedish research council Formas.

AUTHORS' CONTRIBUTIONS

AL designed and coordinated the study, EE participated in the design of the study and coordinated the laboratory work, AO was responsible for selection of clinics and the sampling process, MH was the main author of the manuscript and performed the data analysis, TR participated in the data analysis and the preparation and editing of the manuscript. All authors have read and approved the final manuscript.

REFERENCES

1. EFSA, ECDC The European Union Summary Report on Trends and Sources of Zoonoses, Zoonotic Agents and Food-borne Outbreaks in 2011. EFSA J. 2013;11(4):3129.
2. Folkhälsomyndigheten. http://folkhalsomyndigheten.se/amnesomraden/statistik-och-undersokningar/sjukdomsstatistik/campylobacterinfektion/. 2014.
3. Hald B, Pedersen K, Waino M, Jorgensen JC, Madsen M. Longitudinal study of the excretion patterns of thermophilic *Campylobacter* spp. in young pet dogs in Denmark. J Clin Microbiol. 2004;42(5):2003–12. doi: 10.1128/JCM.42.5.2003-2012.2004.
4. Engvall EO, Brandstrom B, Andersson L, Baverud V, Trowald-Wigh G, Englund L. Isolation and identification of thermophilic *Campylobacter* species in faecal samples from Swedish dogs. Scand J Infect Dis. 2003;35(10):713–8. doi: 10.1080/00365540310014558.
5. Chaban B, Ngeleka M, Hill JE. Detection and quantification of 14 *Campylobacter* species in pet dogs reveals an increase in species richness in feces of diarrheic animals. BMC Microbiol. 2010;10:73. doi: 10.1186/1471-2180-10-73.
6. Parsons BN, Porter CJ, Ryvar R, Stavisky J, Williams NJ, Pinchbeck GL, et al. Prevalence of *Campylobacter* spp. in a cross-sectional study of dogs attending veterinary practices in the UK and risk indicators associated with shedding. Veterinary J. 2010;184(1):66–70. doi: 10.1016/j.tvjl.2009.01.009.
7. Weese JS. Bacterial Enteritis in Dogs and Cats: Diagnosis, Therapy, and Zoonotic Potential. Vet Clin North Am Small Anim Pract. 2011;41(2):287–309. doi: 10.1016/j.cvsm.2010.12.005.
8. Sandberg M, Bergsjo B, Hofshagen M, Skjerve E, Kruse H. Risk factors for *Campylobacter* infection in Norwegian cats and dogs. Prev Vet Med. 2002;55(4):241–53. doi: 10.1016/S0167-5877(02)00095-8.
9. Acke E, McGill K, Golden O, Jones BR, Fanning S, Whyte P. Prevalence of thermophilic *Campylobacter* species in household cats and dogs in Ireland. Vet Rec. 2009;164(2):44–7. doi: 10.1136/vr.164.2.44.
10. Workman SN, Mathison GE, Lavoie MC. Pet dogs and chicken meat as reservoirs of *Campylobacter* spp. in Barbados. J Clin Microbiol. 2005;43(6):2642–50. doi: 10.1128/JCM.43.6.2642-

2650.2005.

11. Sandstedt K, Wierup M. Concomitant occurrence of *Campylobacter* and parvoviruses in dogs with gastroenteritis. Vet Res Commun. 1981;4(4):271–3.
12. Leonard EK, Pearl DL, Janecko N, Weese JS, Reid-Smith RJ, Peregrine AS, et al. Factors related to *Campylobacter* spp. carriage in client-owned dogs visiting veterinary clinics in a region of Ontario, Canada. Epidemiology & Infection. 2011;139(Special Issue 10):1531–41. doi: 10.1017/S0950268810002906.
13. Man SM. The clinical importance of emerging *Campylobacter* species. Nat Rev Gastroenterol Hepatol. 2011;8(12):669–85. doi: 10.1038/nrgastro.2011.191.
14. Folkhälsomyndigheten. http://www.folkhalsomyndigheten.se/documents/statistik-uppfoljning/smittsamma-sjukdomar/2012/brucellos-epidemiologisk-arsrapport-2012-2013-101-8.pdf. 2013.
15. Neimann J, Engberg J, Molbak K, Wegener HC. A case–control study of risk factors for sporadic campylobacter infections in Denmark. Epidemiol Infect. 2003;130(3):353–66.
16. Studahl A, Andersson Y. Risk factors for indigenous campylobacter infection: a Swedish case–control study. Epidemiol Infect. 2000;125(2):269–75. doi: 10.1017/S0950268899004562.
17. Carrique-Mas J, Andersson Y, Hjertqvist M, Svensson A, Torner A, Giesecke J. Risk factors for domestic sporadic campylobacteriosis among young children in Sweden. Scand J Infect Dis. 2005;37(2):101–10. doi: 10.1080/00365540510027165.
18. Damborg P, Olsen KE, Moller Nielsen E, Guardabassi L. Occurrence of *Campylobacter jejuni* in pets living with human patients infected with *C. jejuni*. J Clin Microbiol. 2004;42(3):1363–4. doi: 10.1128/JCM.42.3.1363-1364.2004.
19. Wolfs TF, Duim B, Geelen SP, Rigter A, Thomson-Carter F, Fleer A, et al. Neonatal sepsis by *Campylobacter jejuni*: genetically proven transmission from a household puppy. Clin Infect Dis. 2001;32(5):E97–9. doi: 10.1086/319224.
20. Tenkate TD, Stafford RJ. Risk factors for campylobacter infection in infants and young children: a matched case–control study. Epidemiol Infect. 2001;127(3):399–404. doi: 10.1017/S0950268801006306.
21. Gras LM, Smid JH, Wagenaar JA, Koene MG, Havelaar AH, Friesema

IH et al. Increased risk for *Campylobacter jejuni* and *C. coli* infection of pet origin in dog owners and evidence for genetic association between strains causing infection in humans and their pets. Epidemiol Infect. 2013:1–10. doi:10.1017/S0950268813000356.

22. Carbonero A, Torralbo A, Borge C, Garcia-Bocanegra I, Arenas A, Perea A. *Campylobacter* spp., *C. jejuni* and *C. upsaliensis* infection-associated factors in healthy and ill dogs from clinics in Cordoba, Spain. Screening tests for antimicrobial susceptibility. Comp Immunol Microbiol Infect Dis. 2012;35(6):505–12. doi: 10.1016/j.cimid.2012.05.002.
23. Bessede E, Solecki O, Sifre E, Labadi L, Megraud F. Identification of *Campylobacter* species and related organisms by matrix assisted laser desorption ionization-time of flight (MALDI-TOF) mass spectrometry. Clin Microbiol Infect. 2011;17(11):1735–9. doi: 10.1111/j.1469-0691.2011.03468.x.
24. Lawson AJ, Linton D, Stanley J, Owen RJ. Polymerase chain reaction detection and speciation of *Campylobacter upsaliensis* and *C. helveticus* in human faeces and comparison with culture techniques. J Appl Microbiol. 1997;83(3):375–80. doi: 10.1046/j.1365-2672.1997.00240.x.
25. Denis M, Soumet C, Rivoal K, Ermel G, Blivet D, Salvat G, et al. Development of a m-PCR assay for simultaneous identification of *Campylobacter jejuni* and *C. coli.* Lett Appl Microbiol. 1999;29(6):406–10. doi: 10.1046/j.1472-765X.1999.00658.x.
26. Dingle KE, Colles FM, Wareing DR, Ure R, Fox AJ, Bolton FE, et al. Multilocus sequence typing system for *Campylobacter jejuni*. J Clin Microbiol. 2001;39(1):14–23. doi: 10.1128/JCM.39.1.14-23.2001.
27. Procter TD, Pearl DL, Finley RL, Leonard EK, Janecko N, Reid-Smith RJ, et al. A Cross-Sectional study examining *Campylobacter* and other zoonotic enteric pathogens in dogs that frequent dog parks in three cities in South-Western Ontario and risk factors for shedding of *Campylobacter* spp. Zoonoses Public Health. 2013. doi:10.1111/zph.12062.
28. Parsons BN, Williams NJ, Pinchbeck GL, Christley RM, Hart CA, Gaskell RM, et al. Prevalence and shedding patterns of *Campylobacter* spp. in longitudinal studies of kennelled dogs. Vet J. 2011;190(2):249–54. doi: 10.1016/j.tvjl.2010.10.006.

29. Wieland B, Regula G, Danuser J, Wittwer M, Burnens AP, Wassenaar TM, et al. *Campylobacter* spp. in dogs and cats in Switzerland: Risk factor analysis and molecular characterization with AFLP. J Vet Med Ser B-Infect Dis Vet Public Health. 2005;52(4):183–9. doi: 10.1111/j.1439-0450.2005.00843.x.

30. Amar C, Kittl S, Spreng D, Thomann A, Korczak BM, Burnens AP, et al. Genotypes and antibiotic resistance of canine *Campylobacter jejuni* isolates. Vet Microbiol. 2014;168(1):124–30. doi: 10.1016/j.vetmic.2013.10.006.

31. ECDC. Annual Epidemiological Report 2012 . Reporting on 2010 surveillance data and 2011 epidemic intelligence data. Stockholm: ECDC; 2013.

32. Patton CM, Shaffer N, Edmonds P, Barrett TJ, Lambert MA, Baker C, et al. Human disease associated with "*Campylobacter upsaliensis*" (catalase-negative or weakly positive *Campylobacter* species) in the United States. J Clin Microbiol. 1989;27(1):66–73.

33. Taylor DE, Hiratsuka K, Mueller L. Isolation and characterization of catalase-negative and catalase-weak strains of *Campylobacter* species, including «*Campylobacter upsaliensis*," from humans with gastroenteritis. J Clin Microbiol. 1989;27(9):2042–5.

34. Labarca JA, Sturgeon J, Borenstein L, Salem N, Harvey SM, Lehnkering E, et al. *Campylobacter upsaliensis*: Another pathogen for consideration in the United States. Clin Infect Dis. 2002;34(11):E59–60. doi: 10.1086/340266.

35. Goossens H, Giesendorf BA, Vandamme P, Vlaes L, Van den Borre C, Koeken A, et al. Investigation of an outbreak of *Campylobacter upsaliensis* in day care centers in Brussels: analysis of relationships among isolates by phenotypic and genotypic typing methods. J Infect Dis. 1995;172(5):1298–305. doi: 10.1093/infdis/172.5.1298.

36. Damborg P, Guardabassi L, Pedersen K, Kokotovic B. Comparative analysis of human and canine *Campylobacter upsaliensis* isolates by amplified fragment length polymorphism. J Clin Microbiol. 2008;46(4):1504–6. doi: 10.1128/JCM.00079-08.

37. Westgarth C, Porter CJ, Nicolson L, Birtles RJ, Williams NJ, Hart CA, et al. Risk factors for the carriage of *Campylobacter upsaliensis* by dogs in a community in Cheshire. Vet Rec. 2009;165(18):526–30. doi: 10.1136/vr.165.18.526.

38. Manning G, Dowson CG, Bagnall MC, Ahmed IH, West M, Newell DG. Multilocus sequence typing for comparison of veterinary and human isolates of *Campylobacter jejuni*. Appl Environ Microbiol. 2003;69(11):6370–9. doi: 10.1128/AEM.69.11.6370-6379.2003.

SECTION 5:
FISH

CHAPTER 17

Taxonomy of Bacterial Fish Pathogens

Brian Austin

Institute of Aquaculture, Pathfoot Building, University of Stirling, Stirling FK9 4LA, Scotland, UK

ABSTRACT

Bacterial taxonomy has progressed from reliance on highly artificial culture-dependent techniques involving the study of phenotype (including morphological, biochemical and physiological data) to the modern applications of molecular biology, most recently 16S rRNA gene sequencing, which gives an insight into evolutionary pathways (= phylogenetics). The latter is applicable to culture-independent approaches, and has led directly to the recognition of new uncultured bacterial groups, i.e. "*Candidatus*", which have been associated as the cause of some fish diseases, including rainbow trout summer enteritic syndrome. One immediate benefit is that 16S rRNA gene sequencing has led to increased confidence in the accuracy of names

Citation: Austin, B. Taxonomy of bacterial fish pathogens. Vet Res 42, 20 (2011). https://doi.org/10.1186/1297-9716-42-20

allocated to bacterial pathogens. This is in marked contrast to the previous dominance of phenotyping, and identifications, which have been subsequently challenged in the light of 16S rRNA gene sequencing. To date, there has been some fluidity over the names of bacterial fish pathogens, with some, for example *Vibrio anguillarum*, being divided into two separate entities *(V. anguillarum* and *V. ordalii)*. Others have been combined, for example *V. carchariae, V. harveyi* and *V. trachuri* as *V. harveyi*. Confusion may result with some organisms recognized by more than one name; *V. anguillarum* was reclassified as *Beneckea* and *Listonella*, with *Vibrio* and *Listonella* persisting in the scientific literature. Notwithstanding, modern methods have permitted real progress in the understanding of the taxonomic relationships of many bacterial fish pathogens.

INTRODUCTION

"What's in a name?" (William Shakespeare; Romeo and Juliet)

The Swedish botanist Carl Linnaeus (1707-1778), who was also known as Carolus Linnaeus and Carl von Linné, is undoubtedly the Father of Taxonomy, and was responsible for developing a system for naming and ranking living organisms. His lasting contribution was the development of a simplified naming system in Latin with consistency across all living organisms, i.e. the binomial system, in which each organism has a unique two-word name - incorporating genus and species. A simplistic view is that Linnaeus made order out of chaos. Yet, for Linnaeus and his contemporaries, the process was comparatively easy, and involved only large organisms, which were clearly visible to the naked eye (= macro-organisms) and easily seen morphological characteristics (= a category of phenotypic characters). Thus, these early classifications (= the process of arranging organisms into groups) were based on limited but easily visible data, and the outcomes were largely obvious, for example a dog is notably different from a horse and would therefore belong in separate species.

The founding father of microbiology, the Dutch textile merchant and lens maker, Antonie van Leeuwenhoek (1632-1723), observed small organisms initially from the proximity of his teeth (= bacteria and protozoa?), and these entities were termed "animalcules", which he wrote about in a letter to the Royal Society of London in 1676. His careful illustrations suggested morphological variation between the cells. Yet, another two centuries were to pass before serious attempts at naming and ordering bacteria started. Thus, bacterial taxonomy has progressed from the simplistic approach involving a small number of readily observable characteristics, such as morphology as deduced from observation using light microscopes, to the modern applications of molecular biology. With improvements in knowledge, there have been refinements in taxonomic processes and an increase in reliability. It should be remembered that taxonomy (= the theory of classification, nomenclature and identification) is a man-made process, i.e. the organisms included in any classification have not chosen to be placed in the groups that have been created by human beings. Nevertheless if done properly, taxonomy has value in:

- Understanding biodiversity, namely the range of organisms in a given habitat
- Communication between scientists, thus enabling exchange of information about similar organisms
- Cataloguing information - the name is the key to a catalogue of information about the organism
- Enabling identification, such that new isolates may be readily and reliably identified
- Providing an insight into evolutionary pathways (=phylogenetics).

To be effective, taxonomy should be

- based on a high information content
- reproducible, and
- stable,

otherwise confusion will surely result.

Since the start of bacterial taxonomic processes in the nineteenth century, there has been a progression in the type of information used in the procedure. It may be argued that early bacteriologists had considerable taxonomic insight judging from the conclusions reached from the comparatively simple data that were available. However, taxonomy is a dynamic science, with new developments/methods being incorporated into processes including the descriptions of bacterial species. Since the 1950 s, bacterial taxonomy has evolved rationally, encompassing numerical methods [1,2], chemotaxonomy (e.g. [3,4]), and molecular techniques [5]. Taxonomy has progressed from a highly artificial process involving limited amounts of phenotypic data to the recognition of more natural relationships between organisms, based on comparatively large amounts of varied and reliable data covering multiple aspects of the biology of an organism, and including phenotypic, chemotaxonomic, genotypic and phylogenetic date, i.e. a polyphasic approach. However, the current dominance of 16S rRNA gene sequencing although revolutionising some aspects of bacterial classification needs to be treated cautiously as overreliance on the approach may lead to erroneous conclusions [5]. Nevertheless, it is apparent that sequencing methods are instrumental with the explosion of new species names, which have greeted the arrival of the twenty-first century. Whereas, the information content of many of the new species descriptions is generally high, an unwelcome trend is that many new taxa (= taxonomic groups) are described solely after the study of only single strains. Therefore, the diversity/variability within the new taxon cannot be adequately assessed. Also, it is impossible to determine whether a single strain is effectively an outlier or a median representative of the group (in future years, will it be regarded as typical or atypical of the group?). However, taxonomy is often ignored by many microbiologists in other specialisms, and there may well be concern that basic principles could be forgotten, e.g. is the purity and authenticity of cultures always verified before use? Where culturing is not possible, there is the possibility of analyzing the nucleic acids, determining species composition, and even proposing new taxa, i.e. by the use of culture-independent approaches.

BACTERIAL FISH PATHOGENS

There has been a steady increase in the numbers of bacterial species associated with fish diseases, with new pathogens regularly recognised in the scientific literature [6]. However, the names of many bacterial fish pathogens have been subjected to taxonomic change over time, with some species split, for example *Vibrio anguillarum* biotype 2 becoming re-classified as a separate species *V. ordalii* [7,8]. In other cases, different nomenspecies have been combined, for example *V. carchariae, V. harveyi* and *V. trachuri* into *V. harveyi*, which had precedence because it was the first name to be proposed, albeit as the luminous *Achromobacter harveyi* [9-11]. The oldest known fish pathogen, *V. anguillarum*, has undergone name changes to *Beneckea* [12] and *Listonella* [13]; neither of which was widely accepted. However, *Listonella* remains a valid name and is mentioned in the current edition of Bergey's Manual of Systematic Bacteriology, and *Beneckea* has been consigned to the history books. A positive aspect of sequencing methods is that there has been a progression towards the Orwellian notion of "Order out of Chaos" even if scientists do not always appreciate the significance of the data.

ISOLATION OF FISH PATHOGENS: THE CULTURE-DEPENDENT APPROACH

With the rapid development in molecular biology, it is not always necessary to culture an organism in order to enable its study, including the allocation of a species name. Thus, the concept of culture-independent techniques was developed and refined. Sensitivity and specificity increased, but without culturing there was an inability to carry out associated studies, such as the determination of pathogenicity factors. The attraction of culture-dependent approaches is that a pure culture may be obtained and deposited in culture collections as reference material for use by others. This raises a concern about the usefulness of cultures. An assumption is made that pathological material may be used for the recovery of a pure culture of the aetiological agent. This will depend on using appropriate media and incubation conditions, and assumes that the organism is in a form that may be cultured and that the microbiologist picks the "correct" colony. If mixed growth occurs or if the pathogen is largely overgrown by opportunists/ secondary invaders/saprophytes, then there is concern that the actual pathogen will be missed. In addition, infections resulting from two or more organisms working synergistically will undoubtedly be mis-diagnosed if the diagnostician chooses only one culture for study. However, there are only a comparatively few indications of disease resulting from multiple species, such as *Aeromonas hydrophila* with *A. salmonicida* [6]. It is speculative how many diagnoses (if any) are made of contaminants rather than the actual pathogen. Moreover, it is surprising that only two species of anaerobic bacteria, namely *Clostridium botulinum* and *Eubacterium tarantellae*, have been implicated as fish pathogens [6]. Of course, this could reflect the general lack of use of appropriate anaerobic procedures by microbiologists rather than the absence of anaerobic pathogens.

APPROACHES TO CHARACTERIZATION

Phenotype

Traditionally, bacteria were characterized phenotypically, and undoubtedly for some groups, e.g. the Enterobacteriaceae, a wealth of knowledge emerged particularly from the 1950's onwards. Currently, emphasis on phenotype has declined with a concomitant move towards molecular-based approaches. Nevertheless, phenotypic data have a role in polyphasic studies whereby many facets of the biology of an organism are studied

[14]. Phenotyping leads the way with diagnoses worldwide; emphasis often being placed on commercial kits and the use of manufacturer's probabilistic databases to achieve an acceptable identification. Although the approach has standardized diagnoses, the weakness is that most identification systems have been developed for medically important bacteria that grow within 24-48 h at 35-37°C. Consequently, the reliability of these kits for use with fish pathogens which need lower incubation temperatures for longer periods needs to be questioned [6,15]. In particular, the API 20E system includes the use of sugar fermentation reactions, which may be influenced by the presence of plasmids [6]. Moreover, there may be confusion over the interpretation of the profiles. For example, some of the profiles of *A. hydrophila* are similar to those of *A. allosaccharophila* and *A. sobria; Yersinia ruckeri* may be confused with *Hafnia alvei;* moreover *Tenacibaculum maritimum* and *Pseudomonas anguilliseptica* are indistinguishable by API 20E [6]. Problems may result when data from rapid commercial kits are used in conjunction with conventional diagnostic schemes, which have been developed for traditional and often laborious phenotypic characters. Also, some of the traditional tests, e.g. the Voges Proskauer reaction, are not noted for their reproducibility and may introduce errors into the taxonomic process and lead to mis-identification [16]. Undoubtedly, selective media have proved useful for the recovery of some fish pathogens, with an example including selective kidney disease medium (SKDM) for *Renibacterium salmoninarum* [17]. However, selective media are only available for a minority of all fish pathogens, therefore recovery is dependent on more general culturing methods. Specially developed diagnostic procedures have aided identification of some group, e.g. the glucose motility deep cultures have benefitted the recovery and identification of *V. anguillarum* [18].

Immunological Methods

The development and availability of standardized immunological (antibodies and kits) reagents have improved diagnoses considerably [6,19,20], and enhanced the reliability of methods for the detection of pathogens, including *Mycobacterium* spp., *Photobacterium damselae* subsp. *piscicida* [21], *Piscirickettsia salmonis* [22], *R. salmoninarum* [23] and *Streptococcus iniae* [24]. Tentative diagnoses, including of asymptomatic infections, may result from use of monospecific polyclonal or monoclonal antibodies in a range of antibody-based procedures, including the indirect fluorescent antibody test (iFAT), whole-cell (slide) agglutination, precipitin reactions, complement fixation, immunodiffusion, antibody-coated latex particles, co-agglutination using antibody-coated staphylococcal cells, passive haemagglutination, immuno-India ink technique (Geck) or enzyme linked immunosorbent assay (ELISA; reviewed by [25]), the latter of which may also be used for serology, i.e. detecting antibodies in the host to specific pathogens [26]. Antibody-based methods are used effectively for detecting exposure to fish viruses, such as Koi herpes virus [27], but bacterial pathogens pose a more complicated picture with cross reactivities likely unless specific known molecules/antigens are used to coat the ELISA plates rather than whole pathogens [19]. Techniques are often sensitive, specific, rapid and reliable, and in some cases may be used in the field [6]. This is in marked contrast to molecular biology, which may be much slower and relies on specialist, well equipped laboratories.

Chemotaxonomy

Chemotaxonomy involves the investigation of chemical constituents of bacteria, and is particularly useful for the study of Gram-positive bacteria. The molecules studied include fatty acids, polar lipids, lipopolysaccharide (nature of the chain length of the fatty acid and the sugar in the Lipid A moiety; [28]), menaquinones, naphthoquinones, ubiquinones, mycolic acids, peptidoglycan, polyamines, teichoic and teichuronic acids and isoprenoid quinones [29]. Mycolic acids, which are useful taxonomic markers, are present in Gram-positive bacteria with high G+C ratios of their DNA [4], and have been reported for a range of fish pathogens, including *Mycobacterium chelonei* subsp. *piscarium* [3] and *M. shottsii* [30]. The length of the mycolate side chain has been correlated to 16S rRNA gene sequence homology [31]. Specific examples for which reliable chemotaxonomic data exist for Gram-positive bacterial fish pathogens are detailed below:

Lactocolccus Piscium

The long chain cellular fatty acids of *Lactococcus piscium* were reported to be straight chain saturated, mono-unsaturated and cyclopropane-ring types. The major acids corresponded to hexadecanoic acid, Δ 11-octadecanoic acid and Δ 11-methylenoctadecanoic acid [32].

Mycobacterium Neoarum

The cell wall chemotype has been given as IVA, with glycolated muramic acids, mycolic acids and MK-9, as the predominant isoprenoid quinone, being present [33].

Nocardia

Nocardia salmonicida contains LL-diaminopimelic acid (DAP) and glycine but not meso-DAP, arabinose or galactose in the cell wall (i.e. Type I). The major cellular fatty acids are hexdecanoic, octadecanoic, octadecanoic and 10-methyloctadecanoic acid [34].

N. seriolae contains *meso*-diaminopimelic acid, arabinose and galactose, indicative of chemotype IVA. Mycolic acids containing 44-58 carbon atoms are present. The cellular fatty acids are dominated by $n\text{-}C_{16:0}$, $n\text{-}C_{16:1}$ and $n\text{-}C_{18:1}$; 10-methyl-$C_{19:0}$ is also present as a major component in some isolates. The predominant isoprenoid quinone is tetrahydrogenated menaquinone with eight isoprene units [34].

Renibacterium Salmoninarum

Chemotaxonomic traits of *R. salmoninarum* have been highlighted in part because of the comparative difficulty with obtaining conventional phenotypic test results. Thus, the cell wall peptidoglycan was deduced to contain D-alanine, D-glutamic acid, glycine and lysine as the diamino acids [35]. The principal cell wall sugar was glucose, although arabinose, mannose and rhamnose were also present; mycolic acids were absent [36]. Methyl-branched fatty acids form > 92% of the total fatty acid component of the cells, with 12-methyltetradecanoic (anteiso-C_{15}), 13-methyldecanoic (iso-C_{15}) and 14-methylhexadecanoic (anteiso-C_{17}) as the major components. Straight chain fatty acids generally account for 1% of the total fatty acids, and unsaturated fatty acids are not detected at all. Over 81% of the total fatty acids are composed of the lower melting point anteiso acids, which may contribute to membrane fluidity at low temperatures. Unsaturated menaquinones with nine isoprene units are present. All strains contain diphosphatydylglycerol, 2 major and 6 or 7 minor glycolipids and two unidentified minor phospholipids [37].

Molecular/Genetic Methods

Molecular/genetic methods involving 16S rRNA gene sequencing [38], reverse transcriptase-sequencing [39,40] and polymerase chain reaction (PCR)-based gene sequencing [41] have been useful additions to the armoury of techniques applicable to bacterial taxonomy [29,31]. DNA hybridization, which is regarded as the "gold standard" for demonstrating the presence or absence of new species, was introduced into bacterial taxonomy during the 1960s (e.g. [42]). Genotypic classification involving sequencing of the 16S and 23S RNA genes (the latter is less popular) is regarded as the definitive standard for determining phylogenetic relationships of bacteria [29,38]. In particular, the genes are regarded as having all the attributes of useful, relevant and stable biological markers being present and homologous in all bacteria. Also, they are not prone to the effects of gene transfer [29]. Yet, the exact homology values have a profound effect on interpretation of the outputs. Thus, homology values of ≤ 98.7% (97% according to [31]) indicates membership of different species, and this correlates well with DNA hybridization results. Yet, occasionally higher homology values may be attributed to distinct species groupings [43]. By themselves, 16S rRNA gene sequences are insufficient to describe a new species, but may be used indicatively and in conjunction with DNA:DNA hybridization [31]. However, sequencing has permitted the recognition of new variants. For example, sequencing revealed a new variant among Israeli isolates of *Streptococcus iniae* [44]. Moreover, 16S rRNA cataloguing has been useful in providing information about the position of species in existing classifications. Thus, small-subunit rRNA sequencing and DNA:DNA hybridization revealed that *Pasteurella piscicida* was related to *Photobacterium damsela* leading to the proposal that the pathogen be re-classified as *Ph. damsela(e)* subsp. *piscicida* [45]. Furthermore, *R. salmoninarum* was deduced to be a member of the actinomycete subdivision, being related to *Arthrobacter, Brevibacterium, Cellulomonas, Jonesia, Micrococcus, Promicromonospora, Stomatococcus* and *Terrabacter* [46,47]. The evolutionary relationship of *R. salmoninarum* to *Arthrobacter* was reinforced as the result of genome sequencing, which suggested that the genome of the former had been reduced significantly since its divergence from a common ancestor [48].

Nucleic acid fingerprinting methods, including amplified fragment-length polymorphism PCR (AFLP), pulsed field gel electrophoresis (PFGE), random amplified polymorphic DNA (RAPD), rep-PCR (repetitive element primed PCR), REP-PCR (repetitive extragenic palindromic-PCR), ERIC-PCR (enterobacterial repetitive intergenic consensus sequences-PCR), BOX-PCR (derived from the boxA element) and ribotyping, provide information at or below the subspecies level [31]. Of these, AFLP and ribotyping are extremely useful and standardized.

It is unfortunate that with the increasing use of molecular methods, the description of bacterial groups has been often met with the use of minimal phenotypic data, which causes problems for diagnostics especially in laboratories, which are not equipped for molecular biology [6]. In these situations where distinguishing phenotypic feature have not been or could not be provided then the species should be referred to as a geno[mo]species. Nevertheless, molecular methods have revolutionized taxonomy, and led to the description of an increasing number of new taxa. The methodologies may be culture-independent, allowing for the study of uncultured organisms but there are issues with genomic fluidity [29]. "*Candidatus*" describes uncultured prokaryotes for which phylogenetic relationships have been determined, and authenticity confirmed by methods such as in situ probing [29].

Sequencing of the 16S rDNA is becoming an accepted procedure for the identification of fish pathogens, for example *V. harveyi* [49] and confirming its synonymy with *V. carchariae* [9,10], and has been instrumental in the recognition of new pathogens, including *Streptococcus dysgalactiae* [50], *S. parauberis* (previously recognised as *S. uberis* genotype II; [51]) and *Vagococcus salmoninarum* [8,52], and confirmed the presence of *Lactococcus garvieae* in Taiwan [53].

DNA:DNA and RNA:DNA hybridization, 16S RNA cataloguing, and 5S and 16S rRNA sequencing techniques have been used with increasing regularity and success. A review of PCR with emphasis on validation of the techniques and problems with diagnosis has been published [54]. PCR has been used successfully to identify hard-to-isolate fish pathogens, such as *Mycobacterium* spp. in sea bass (*Dicentrarchus labrax)* [55] and *M. chelonei* in a cichlid oscar *(Astronotus ocellatus)* [56]. Moreover, PCR has been useful with distinguishing different species from within the same genus, such as *Lactococcus garvieae* from *L. lactis* [57], from related genera, i.e. *L. garvieae, S. difficilis, S. iniae* and *S. parauberis* [58], and with an admirable level of specificity [59].

The sensitivity of PCR is clearly a positive attribute particularly with slow growing and/or nutritionally fastidious pathogens that are otherwise difficult to study in the laboratory. Of relevance, a PCR was developed [60,61], which detected only 22 cells of *R. salmoninarum;* a sensitivity of 10 cells was reported by others [62]. Similarly, PCR detected only 10^2 colony forming units (CFU) of *N. seriolae* in yellowtail [63].

A recent development is multi locus sequence analysis (MLSA), which permits the genotypic examination of micro-organisms by comparison of the sequences of multiple, i.e. 12 or more, house-keeping genes. The benefit of using multiple genes is that the outputs are more informative and less likely to generate results that are distorted by recombination of single loci

[29]. The resulting phylogenetic trees are capable of recognizing deeply branching clusters and permit the delineation of groups within a species or genus [64].

New Species of Fish Pathogens Recognized by 16S rRNA Sequencing

16S rRNA sequencing has helped the description of fish pathogens where phenotypic characterization alone does not permit their incorporation in classifications. For example, a new disease of Atlantic salmon (*Salmo salar*) was linked to the *Streptobacillus moniliformis* and the fusobacteria group on the basis of sequence homology; biochemical traits did not permit identification [65]. The newly described cause of a mycobacteriosis in Chesapeake Bay (USA) striped bass *(Morone saxatilis)* was equated to a new species, *M. shottsii*, with confirmation by 16S rRNA sequence homology in which the pathogen was linked most closely to *M. marinum* and *M. ulcerans* (similarity = 99.2%) [30]. In one study, *M. gordonae* was identified by 16S rRNA sequencing [66]. Furthermore, phylogenetic analysis based on 16S rRNA gene sequencing together with partial sequences from the 65 kDa heat-shock protein (hsp65) and the beta-subunit of the bacterial RNA polymerase (*rpoB*) genes and the 16S-23S internal transcribed spacer 1 (ITS 1) region named other novel mycobacteria as *M. stomatepiae* and *M. barombii* [67].

During an examination of 16S rRNA sequences, two isolates of motile aeromonads from diseased elvers in Spain were described as a new species, *Aeromonas allosaccharophila* [68], albeit phenotypically heterogeneous [69]. This heterogenicity has caused problems for reliable phenotypic-based diagnoses.

Francisella became recognized as the cause of a new disease of Atlantic cod (*Gadus morhua*) in Norway in which the affected fish displayed white granuloma in the viscera and skin. Isolates were recovered, and determined to possess the key phenotypic characters of *Francisella*, viz. non-motile, strictly aerobic Gram-negative intracellular coccobacilli which produced H_2S from cysteine-containing media [70]. 16S rRNA sequencing revealed a 99.17% homology to *Francisella philomiragia* [71], although a slightly higher value of 99.3% was published [70] with the proposal for a new subspecies, i.e. *Francisella philomiragia* subsp. *noatunensis*, to accommodate the organisms. There was 92.2-99.0% homology with *Francisella philomiragia* housekeeping genes, *groEL, shdA, rpoB, rpoA, pgm* and *atpA*. A DNA:DNA hybridization of 68% was recorded between the fish pathogen and *Francisella philomiragia* [70].

Pasteurella skyensis was recovered from diseased Atlantic salmon in Scotland, linked to the family Pasteurellaceae by phenotypic analysis, and elevated to a new species largely as a result of 16S rRNA sequencing that identified the closest neighbour as *Pasteurella phocoenarum* (homology = 97.1%; [72]).

Piscirickettsia salmonis was named to accommodate isolates from diseased salmon in Chile, of which LF-89 was studied in detail [73] with 16S rRNA conforming to the gamma subdivision of the Proteobacteria with similarities to the family Rickettsiales, and in particular *Wolbachia persica* (similarity = 86.3%) and *Coxiella burnetii* (similarity = 87.5%) more than to representatives of *Ehrlichia, Rickettsia* or *Rochalimaea* leading to the description of a new genus and species [73]. Other rickettsias not conforming exactly with *Piscirickettia salmonis* have been described. For example, an organism recovered from white sea bass (*Atractascion nobilis*) was reported to have a 96.3-98.7% 16S rDNA homology with *Piscirickettsia salmonis* [74], which was considered by the authors to be too low for a confirmed identity. A Tasmanian isolate from Atlantic salmon was distinct from *Piscirickettsia* in terms of sequence alignment of the 16S rRNA, and for the present regarded as a rickettsial-like organism (RLO; [75]).

Pseudomonas plecoglossicida, the causal agent of bacterial ascites of ayu *(Plecoglossus altivelis)*, was described as a new species as a result of 16S rRNA gene sequence analysis confirming distinctiveness from *P. putida* biovar A. DNA:DNA hybridization confirmed the isolates to be a new centre of variation insofar as < 50% homology was recorded with other pseudomonads, including *P. putida* [76].

Streptococcus phocae was recognized as a cause of systemic disease in Atlantic salmon farmed in Chile. Phenotypic testing linked the pathogen with the streptococci, notably *Gemella*, but analysis of 16S rRNA genes provided a link to *S. phocae* [77].

Tenacibaculum soleae was recovered from diseased sole *(Solea senegalensis)* in Spain, and confirmed as a new species largely on account of 16S rRNA homology values of 94.8-96.7% with other members of the genus [78].

Two groups of bacteria were recovered from Atlantic salmon with winter ulcer disease/syndrome [79], of which one cluster was found to be closest to *Moritella marina* (43% re-association by DNA:DNA hybridization), and was named as *V. viscosus*. By 16S rDNA sequencing, the closest match was with *Moritella* [79] and *M. marina* (99.1% sequence homology) so that the organism was re-classified to *Moritella*, but as a new species, as *M. viscosa* [80], despite the high sequence homology [29]. Separately, 19 Icelandic and one Norwegian isolate from shallow skin lesions on Atlantic salmon, and the type strain of *V. marinus* NCIMB 1144 were identified as *V. marinus* after an examination of phenotypic data and analyses by numerical taxonomy [81]. On the basis of 16S rRNA sequencing, the species was transferred to *Moritella* as *M. marina* [82].

New and Uncultured Fish Pathogens: "*Candidatus*"

Molecular techniques have permitted the recognition of uncultured pathogens belonging to new groupings for which the name of "*Candidatus*" has been used. "*Candidatus* arthromitus" has been recovered from rainbow trout (*Oncorhynchus mykiss*) with summer enteritic syndrome, which is a gastro-enteritis [83,84]. The organism was observed in histological preparations to which nested polymerase chain reaction was used, with confirmation by sequencing [85]. "*Candidatus* piscichlamydia salmonis" was detected by RT-DGGE in intracellular inclusions, i.e. epitheliocysts, in Atlantic salmon with proliferative gill inflammation [86]. "*Candidatus* clavochlamydia salmonicola" is an intracellular organism, causing epitheliocystis in Atlantic salmon, which was recognized as novel as a result of 16S rRNA sequencing [87].

TAXONOMIC DEVELOPMENTS ASSOCIATED WITH SPECIFIC BACTERIAL FISH PATHOGENS

From the early literature, a question-mark has hung over the reliability of some bacterial names insofar as there was often negligible evidence to support the use of those names. Concern may also be expressed about the value of studies based on only single isolates where concern about the reasons for choice of the culture may be aired. Some of the controversy surrounding specific diseases/pathogens follows:

Motile Aeromonas Septicaemia

Aeromonas hydrophila (= *A. formicans* and *A. liquefaciens)* would appear to have worldwide distribution and to be a pathogen, causing motile aeromonas septicaemia, of many species of freshwater fish. Indeed, there are reports of a spread into marine fish, notably ulcer disease of cod [88]. Since its initial recognition in the literature, a wealth of knowledge has been accumulated about many facets of its biology (see [6]). A new variant *A. hydrophila* subsp. *dhakenis*, which was originally recovered from children with diarrhoeae in Bangladesh, was determined to be pathogenic to rainbow trout [89]. However overall, there has been some doubt about the role of *A. hydrophila* as a pathogen, and in some cases it may well be present in fish tissue only as a secondary invader [6]. Moreover with developments in the taxonomy of motile aeromonads [90], the accuracy of some of the early published identifications may be justifiably questioned. Could other motile aeromonads be associated with fish disease and may have been confused previously with *A. hydrophila* [89]?

It is clear that there is phenotypic, serological and genotypic heterogeneity within the descriptions of fish pathogenic *A. hydrophila* [e.g. 91, 92], and other motile aeromonads have been implicated as the aetiological agents of (fish) diseases. Thus, 8 isolates reported as pathogenic to eel in Spain were identified by numerical taxonomy as *A. jandaei* [93,94]. Certainly, the current approach of allocating species names as a result of the examination of 16S rRNA gene sequences has encompassed fish pathogenic motile aeromonads. For example, isolates from diseased fish which were recovered in *Aeromonas* DNA Hybridization Group 2 (= *A. hydrophila)* were equated with a new group, *A. bestiarum* [95]. Subsequently, *A. sobria* (*A. sobria* biovar sobria and *A. veronii* biovar sobria were reported as pathogenic to rainbow trout [89]. Indeed, *A. sobria* has been previously found to have a role as a fish pathogen, with isolates recovered from wild spawning gizzard shad (*Dorosoma cepedianum)* in Maryland, USA during 1987 [15,96]. Also, *A. veronii* has been recovered from Siberian sturgeon (*Acipenser baerii*) with identification of the pathogen resulting from phenotyping and 16S rRNA gene sequencing [97].

Aeromonas Salmonicida

Aeromonas salmonicida is one of the oldest described fish pathogens, being isolated initially from diseased hatchery-maintained brown trout (*Salmo trutta*) in Germany, and named as "Bacillus der Forellenseuche" or bacillus of trout contagious disease. The history of the organism reveals a plethora of synonyms including *Bacillus devorans*, *Bacterium salmonica*, *Bacterium salmonicida*, *Bacillus truttae* and *Bacillus salmonicida* [6]. The 7th edition of *Bergey's Manual of Determinative Bacteriology* (1957) placed the pathogen in the genus *Aeromonas* within the family Pseudomonadaceae

[98]. Later, there was a transfer to the family Vibrionaceae and subsequently to its own family, i.e. the Aeromonadaceae [99]. Re-classification was based primarily on phenotyping [100]. Thus in 1953, the first detailed description of the pathogen was published, and from an examination of 10 isolates, it was concluded that *Bacterium salmonicida* was homogeneous in cultural and biochemical characteristics [100]. Numerous studies have addressed the homogeneity of the species (e.g. [101]). The basic description is of an organism, which comprises non-motile (motility and *flaA* and *flaB* flagellar genes have been reported [52,82]), fermentative, Gram-negative rods, which produce a brown water-soluble pigment on tryptone-containing agar, do not grow at 37°C, and produce catalase and oxidase [6]. Cultures have the ability to dissociate into rough, smooth and G-phase (= intermediate) colonies [102]. The pathogen has spread from its dominance in salmonids to cyprinids and marine flatfish [6]. An ongoing issue surrounds the intraspecies structure, i.e. the validity of subspecies *achromogenes, masoucida, pectinolytica, salmonicida* and *smithia*, and the status of so-called atypical isolates.

A. salmonicida subsp. *salmonicida*, isolates of which have been obtained almost exclusively from outbreaks of furunculosis in salmonids, is regarded as homogeneous, and is referred to as "typical" [103]; all other isolates are considered as heterogeneous and "atypical" [6]. So called atypical strains may demonstrate weak, slow or non-pigment production [104,105], catalase [106] or oxidase-negativity [e.g. 106, 107], nutritional fastidiousness for blood products [108], slow growth, i.e. ≥ 5 days compared with 1-2 days for typical isolates [106,108], and be pathogenic for fish other than salmonids, e.g. cyprinids (e.g. [106,108,109]) and marine flatfish, namely dab (*Limanda limanda*), plaice (*Pleuronectes platessa*), flounder (*Platichthys flesus*) and turbot (*Scophthalmus maximus*) [110-112], and cause ulceration rather than furunculosis [6,113]. The deviation in characteristics from the typical to atypical isolates has made typing difficult [114-116]. Even 16S rDNA sequencing has not helped with the clustering of atypical forms (e.g. [117]).

Smith [118] recognized heterogeneity in the species description of *A. salmonicida*. She examined six isolates of non-pigmented *A. salmonicida*, which were clustered as Group I in her numerical taxonomy study, for which a separate new species name was proposed, i.e. *A. achromogenes*, but the proposal was not adopted widely. A second non-pigmented group was described by Kimura [119], and named as *A. salmonicida* subsp. *masoucida*. Schubert [120] considered these non-pigmented isolates as subspecies, and coined the names of *A. salmonicida* subsp. *achromogenes* and *A. salmonicida* subsp. *masoucida*, respectively. Pigmented strains (= typical) were classified as *A. salmonicida* subsp. *salmonicida* [120]. The precise relationship of the subspecies has been the subject of detailed discussion. In particular, it was contended that subsp. *achromogenes* and *masoucida* were more closely related to *A. hydrophila* than to *A. salmonicida* [121]. Later, it was mooted that subsp. *masoucida* bridged typical *A. salmonicida* and *A. hydrophila* [122]. Yet, *A. salmonicida* subsp. *masoucida* is non-motile, sensitive to *A. salmonicida* bacteriophages, possesses an antigenic profile specific to *A. salmonicida*, and shares a DNA homology of 103% with *A. salmonicida* [92]. By PCR, a combination of *achromogenes* with *masoucida* could be justified, but this was not substantiated by ribotyping and RAPD analyses [114]. Phenotypic data suggest a case for combining subsp. *masoucida* with *salmonicida*, and subsp. *achromogenes* with *Haemophilus piscium*, which is the causal agent of ulcer disease of trout [123]. Examination of the small subunit rRNA gene sequences revealed 99.9% homology of an authentic strain of *H. piscium* with *A. salmonicida* subsp. *salmonicida* [124]. So far, the comparative uniqueness of subsp. *smithia* has been indicated from several studies (e.g. [114]). The complication is with aberrant strains of *A. salmonicida* from fish species other than salmonids.

DNA homology was used to reveal that all isolates of *A. salmonicida* (including *A. salmonicida* subsp. *masoucida)* were highly related, i.e. 96-106% homology, when hybridized against *A. salmonicida* subsp. *salmonicida* [92]. It was opined that *A. salmonicida* subsp. *masoucida* and some atypical isolates did not warrant separate subspecies status, because they were regarded as variants of other well-recognized groups. Also as a result of genotypic analyses, it was reported that typical and atypical isolates of *A. salmonicida* were very closely related, with minimal divergence [125]. Using DNA:DNA re-association, it was concluded that typical *A. salmonicida* were recovered in a homogeneous group, whereas the atypical representatives were more diverse [126]. From numerical taxonomy and DNA:DNA hybridization, similar conclusions resulted regarding the homogeneity of typical isolates of *A. salmonicida* [101]. However using 16S rRNA sequencing techniques, it was reported that subspecies *achromogenes* and *masoucida* were indistinguishable, and only differed from subspecies *salmonicida* by two bases [68].

The relation of *A. salmonicida* to other aeromonads has been discussed. Eddy [127] focused on the inability of *A. salmonicida* to produce 2,3-butanediol from glucose, and the absence of motility, which were in contrast to the genus description [128]. A new genus, i.e. *Necromonas*, was proposed with two species, namely *N. salmonicida* for the typical isolates and *N. achromogenes* for the non-pigmented strains [118]. This proposal was not formally widely accepted, although Cowan [129] used the suggestion in his landmark identification scheme for medically important bacteria. Subsequent serological and bacteriophage sensitivity data supported the relationship between *A. salmonicida* and the motile aeromonads. Common antigens between *A. hydrophila* and *A. salmonicida* subsp. *masoucida* and other isolates of *A. salmonicida* were reported [122,130]. Furthermore, serological cross-reactions between *A. salmonicida* and motile aeromonads were discussed [131]. Moreover, *A. hydrophila* cultures were found to be sensitive to *A. salmonicida* bacteriophages [132,133]. The outcome of all the studies is that DNA homology supports the classification of *A. salmonicida* in the genus *Aeromonas* (e.g. [92,122,126]).

There are certainly outstanding questions about the validity and taxonomic placing of *Haemophilus piscium* [123], but an authentic reference strain was not deposited any in any recognized culture collection at the time of its first isolation. Later, it was concluded that the organism was not a *bona fide Haemophilus* because of the lack of requirement for haemin or NAD [134]. In particular *H. piscium* differed from the type species of the genus, *H. influenzae*, in the inability to reduce nitrate or alkaline phosphatase and to grow at 37°C, in conjunction with a higher G+C ratio of the DNA. It was commented that there was only a low similarity between *H. piscium* and other *Haemophilus* spp. in a numerical taxonomic study [135]. A link with atypical, achromogenic *A. salmonicida* was made [122]. This link was reinforced by bacteriophage sensitivity, when it was concluded that *H. piscium* is an atypical *A. salmonicida* [136]. Other workers have supported this view (e.g. [114]). However with the absence of an authentic, original type strain, the definitive taxonomic position of *H. piscium* is only speculative.

A lack of congruence has been reported between the results of molecular (PCR, RAPD and ribotyping) and phenotypic methods in taxonomy of aeromonads [114]. Moreover, there are problems of inter-laboratory differences and lack of standardisation in test methods [137]. The outcome is that the definitive classification of *A. salmonicida* has not been achieved, to date.

Enteric Redmouth (ERM)

There has been discussion about the taxonomic position of the aetiological agent of ERM. Strong agglutination with *Salmonella enterica* subsp *arizonae* O group 26, and a weak reaction with O group 29 was reported [138]. In addition, biochemical similarities with enterics, notably *Enterobacter liquefaciens, Serratia marcescens* subsp. *kiliensis* as well as *Salmonella enterica* subsp. *arizonae* were mentioned [138]. Serological cross-reactions were also recorded with *Hafnia alvei* [139]. Nevertheless, a new species, i.e. *Yersinia ruckeri* was described although there was only a 30-31% DNA homology with *Y. enterocolitica* and *Y. pseudotuberculosis* [140]. This compares to DNA homologies of 24-28% and 31% with *Serratia marcescens* and *Serratia liquefaciens*, respectively [141]. Indeed, it has been suggested that the causal agent of ERM should belong in a new genus of the Enterobacteriaceae [142]. A complication developed when a new non-motile form of the pathogen was recovered from rainbow trout. By 16S rRNA sequencing and a homology of 100%, the organisms were linked to *Y. ruckeri* but regarded as a new biogroup [143]. Similar non-motile variants were also recovered from previously vaccinated rainbow trout in Spain [144].

Vibriosis

The causal agent of "red-pest" in eels was first designated as *Bacterium anguillarum* [6]. Subsequently, an outbreak among eels in Sweden led to the use of the name *Vibrio anguillarum*. Numerous studies have pointed to heterogeneity in *V. anguillarum* initially with the delineation of two sub-groupings/biotypes (e.g. [12]). This increased to 3 [145] and then 4 sub-groups/phena within the species definition [146,147]. Ribotyping has confirmed the heterogeneity [148], although a single taxon, homogeneous by ribotyping but heterogeneous by LPS profiles, plasmid composition, serogrouping, and BIOLOG-GN fingerprints and API 20E profiles was described [149,150]. Biotype II became recognized as a separate species, i.e. *V. ordalii* [151], which is homogeneous by plasmid profiling, ribotyping and serogrouping, accommodates two LPS groups, but is heterogeneous by BIOLOG-GN fingerprints and API 20E profiles [150].

Serology has been widely used for diagnosis, but has complicated the understanding of *V. anguillarum* [151], and the establishment of serotypes has to some extent traversed species boundaries. With *V. anguillarum*, serogroup 02 and 05, there are common antigens with *V. ordalii* [152] and *V.*

harveyi [149], respectively. Initially, three serotypes were recognised for isolates from salmonids from the northwest USA, Europe, and the Pacific-northwest (USA) [103]. This number increased to 6 [128], and then 10 [153] and upwards [148,154]. Serogroup O1 dominates the number of isolates available for study and the relative importance to fish pathology [149,155-157]. Serogroup O2 has been further subdivided into serogroup O2a and O2b [158].

V. anguillarum was re-classified initially to *Beneckea* [12] and then to a newly proposed genus *Listonella* [13], but the changes were not widely accepted.

THE ROLE OF PHYLOGENETICS IN BACTERIAL FISH PATHOLOGY

The techniques described above are relevant for the taxonomy of bacterial fish pathogens. Yet, molecular methods, namely sequencing of the 16S rRNA gene, permit the study of evolutionary relationships, i.e. phylogenetics, which may be viewed as phylogenetic trees, which are interpreted by cladistics and used in defining taxa. The approach is essential in the study of the evolutionary tree of life, but is it strictly necessary for fish pathology and the recognition of species? One concern is the comparative fluidity by which genes may be exchanged, such as by horizontal gene transfer, and the impact of this movement on the outcome of the taxonomic/phylogenetic process.

CONCLUSIONS

There has been a resurgence of interest in bacterial taxonomy partially because of the current focus on biodiversity and the development of reliable molecular methods [159], notably 16S rRNA sequencing. Undoubtedly, these molecular approaches have led to greater confidence and accuracy in the reporting of bacterial names. Nevertheless, it is conceded that bacterial taxonomy is a specialist subject, which is not of interest to all fish pathologists. However, it cannot be overstated that there is a real value for good taxonomy as a means of communication. In terms of fish pathology, taxonomy enables the recognition of new pathogens, improvements in the understanding of relationships between taxa, an appreciation of variation within existing nomenspecies including the recognition of new subspecies and biogroups, and facilitates accurate commentary about all aspects of biology from epizootiology to pathogenicity (e.g. [160]), although the position of so-called atypical isolates in taxonomic hierarchies is often difficult to determine [161].

For the future, a range of new techniques, including in situ hybridization, probe hybridization, microarray techniques and restriction enzyme digestion, are entering taxonomic use, and are likely to be used in fish pathology. The impact of these new approaches is difficult to predict, but will undoubtedly be incorporated in some fish bacteriology laboratories.

REFERENCES

1. Sneath PHA. The application of computers to taxonomy. J Gen Microbiol. 1957;17:201–226.
2. Sokal RR, Michener CD. Numerical Taxonomy. WH Freeman, London; 1958.
3. Arakawa CK, Fryer JL. Isolation and characterization of a new subspecies of *Mycobacterium chelonei* infections for salmonid fish. Helgoländer Meeresuntersuch. 1984;37:329–342.
4. Brennan PJ. In: Microbial Lipids. Ratledge C, Wilkinson SG, editor. Vol. 1. Academic Press, London; 1988. *Mycobacterium* and other actinomycetes; pp. 203–298.
5. Kämpfer P. The characterization of prokaryote strains for taxonomic purposes. Int J Syst Evol Microbiol. 2010;60:7.
6. Austin B, Austin DA. Bacterial Fish Pathogens: Disease in Farmed and Wild Fish. 4. Springer-Praxis, Goldalming; 2007.
7. Schiewe MH, Trust TJ, Crosa JH. *Vibrio ordalii* sp. nov.: a causative agent of vibriosis in fish. Curr Microbiol. 1981;6:343–348. doi: 10.1007/BF01567009.
8. Schmidtke LM, Carson J. Characteristics of *Vagococcus salmoninarum* isolated from diseased salmonid fish. J Appl Bacteriol. 1994;77:229–236.
9. Gauger E, Gómez-Chiarri M. 16S ribosomal DNA sequencing confirms the synonymy of *Vibrio harveyi* and *V. carchariae*. Dis Aquat Org. 2002;52:39–46. doi: 10.3354/dao052039.
10. Pedersen K, Verdonck L, Austin B, Austin DA, Blanch AR, Grimont PAD, Jofre J, Koblavi S, Larsen JL, Tiainen T, Vigneulle M, Swings J. Taxonomic evidence that *Vibrio carchariae* Grimes et al. 1985 is a junior synonym of *Vibrio harveyi* (Johnson and Shunk 1936) Baumann et al. 1981. Int J Syst Bacteriol. 1998;48:749–758. doi: 10.1099/00207713-48-3-749.
11. Thompson FL, Hoste B, Vandemeulebroecke K, Engelbeen K, Denys R, Swings J. *Vibrio trachuri* Iwamoto et al., 1995 is a junior synonym of *Vibrio harveyi* (Johnson and Shunk 1936) Baumann et al. 1981. Int J Syst Evol Microbiol. 2002;52:973–976. doi: 10.1099/ijs.0.02046-0.
12. Baumann P, Bang SS, Baumann L. Phenotypic characterization of *Beneckea anguillara* biotypes I and II. Curr Microbiol. 1978;1:85–88. doi: 10.1007/BF02605421.

13. MacDonell MT, Colwell RR. Phylogeny of the Vibrionaceae and recommendations for two new genera, *Listonella* and *Shewanella*. Syst Appl Microbiol. 1985;6:171–182.
14. Vandamme P, Pot B, Gillis M, de Vos P, Kersters K, Swings J. Polyphasic taxonomy, a consensus approach to bacterial systematics. Microbiol Rev. 1996;60:407–438.
15. Toranzo AE, Santos Y, Nieto TP, Barja JL. Evaluation of different assay systems for identification of environmental *Aeromonas* strains. Appl Environ Microbiol. 1986;51:652–656.
16. Sneath PHA, Johnson R. The influence of numerical taxonomic similarities of errors in microbiological tests. J Gen Microbiol. 1972;72:377–392.
17. Austin B, Embley TM, Goodfellow M. Selective isolation of *Renibacterium salmoninarum*. FEMS Microbiol Lett. 1983;17:111–114. doi: 10.1111/j.1574-6968.1983.tb00383.x.
18. Walters GR, Plumb JA. Modified oxidation/fermentation medium for use in the identification of bacterial fish pathogens. J Fish Res Bd Can. 1978;35:1629–1630.
19. Adams A, Thompson KD. Recent applications of biotechnology to novel diagnostics for aquatic animals. Rev Sci Tech. 2008;27:197–209.
20. Goerlich R, Schlüsener HJ, Lehmann J, Greuel E. The application of monoclonal antibodies to diagnosis of *Aeromonas salmonicida* infections in fishes. Bull Europ Assoc Fish Pathol. 1984;4:66.
21. Jung TS, Thompson KD, Adams A, Morris DJ, Snedden K. The production of monoclonal antibodies against *Photobacterium damselae* subsp. *piscicida* and their application in immunohistochemistry. J Fish Dis. 2001;24:64–67. doi: 10.1046/j.1365-2761.2001.00260.x.
22. Steiropoulos NA, Yuksel SA, Thompson KD, Adams A, Ferguson HW. Detection of *Rickettsia*-like organisms (RLOs) in European sea bass *(Dicentrarchus labrax*, L) by immunohistochemistry, using rabbit anti-*Rickettsia salmonis* serum. Bull Eur Assoc Fish Pathol. 2002;22:428–432.
23. Adams A, Thompson KD, Morris D, Farias C, Chen SC. Development and use of monoclonal antibody probes for immunohistochemistry, ELISA and IFAT to detect bacterial and parasitic fish pathogens. Fish Shellfish Immunol. 1995;5:537–547. doi: 10.1016/S1050-4648(95)80040-9.

24. Klesius P, Evans J, Shoemaker C, Yeh H, Goodwin AE, Adams A, Thompson KD. Rapid detection and identification of *Streptococcus iniae* using a monoclonal antibody-based indirect fluorescent antibody technique. Aquaculture. 2005;258:180–186. doi: 10.1016/j.aquaculture.2005.06.040.

25. Adams A. In: Advances in disease diagnosis, vaccine development and other emerging methods to control pathogens. Burnell G, Allan G, editor. Chapter 7. New Technologies in Aquaculture, Improving Production, Efficiency, Quality and Environmental Management, Woodhead, England; 2009. pp. 197–211.

26. Saeed MO, Plumb JA. Serological detection of *Edwardsiella ictaluri* Hawke lipopolysaccharide antibody in serum of channel catfish *Ictalurus punctatus* Rafinesque. J Fish Dis. 1987;10:205–209. doi: 10.1111/j.1365-2761.1987.tb01062.x.

27. Adkinson MA, Gilad O, Hedricks RP. An enzyme linked immunosorbent assay (ELISA) for detection of antibodies to the Koi herpes virus (KHV) in the serum of Koi *Cyprinus carpio*. Fish Pathol. 2005;40:53–62.

28. Mayer H, Masoud H, Urbanik-Sypniewska T, Weckesser J. Lipid A composition and phylogeny of Gram-negative bacteria. Bull Jap Fed Cult Collect. 1989;5:19–25.

29. Scheifer KH. Classification of bacteria and Archaea: past, present and future. Syst Appl Microbiol. 2009;32:533–542. doi: 10.1016/j.syapm.2009.09.002.

30. Rhodes MW, Kator H, Kotob S, van Berkum P, Kaattari I, Vogelbein W, Quinn F, Floyd MM, Butler WR, Ottinger CA. *Mycobacterium shottsii* sp nov., a slowly growing species isolated from Chesapeake Bay striped bass (*Morone saxatilis)* Int J Syst Evol Microbiol. 2003;53:421–424. doi: 10.1099/ijs.0.02299-0.

31. Tindall BJ, Rosselló-Móra R, Busse HJ, Ludwig W, Kämpfer P. Notes on the characterization of prokaryote strains for taxonomic purposes. Int J Syst Evol Microbiol. 2010;60:249–266. doi: 10.1099/ijs.0.016949-0.

32. Williams AM, Fryer JL, Collins MD. *Lactococcus piscium* sp. nov. a new *Lactococcus* species from salmonid fish. FEMS Microbiol Lett. 1990;68:109–114. doi: 10.1111/j.1574-6968.1990.tb04132.x.

33. Backman S, Ferguson HW, Prescott JF, Wilcock BP. Progressive panophthalmitis in chinook salmon, *Oncorhynchus tshawytscha* (Walbaum): a case report. J Fish Dis. 1990;13:345–353.

doi: 10.1111/j.1365-2761.1990.tb00793.x.

34. Isik K, Chun J, Hah YC, Goodfellow M. *Nocardia salmonicida* nom. rev., a fish pathogen. Int J Syst Bacteriol. 1999;49:833–837. doi: 10.1099/00207713-49-2-833.
35. Fiedler F, Draxl R. Biochemical and immunochemical properties of the cell surface of *Renibacterium salmoninarum*. J Bacteriol. 1986;168:799–804.
36. Sanders JE, Fryer JL. *Renibacterium salmoninarum* gen. nov., sp. nov., the causative agent of bacterial kidney disease in salmonid fishes. Int J Syst Bacteriol. 1980;30:496–502. doi: 10.1099/00207713-30-2-496.
37. Embley TM, Goodfellow M, Minnikin DE, Austin B. Fatty acid, isoprenoid quinone and polar lipid composition in the classification of *Renibacterium salmoninarum*. J Appl Bacteriol. 1983;55:31–37.
38. Fox GE, Pechman KR, Woese CR. Comparative cataloging of 16S ribosomal ribonucleic acid: molecular approach to procaryotic systematics. Int J Syst Bacteriol. 1977;27:44–57. doi: 10.1099/00207713-27-1-44.
39. Lane DJ, Field KG, Olsen GJ, Pace NR. Reverse transcriptase sequencing of ribosomal RNA for phylogenetic analysis. Meth Enzymol. 1988;167:138–144. full_text.
40. Sanger F, Nicklen S, Coulson AR. DNA sequencing with chain-terminating inhibitors. Proc Natl Acad Sci USA. 1977;74:5463–5467. doi: 10.1073/pnas.74.12.5463.
41. Saiki RK, Gelfand DH, Stoffel S, Scharf SJ, Higuchi R, Horn GT, Mullis KB, Erlich HA. Primer-directed enzymatic amplification of DNA with a thermostable DNA polymerase. Science. 1988;239:487–491. doi: 10.1126/science.2448875.
42. De Ley J, Park IW, Tijtgat R, van Ermengem J. DNA homology and taxonomy of *Pseudomonas* and *Xanthomonas*. J Gen Microbiol. 1966;42:43–56.
43. Fox GE, Wisotzkey JD, Jurtshuk P Jr. How close is close: 16S rRNA sequence identity may not be sufficient to guarantee species identity. Int J Syst Bacteriol. 1992;42:166–170. doi: 10.1099/00207713-42-1-166.
44. Kvitt H, Colorni A. Strain variation and geographic endemism in *Streptococcus iniae*. Dis Aquat Org. 2004;61:67–73. doi: 10.3354/dao061067.
45. Gauthier G, Lafay B, Ruimy R, Breittmayer V, Nicolas JL, Gauthier M,

Christen R. Small-subunit rRNA sequences and whole DNA relatedness concur for the re-assignment of *Pasteurella piscicida* (Snieszko et al.) Janssen and Surgalla to the genus *Photobacterium* as *Photobacterium damsela* subsp. *piscicida* comb. nov. Int J Syst Bacteriol. 1995;45:139–144. doi: 10.1099/00207713-45-1-139.

46. Gutenberger SK, Giovannoni SJ, Field KG, Fryer JL, Rohovec JS. A phylogenetic comparison of the 16S rRNA sequence of the fish pathogen, *Renibacterium salmoninarum*, to Gram-positive bacteria. FEMS Microbiol Lett. 1991;77:151–156. doi: 10.1111/j.1574-6968.1991.tb04339.x.
47. Stackebrandt E, Wehmeyer U, Nader H, Fiedler F. Phylogenetic relationship of the fish pathogenic *Renibacterium salmoninarum* to *Arthrobacter, Micrococcus* and related taxa. FEMS Microbiol Lett. 1988;50:117–120. doi: 10.1111/j.1574-6968.1988.tb02922.x.
48. Wiens GD, Rockey DD, Wu ZN, Chang J, Levy R, Crane S, Chen DS, Capri GR, Burnett JR, Sudheesh PS, Schipma MJ, Burd H, Bhattacharyya A, Rhodes LD, Kaul R, Strom MS. Genome sequence of the fish pathogen *Renibacterium salmoninarum* suggests reductive evolution away from an environmental *Arthrobacter* ancestor. J Bacteriol. 2008;190:6970–6982. doi: 10.1128/JB.00721-08.
49. Ransangan J, Mustafa S. Identification of *Vibrio harveyi* isolated from diseased Asian sea bass *Lates calcarifer* by use of 16S ribosomal DNA sequencing. J Aquat Anim Health. 2009;21:150–155. doi: 10.1577/H09-002.1.
50. Nomoto R, Munasinghe LI, Shimahara Y, Yasuda H, Nakamura A, Misawa N, Itami T, Yoshida T. Lancefield group C *Streptococcus dysgalactiae* infection responsible for fish mortalities in Japan. J Fish Dis. 2004;27:679–686. doi: 10.1111/j.1365-2761.2004.00591.x.
51. Doménech A, Fernández-Garayzábal JF, Pascual C, Garcia JA, Cutuli MT, Moreno MA, Collins MD, Dominguez L. Streptococcosis in cultured turbot, *Scophthalmus maximus* (L.), associated with *Streptococcus parauberis*. J Fish Dis. 1996;19:33–38.
52. Wallbanks S, Martinez-Murcia AJ, Fryer JL, Phillips BA, Collins MD. 16S rRNA sequence determination for members of the genus *Carnobacterium* and related lactic acid bacteria and description of *Vagococcus salmoninarum* sp. nov. Int J Syst Bacteriol. 1990;40:224–230. doi: 10.1099/00207713-40-3-224.

53. Chen SC, Liaw LL, Su HY, Ko SC, Wu CY, Chaung HC, Tsai YH, Yang KL, Chen YC, Chen TH, Lin GR, Cheng SY, Lin YD, Lee JL, Lai CC, Weng YJ, Chu SY. *Lactococcus garvieae*, a cause of disease in grey mullet, *Mugil cephalus* L., in Taiwan. J Fish Dis. 2002;25:727–732. doi: 10.1046/j.1365-2761.2002.00415.x.
54. Hiney MP, Smith PR. Validation of polymerase chain reaction-based techniques for proxy detection of bacterial fish pathogens: framework, problems and possible solutions for environmental applications. Aquaculture. 1998;162:41–68. doi: 10.1016/S0044-8486(98)00207-5.
55. Knibb W, Colorni A, Ankaoua M, Lindell D, Diamant A, Gordin H. Detection and identification of a pathogenic marine mycobacterium from the European seabass *Dicentrarchus labrax* using polymerase chain reaction and direct sequencing of 16S rRNA sequences. Mol Mar Biol Biotechnol. 1993;2:225–232.
56. McCormick JI, Hughes MS, McLoughlin MF. Identification of *Mycobacterium chelonae* in a cichlid oscar *Astronotus ocellatus* Cuvier, by direct cycle sequencing of polymerase chain reaction amplified 16S rRNA gene sequences. J Fish Dis. 1995;18:459–461. doi: 10.1111/j.1365-2761.1995.tb00337.x.
57. Zlotkin A, Eldar A, Ghittino C, Bercovier H. Identification of *Lactococcus garvieae* by PCR. J Clin Microbiol. 1998;36:983–985.
58. Mata AI, Gibello A, Casamayor A, Blanco MM, Domínquez L, Fernández-Garayzábal JF. Multiplex PCR assay for the detection of bacterial pathogens associated with warm-water streptococcosis in fish. Appl Environ Microbiol. 2004;70:3183–3187. doi: 10.1128/AEM.70.5.3183-3187.2004.
59. Aoki T, Park CI, Yamashita H, Hirono I. Species-specific polymerase chain reaction primers for *Lactococcus garvieae*. J Fish Dis. 2000;23:1–6. doi: 10.1046/j.1365-2761.2000.00207.x.
60. León G, Maulén N, Figueroa J, Villanueva J, Rodríguez C, Vera MI, Krauskopf M. A PCR based assay for the identification of the fish pathogen *Renibacterium salmoninarum*. FEMS Microbiol Lett. 1994;115:131–136.
61. León G, Martinez MA, Etchegaray JP, Vera MI, Figueroa J, Krauskopf M. Specific DNA probes for the identification of the fish pathogen, *Renibacterium salmoninarum*. Wld J Microbiol Biotechnol. 1994;10:149–153.

62. Magnüssen HB, Fridjónsson ÓH, Andrésson ÓS, Benediktsdóttir E, Gudmundsdóttir S, Andrésdottír V. *Renibacterium salmoninarum*, the causal agent of bacterial kidney disease in salmonid fish, detected by nested reverse transcription-PCR of 16S rRNA sequences. Appl Environ Microbiol. 1994;60:4580–4583.

63. Miyoshi Y, Suzuki S. PCR method to detect *Nocardia seriolae* in fish samples. Fish Pathol. 2003;38:93–97.

64. Soria-Carrasco V, Talavera G, Igea J, Castresana J. The K tree score: quantification of differences in the relative branch length and topology of phylogenetic trees. Bioinformatics. 2007;23:2954–2956. doi: 10.1093/bioinformatics/btm466.

65. Maher M, Palmer R, Gannon F, Smith T. Relationship of a novel bacterial fish pathogen to *Streptobacillus moniliformis* and the fusobacteria group, based on 16S ribosomal RNA analysis. Syst Appl Microbiol. 1995;18:79–84.

66. Sakai M, Konno T, Tassakka ACMAR, Ponpornpisit A, Areechon N, Katagiri T, Yoshida T, Endo M. Characterization of a *Mycobacterium* sp. isolated from guppy *Poecilia reticulata*, using 16S ribosomal RNA and its internal transcribed spacer sequences. Bull Eur Assoc Fish Pathol. 2005;25:64–69.

67. Pourahmad F, Cervellione F, Thompson KD, Taggart JB, Adams A, Richards RH. *Mycobacterium stomatepiae* sp. nov., a slowly growing non-chromogenic species isolated from fish. Int J Syst Evol Microbiol. 2008;58:2821–2827. doi: 10.1099/ijs.0.2008/001164-0.

68. Martinez-Murcia AJ, Esteve C, Garay E, Collins MD. *Aeromonas allosaccharophila* sp. nov., a new mesophilic member of the genus *Aeromonas*. FEMS Microbiol Lett. 1992;91:199–206. doi: 10.1111/j.1574-6968.1992.tb05209.x.

69. Huys G, Kämpfer P, Swings J. New DNA-DNA hybridization and phenotypic data on the species *Aeromonas ichthiosmia* and *Aeromonas allosaccharophila*: A. *ichthiosmia is a later synonym of A. veronii* Hickman-Brenner et al. 1987. Syst Appl Microbiol. 2001;24:177–182. doi: 10.1078/0723-2020-00038.

70. Mikalsen J, Olsen AB, Tengs T, Colquohoun DJ. *Francisella philomiragia* subsp *noatunensis* subsp nov., isolated from farmed Atlantic cod (*Gadus morhua* L.) Int J Syst Evol Microbiol. 2007;57:1960–1965. doi: 10.1099/ijs.0.64765-0.

71. Nylund A, Ottem KF, Watanabe K, Karlsbakk E, Krossoy B. *Francisella* sp (Family Francisellaceae) causing mortality in Norwegian cod *(Gadus morhua)* farming. Arch Microbiol. 2006;185:383–392. doi: 10.1007/s00203-006-0109-5.

72. Birkbeck TH, Laidler LA, Grant AN, Cox DI. *Pasteurella skyensis* sp nov., isolated from Atlantic salmon *(Salmo salar* L.) Int J Syst Evol Microbiol. 2002;52:699–704. doi: 10.1099/ijs.0.01884-0.

73. Fryer JL, Lannan CN, Giovannoni SJ, Wood ND. *Piscirickettsia salmonis* gen. nov., the causative agent of an epizootic disease in salmonid fishes. Int J Syst Bacteriol. 1992;42:120–126. doi: 10.1099/00207713-42-1-120.

74. Arkush KD, Mcbride AM, Mendonca HL, Okihiro MS, Andree KB, Marshall S, Henriquez V, Hedrick RP. Genetic characterization and experimental pathogenesis of *Piscirickettsia salmonis* isolated from white seabass *Atractascion nobilis*. Dis Aquat Org. 2005;63:139–149. doi: 10.3354/dao063139.

75. Corbeil S, Hyatt AD, Crane MStJ. Characterisation of an emerging rickettsia-like organism in Tasmanian farmed Atlantic salmon *Salmo salar*. Dis Aquat Org. 2005;64:37–44. doi: 10.3354/dao064037.

76. Nishimori E, Kita-Tsukamoto K, Wakabayashi H. *Pseudomonas plecoglossicida* sp nov., the causative agent of bacterial haemorrhagic ascites of ayu, *Plecoglossus altivelis*. Int J Syst Evol Microbiol. 2000;50:83–89.

77. Romalde JL, Ravelo C, Valdes I, Magariños B, de la Fuente E, Martin CS, Avendano-Herrera R, Toranzo AE. *Streptococcus phocae*, an emerging pathogen for salmonid culture. Vet Microbiol. 2008;130:198–207. doi: 10.1016/j.vetmic.2007.12.021.

78. Pineiro-Vidal M, Carballas C, Gomez-Barreiro O, Riaza A, Santos Y. *Tenacibaculum soleae* sp nov., isolated from diseased sole *(Soleae senegalensis* Kaup) Int J Syst Evol Microbiol. 2008;58:881–885. doi: 10.1099/ijs.0.65539-0.

79. Lunder T, Sorum H, Holstad G, Steigerwalt AG, Mowinckel P, Brenner DJ. Phenotypic and genotypic characterization of *Vibrio viscosus* sp nov and *Vibrio wodanis* sp nov isolated from Atlantic salmon *(Salmo salar)* with "winter ulcer". Int J Syst Evol Microbiol. 2000;50:427–450.

80. Benediktsdóttir E, Verdonck L, Spröer C, Helgason S, Swings J.

Characterization of *Vibrio viscosus* and *Vibrio wodanis* isolated from different geographical locations: a proposal for re-classification of *Vibrio viscosus* as *Moritella viscose* comb. nov. Int J Syst Evol Microbiol. 2000;50:479–488.

81. Benediktsdóttir E, Helgason S, Sigurjónsdóttir H. *Vibrio* spp. isolated from salmonids with shallow skin lesions and reared at low temperature. J Fish Dis. 1988;21:19–28.
82. Urakawa H, Kita-Tsukamoto K, Stevens SE, Ohwada K, Colwell RR. A proposal to transfer *Vibrio marinus* (Russell 1891) to a new genus *Moritella* gen. nov. as *Moritella marina* comb. nov. FEMS Microbiol Lett. 1998;165:373–378. doi: 10.1111/j.1574-6968.1998.tb13173.x.
83. Del-Pozo J, Crumlish M, Ferguson HW, Turnbull JF. Retrospective cross-sectional study on "*Candidatus arthromitus*" associated rainbow trout gastroenteritis (RTGE) in the UK. Aquaculture. 2009;290:22–27. doi: 10.1016/j.aquaculture.2009.02.009.
84. Michel C, Bernardet JF, Daniel P, Chilmonczyk S, Urdaci M, de Kinkelin P. Clinical and aetiological aspects of a summer enteritic syndrome associated with the sporulating segmented filamentous bacterium "*Candidatus Arthromitus*" in farmed rainbow trout, *Oncorhynchus mykiss* (Walbaum) J Fish Dis. 2002;25:533–543. doi: 10.1046/j.1365-2761.2002.00400.x.
85. Del-Pozo J, Turnbull J, Ferguson H, Crumlish M. A comparative molecular study of the presence of "*Candidatus arthromitus*" in the digestive system of rainbow trout, *Oncorhynchus mykiss* (Walbaum), healthy and affected with rainbow trout gastroenteritis. J Fish Dis. 2010;33:241–250. doi: 10.1111/j.1365-2761.2009.01117.x.
86. Stevenson RMW, Daly JG. Biochemical and serological characteristics of Ontario isolates of *Yersinia ruckeri*. Can J Fish Aquat Sci. 1982;39:870–876.
87. Karlsen M, Nylund A, Watanabe K, Helvik JV, Nylund S, Plarre H. Characterization of "*Candidatus Clavochlamydia salmonicola*": an intracellular bacterium infecting salmonid fish. Environ Microbiol. 2008;10:208–218.
88. Larsen JL, Jensen NJ. An *Aeromonas* species implicated in ulcer-disease of the cod (*Gadus morhua*) Nord Veterinaermed. 1977;29:199–211.

89. Orozova P, Barker M, Austin DA, Austin B. Identification and pathogenicity to rainbow trout, *Oncorhynchus mykiss* (Walbaum), of some aeromonads. J Fish Dis. 2009;32:865–871. doi: 10.1111/j.1365-2761.2009.01065.x.

90. Carnahan AM, Altwegg M. In: The Genus Aeromonas. Austin B, Altwegg M, Gosling PJ, Joseph S, Wiley, Chichester, editor. 1996. Taxonomy; pp. 1–38.

91. Allen DA, Austin B, Colwell RR. *Aeromonas media*, a new species isolated from river water. Int J Syst Bacteriol. 1983;33:599–604. doi: 10.1099/00207713-33-3-599.

92. MacInnes JI, Trust TJ, Crosa JH. Deoxyribonucleic acid relationships among members of the genus *Aeromonas*. Can J Microbiol. 1979;25:579–586. doi: 10.1139/m79-083.

93. Esteve C. Numerical taxonomy of *Aeromonadaceae* and *Vibrionaceae* associated with reared fish and surrounding fresh and brackish water. System Appl Microbiol. 1995;18:391–402.

94. Esteve C, Biosca EG, Amaro C. Virulence of *Aeromonas hydrophila* and some other bacteria isolated from European eels *Anguilla anguilla* reared in fresh water. Dis Aquat Org. 1993;16:15–20. doi: 10.3354/dao016015.

95. Ali A, Carnahan AM, Altwegg M, Lüthy-Hotenstein J, Joseph SW. *Aeromonas bestiarum* sp. nov. (formerly genomospecies DNA group 2 *A. hydrophila)* a new species isolated from non-human sources. Med Microbiol Lett. 1996;5:156–165.

96. Toranzo AE, Baya AM, Romalde JJ, Hetrick FM. Association of *Aeromonas sobria* with mortalities of adult gizzard shad, *Dorosoma cepedianum* Lesueur. J Fish Dis. 1989;12:439–448. doi: 10.1111/j.1365-2761.1989.tb00555.x.

97. Ma Z, Yang H, Li T, Luo L, Gao JL. Isolation and identification of pathogenic *Aeromonas veronii* isolated from infected Siberian sturgeon (*Acipenser baerii*) Wei Sheng Wu Xue Bao. 2009;49:1289–1294. (article in Chinese)

98. Snieszko SF. In: Genus IV. *Aeromonas* Kluyver and van Niel 1936, Bergey's Manual of Determinative Bacteriology. 7. Breed RS, Murray EGD, Smith NR, editor. Williams and Wilkins, Baltimore; 1957. pp. 189–193.

99. Colwell RR, MacDonell MT, De Ley J. Proposal to recognize the family Aeromonadaceae fam. nov. Int J Syst Bacteriol. 1986;36:473–477. doi: 10.1099/00207713-36-3-473.

100. Griffin PJ, Snieszko SF, Friddle SB. A more comprehensive description of *Bacterium salmonicida*. Trans Am Fish Soc. 1953;82:129–138. doi: 10.1577/1548-8659(1952)82[129:AMCDOB]2.0.CO;2.

101. Austin DA, McIntosh D, Austin B. Taxonomy of fish associated *Aeromonas* spp., with the description of *Aeromonas salmonicida* subsp. *smithia* subsp. nov. Syst Appl Microbiol. 1989;11:277–290.

102. Duff DCB. Dissociation in *Bacillus salmonicida*, with special reference to the appearance of the G form of culture. J Bacteriol. 1937;34:49–67.

103. Pacha RE, Kiehn ED. Characterization and relatedness of marine vibrios pathogenic to fish: physiology, serology and epidemiology. J Bacteriol. 1969;100:1242–1247.

104. Koppang EO, Fjølstad M, Melgård B, Vigerust M, Sørum H. non-pigment-producing isolates of *Aeromonas salmonicida* subspecies *salmonicida:* isolation, identification, transmission and pathogenicity in Atlantic salmon, *Salmo salar* L. J Fish Dis. 2000;23:39–48. doi: 10.1046/j.1365-2761.2000.00205.x.

105. Nakatsugawa T. Atypical *Aeromonas salmonicida* isolted from cultured shotted halibut. Fish Pathol. 1994;29:193–198.

106. Kaku Y, Yamada Y, Wakabayashi H. Characterization of atypical *Aeromonas salmonicida* isolated from an epizootic ulcerative disease in carp (*Cyprinus carpio*) Fish Pathol. 1999;34:155–162.

107. Wiklund T, Bylund G. Skin ulcer disease of flounder *Platichthys flesus* in the northern Baltic Sea. Dis Aquat Org. 1993;17:165–174. doi: 10.3354/dao017165.

108. Austin B. Recovery of "atypical" isolates of *Aeromonas salmonicida*, which grow at 37°C, from ulcerated non-salmonids in England. J Fish Dis. 1993;16:165–168. doi: 10.1111/j.1365-2761.1993.tb00862.x.

109. Goodwin AE, Merry GE. Are all Koi ulcer cases associated with infection by atypical *Aeromonas salmonicida*? Polymerase chain reaction assays of Koi carp skin swabs submitted by hobbyists. J Aquat Anim Health. 2009;21:98–103. doi: 10.1577/H08-042.1.

110. Pedersen K, Kofod H, Dalsgaard I, Larsen JL. Isolation of oxidase-negative *Aeromonas salmonicida* from disease turbot *Scophthalmus*

maximus. Dis Aquat Org. 1994;18:149–154. doi: 10.3354/dao018149.

111. Wiklund T, Dalsgaard I, Eerola E, Olivier G. Characteristics of "atypical" cytochrome-oxidase negative *Aeromonas salmonicida* isolated from ulcerated flounder *(Platichthys flesus* (L.) J Appl Bacteriol. 1994;76:511–520.

112. Wiklund T, Dalsgaard I. Atypical *Aeromonas salmonicida* associated with ulcerated flatfish species in the Baltic Sea and the North Sea. J Aquat Anim Health. 1995;7:218–224. doi: 10.1577/1548-8667(1995)007<0218:AASAWU>2.3.CO;2.

113. Goldschmidt-Clermont E, Hochwartner O, Demarta A, Caminada AP, Frey J. Outbreaks of an ulcerative and haemorrhagic disease in Arctic char *Salvelinus alpinus* caused by *Aeromonas salmonicida* subsp *smithia*. Dis Aquat Org. 2009;86:81–86. doi: 10.3354/dao02110.

114. Austin B, Austin DA, Dalsgaard I, Gudmundsdóttir BK, Høie S, Thornton JM, Larsen JL, O'Hici B, Powell R. Characterization of atypical *Aeromonas salmonicida* by different methods. Syst Appl Microbiol. 1998;21:50–64.

115. Høie S, Dalsgaard I, Aase IL, Heum M, Thornton JM, Powell R. Polymerase chain reaction (PCR) based typing analysis of atypical isolates of the fish pathogen *Aeromonas salmonicida*. Syst Appl Microbiol. 1999;22:403–411.

116. Lund V, Jenssen LM, Wesmajervi MS. Assessment of genetic variability and relatedness among atypical *Aeromonas salmonicida* from marine fishes, using AFLP fingerprinting. Dis Aquat Org. 2002;50:119–126. doi: 10.3354/dao050119.

117. Yamada Y, Kaku Y, Wakabayashi H. Phylogenetic intrarelationships of atypical *Aeromonas salmonicida* isolated in Japan as determined by 16S rDNA sequencing. Fish Pathol. 2000;35:35–40.

118. Smith IW. The classification of *Bacterium salmonicida*. J Gen Microbiol. 1963;33:263–274.

119. Kimura T. A new subspecies of *Aeromonas salmonicida* as an etiological agent of furunculosis on "Sakuramasu" *(Oncorhynchus masou)* and pink salmon *(O. gorbuscha)* rearing for maturity. Fish Pathol. 1969;3:34–44. Part 1. On the serological properties.

120. Schubert RHW. In: Genus II. *Aeromonas* Kluyver and van Niel 1936, 398, Bergey's Manual of Determinative Bacteriology. 8th. Buchanan

RE, Gibbons NE, editor. Williams and Wilkins, Baltimore; 1974. pp. 345–348.

121. Popoff M. Bacterioses, generalities - *Aeromonas* et Aeromonose. Mimeograph. Institut National de la Recherche Agronomique; 1970. pp. 1–38.

122. Paterson WD, Douey D, Desautels D. Relationships between selected strains of typical and atypical *Aeromonas salmonicida, Aeromonas hydrophila*, and *Haemophilus piscium*. Can J Microbiol. 1980;26:588–598. doi: 10.1139/m80-104.

123. Snieszko SF, Griffin P, Friddle SB. A new bacterium *(Haemophilus piscium* n. sp.) from ulcer disease of trout. J Bacteriol. 1950;59:699–710.

124. Thornton JM, Austin DA, Austin B, Powell R. Small subunit rRNA gene sequences of *Aeromonas salmonicida* subsp. *smithia* and *Haemophilus piscium* reveal pronounced similarities with *A. salmonicida* subsp. *salmonicida.* Dis Aquat Org. 1999;35:155–158. doi: 10.3354/dao035155.

125. McCarthy DH. Some ecological aspects of the bacterial fish pathogen - *Aeromonas salmonicida.* Aquatic Microbiology Symp Soc Appl Bacteriol. 1980;6:299–324.

126. Belland RJ, Trust TJ. DNA:DNA reassociation analysis of *Aeromonas salmonicida.* J Gen Microbiol. 1988;134:307–315.

127. Eddy BP. Cephalotrichous, fermentative Gram-negative bacteria: the genus *Aeromonas.* J Appl Bacteriol. 1960;23:216–248.

128. Kluyver AJ, Van Niel CB. Prospects for a natural system of classification of bacteria. Z Bakteriol Parasiten, Infekt Hyg, Orig. 1936;94:369–403. 2 Abt.

129. Cowan ST. Cowan and Steel's Manual for the Identification of Medical Bacteria. 2. Cambridge University Press, Cambridge; 1974.

130. Kimura T. A new subspecies of *Aeromonas salmonicida* as an etiological agent of furunculosis on "Sakuramasu" *(Oncorhynchus masou)* and pink salmon *(O. gorbuscha)* rearing for maturity. Part 2. On the morphological and physiological properties. Fish Pathol. 1969;3:45–52.

131. Liu PV. Observations on the specificities of extra-cellular antigens of the genera *Aeromonas* and *Serratia.* J Gen Microbiol. 1961;24:145–153.

132. Popoff M. Étude sur les *Aeromonas salmonicida*. II. Caracterisation des bactériophages actifs sur les "*Aeromonas salmonicida*" et lysotypie. Ann Recherch Vét. 1971;2:33–45.

133. Popoff M. Intérêt diagnostique d'un bactériophage spécifique des *Aeromonas salmonicida*. Ann Rech Vét. 1971;2:137–139.

134. Kilian M. A taxonomic study of the genus *Haemophilus*, with the proposal of a new species. J Gen Microbiol. 1976;93:9–62.

135. Broom AK, Sneath PHA. Numerical taxonomy of *Haemophilus*. J Gen Microbiol. 1981;126:123–149.

136. Trust TJ, Ishiguro EE, Atkinson HM. Relationship between *Haemophilus piscium* and *Aeromonas salmonicida* revealed by *Aeromonas hydrophila* bacteriophage. FEMS Microbiol Lett. 1980;9:199–201. doi: 10.1111/j.1574-6968.1980.tb05636.x.

137. Dalsgaard I, Gudmundsdóttir BK, Helgason S, Høie S, Thoresen OF, Wichardt UP, Wiklund T. Identification of atypical *Aeromonas salmonicida*: inter-laboratory evaluation and harmonization of methods. J Appl Microbiol. 1998;84:999–1006. doi: 10.1046/j.1365-2672.1998.00435.x.

138. Ross AJ, Rucker RR, Ewing WH. Description of a bacterium associated with redmouth disease of rainbow trout (*Salmo gairdneri*) Can J Microbiol. 1966;12:763–70. doi: 10.1139/m66-103.

139. Steinum T, Sjastad K, Falk K, Kvellestad A, Colquhuin DJ. An RT PCR-DGGE survey of gill-associated bacteria in Norwegian seawater-reared Atlantic salmon suffering proliferative gill inflammation. Aquaculture. 2009;293:172–179. doi: 10.1016/j.aquaculture.2009.05.006.

140. Ewing WH, Ross AJ, Brenner DJ, Fanning GR. *Yersinia ruckeri* sp. nov., the redmouth (RM) bacterium. Int J Syst Bacteriol. 1978;28:37–44. doi: 10.1099/00207713-28-1-37.

141. Steigerwalt AG, Fanning GR, Fife-Asbury MA, Brenner DJ. DNA relatedness among species of *Enterobacter* and *Serratia*. Can J Microbiol. 1976;22:121–137. doi: 10.1139/m76-018.

142. Bercovier H, Mollaret HH. In: Bergey's Manual of Systematic Bacteriology. Krieg NR, Holt JG, editor. I. Williams and Wilkins, Baltimore; 1984. Genus XIV. *Yersinia* van Loghem 1944, 15AL; pp. 489–506.

143. Austin DA, Robertson PAW, Austin B. Recovery of a new biogroup of *Yersinia ruckeri* from diseased rainbow trout *(Oncorhynchus mykiss*, Walbaum) Syst Appl Microbiol. 2005;26:127–131. doi: 10.1078/072320203322337416.

144. Fouz B, Zarza C, Amaro C. First description of non-motile *Yersinia ruckeri* serovar I strains causing disease in rainbow trout, *Oncorhynchus mykiss* (Walbaum), cultured in Spain. J Fish Dis. 2006;29:339–346. doi: 10.1111/j.1365-2761.2006.00723.x.

145. Smith IW. A disease of finnock due to *Vibrio anguillarum.* J Gen Microbiol. 1961;74:247–252.

146. Kaper JB, Lockman H, Remmers EF, Kristensen K, Colwell RR. Numerical taxonomy of vibrios isolated from estuarine environments. Int J Syst Bacteriol. 1983;33:229–255. doi: 10.1099/00207713-33-2-229.

147. Pazos F, Santos Y, Magariños B, Bandín I, Núñez S, Toranzo AE. Phenotypic characteristics and virulence of *Vibrio anguillarum* - related organisms. Appl Environ Microbiol. 1993;59:2969–2976.

148. Olsen JE, Larsen JL. Ribotypes and plasmid content of *Vibrio anguillarum* strains in relation to serovar. Appl Environ Microbiol. 1993;59:3863–3870.

149. Austin B, Alsina M, Austin DA, Blanch AR, Grimont F, Grimont PAD, Jofre J, Koblavi S, Larsen JL, Pedersen K, Tiainen T, Verdonck L, Swings J. Identification and typing of *Vibrio anguillarum* a comparison of different methods. Syst Appl Microbiol. 1995;18:285–302.

150. Austin B, Austin DA, Blanch AR, Cerdà M, Grimont F, Grimont PAD, Jofre J, Koblavi S, Larsen JL, Pedersen K, Tiainen T, Verdonck L, Swings J. A comparison of methods for the typing of fish-pathogenic *Vibrio* spp. Syst Appl Microbiol. 1997;20:89–101.

151. Schiewe MH. Taxonomic status of marine vibrios pathogenic for salmonid fish. Dev Biol Stand. 1981;49:149–158.

152. Muthiara LW, Raymond BT, Dekievit TR, Stevenson RHW. Antibody specificities of polyclonal rabbit and rainbow trout antisera against *Vibrio ordalii* and serotype O:2 strains of *Vibrio anguillarum.* Can J Microbiol. 1993;39:492–499. doi: 10.1139/m93-070.

153. Sørensen UBS, Larsen JL. Serotyping of *Vibrio anguillarum.* Appl Environ Microbiol. 1986;51:593–597.

154. Grisez L, Ollevier F. Comparative serology of the marine fish pathogen *Vibrio anguillarum*. Appl Environ Microbiol. 1995;61:4367–4373.
155. Pedersen K, Dalsgaard I, Larsen JL. Characterization of atypical *Aeromonas salmonicida* isolates by ribotyping and plasmid profiling. J Appl Bacteriol. 1996;80:37–44.
156. Pedersen K, Tiainen T, Larsen JL. Plasmid profiles, restriction length polymorphisms and O-serotypes among *Vibrio anguillarum* isolates. Epidemiol Infect. 1996;117:471–478. doi: 10.1017/S0950268800059136.
157. Kitao T, Aoki T, Fukodome M, Kawano K, Wada Y, Mizuno Y. Serotyping of *Vibrio anguillarum* isolated from diseased freshwater fish in Japan. J Fish Dis. 1983;6:175–181. doi: 10.1111/j.1365-2761.1983.tb00064.x.
158. Tiainen T, Pedersen K, Larsen JL. Immunological reactivity of *Vibrio anguillarum* sero-subgroups O2a and O2b, and comparison of their lipopolysaccharide profiles. Curr Microbiol. 1997;34:38–42. doi: 10.1007/s002849900141.
159. O'Hici B, Olivier G, Powell R. Genetic diversity of the fish pathogen *Aeromonas salmonicida* demonstrated by random amplified polymorphic DNA and pulsed-field gel electrophoresis analyses. Dis Aquat Org. 2000;39:109–119.
160. Umelo E, Trust TJ. Identification and molecular characterization of two tandomly located flagellin genes from *Aeromonas salmonicida* A449. J Bacteriol. 1997;179:5292–5299.
161. McIntosh D, Austin B. Atypical characteristics of the salmonid pathogen *Aeromonas salmonicida*. J Gen Microbiol. 1991;137:1341–1343.

CHAPTER 18

Flavobacterium Plurextorum sp. nov. Isolated from Farmed Rainbow Trout (Oncorhynchus Mykiss)

Leydis Zamora[1] , Jose´ F. Ferna´ndez-Garayza´bal[1,2], Cristina Sa´nchez-Porro[4] , Mari Angel Palacios[3] , Edward R. B. Moore[5] , Lucas Domı´nguez[1] , Antonio Ventosa[4] , Ana I. Vela[1,2]

[1] Centro de Vigilancia Sanitaria Veterinaria (VISAVET), Universidad Complutense, Madrid, Spain

[2]Departamento de Sanidad Animal, Facultad de Veterinaria, Universidad Complutense, Madrid, Spain

[3] Piszolla, S.L., Alba de Tormes, Salamanca, Spain

[4] Departamento de Microbiologı´a y Parasitologı´a, Facultad de Farmacia, Universidad de Sevilla, Sevilla, Spain

[5] Culture Collection University of Gothenburg (CCUG) and Department of Infectious Disease, Sahlgrenska Academy of the University of Gothenburg, Go¨teborg, Sweden

ABSTRACT

Five strains (1126-1H-08[T], 51B-09, 986-08, 1084B-08 and 424-08) were isolated from diseased rainbow trout. Cells were Gram-negative rods, 0.7 μm wide and 3 μm long, non-endospore-forming, catalase and oxidase positive.

Citation: Zamora L, Fernández-Garayzábal JF, Sánchez-Porro C, Palacios MA, Moore ERB, Domínguez L, et al. (2013) Flavobacterium plurextorum sp. nov. Isolated from Farmed Rainbow Trout (Oncorhynchus mykiss). PLoS ONE 8(6): e67741. https://doi.org/10.1371/journal.pone.0067741

Colonies were circular, yellow-pigmented, smooth and entire on TGE agar after 72 hours incubation at 25°C. They grew in a temperature range between 15°C to 30°C, but they did not grow at 37°Cor 42°C. Based on 16S rRNA gene sequence analysis, the isolates belonged to the genus *Flavobacterium*. Strain 1126-1H-08^{T} exhibited the highest levels of similarity with *Flavobacterium oncorhynchi* CECT 7678^{T} and *Flavobacterium pectinovorum* DSM 6368^{T} (98.5% and 97.9% sequence similarity, respectively). DNA–DNA hybridization values were 87 to 99% among the five isolates and ranged from 21 to 48% between strain 1126-1H-08^{T}, selected as a representative isolate, and the type strains of *Flavobacterium oncorhynchi* CECT 7678^{T} and other phylogenetic related *Flavobacterium* species. The DNA G+C content of strain 1126-1H-08^{T} was 33.2 mol%. The predominant respiratory quinone was MK-6 and the major fatty acids were iso-$C_{15:0}$ and $C_{15:0}$. These data were similar to those reported for *Flavobacterium* species. Several physiological and biochemical tests differentiated the novel bacterial strains from related *Flavobacterium* species. Phylogenetic, genetic and phenotypic data indicate that these strains represent a new species of the genus *Flavobacterium,* for which the name *Flavobacterium plurextorum* sp. nov. was proposed. The type strain is 1126-1H-08^{T} (=CECT 7844^{T}=CCUG 60112^{T}).

INTRODUCTION

The genus *Flavobacterium* is the type genus of the family *Flavobacteriaceae* accommodating Gram-negative, non-endospore-forming, aerobic, oxidase-positive, non-fermenting, predominantly gliding, yellow-pigmented bacteria [1], [2]. The genus, initially described to accommodate seven species, has considerably expanded with the description of many new species. Currently it includes 99 species, many of them described during the last five years. [3]. Members of the genus *Flavobacterium* can been isolated from a number of diverse habitats such as soil, water, sludge, plants, food products such as fish, meat, poultry, milk or lactic acid beverages [2], [4]. Most species are non-pathogenic, although a number of species have been associated with different clinical infections, being freshwater fish the animals most prone to flavobacterial infections [5]. Some *Flavobacterium* species, mainly *Flavobacterium columnare, Flavobacterium branchiophilum* and *Flavobacterium psychrophilum,* are well-recognized fish pathogens responsible for important economic losses in the fish farming industry [6], [7]. However, several other species such as *Flavobacterium hydatis*, *Flavobacterium jhonsoniae*, *Flavobacterium succinicans*, *Flavobacterium chilense*, *Flavobacterium araucananum* or *Flavobacterium oncorhynchi* have been also associated with infections in fish [1], [4], [5], [8]–[10]. Additionally, a number of new *Flavobacterium* species also have been described from the water of aquaculture facilities [11]–[13]. This plethora of *Flavobacterium* species could reproduce the diversity of flavobacteria associated with fish or fish surrounding environments. Some of these species could be considered commensal and opportunistic pathogenic bacteria [4], which point out the necessity for an accurate identifications of those strains of *Flavobacterium* spp. isolated from fish or fish farm environments. However, such identifications are extremely difficult based exclusively on biochemical criteria [4], [8], [14] and must be complemented with chemotaxomic and genetic methods [4], [5].

In this article, we report the phenotypic, genotypic and phylogenetic characterization of five novel *Flavobacterium*-like strains isolated from diseased trout. Based on the presented findings, a new species of the genus *Flavobacterium*, *Flavobacterium plurextorum* sp. nov., is proposed.

MATERIALS AND METHODS

The present work does not include any experimental infections trial with farmed trout, just trout exclusively were used to identify microbiologically the etiological agent of the bacterial septicemia. Therefore, we did not consult with the IACUC and no specific national regulations for these procedures are available. Nevertheless, in order to ensure the welfare and ameliorate suffering of trout during transportation to the laboratory and euthanasia, trout were handled according to guidelines of relevant international organisms such as OIE (http://www.oie.int/doc/ged/D7821.PDF) and AVMA (https://www.avma.org/KB/Policies/Documents/euthanasia.pdf) and they were further necropsied under aseptic conditions. In addition, these procedures were approved by the responsible of animal welfare of the UCM Animal Health Department. The trout were sacrificed for the purpose of the study and the sacrifice was approved by the Technical Manager (Mari Angel Palacios, DVM, PhD) of the fish farm located in the west of Spain.

Trout and Strain Isolation

A clinical episode of septicemia occurred in a rainbow trout (*Oncorhynchus mykiss*) farm located in the central region of Spain. Affected trout were submitted by the Technical Manager of the fish farm to the Animal Health Surveillance Centre (VISAVET) of the Universidad Complutense (Madrid, Spain) for a confirmatory microbiological diagnosis.

Five Gram-negative, rod-shaped bacteria were isolated from liver (strains 986-08 and 424-08), gills (strains 1084B-08 and 51B-09) and eggs (1126-1H-08^T) of five different trout. The strains were recovered in two different years (2008 and 2009) and they were isolated on tryptone glucose extract agar (TGE; Difco) after incubation at 25°C for 72 hours under aerobic conditions.

Phylogenetic Analysis

A large continuous sequence (approximately 1,400 bases) of the 16S rRNA gene of five strains was determined bidirectionally using universal primers pA (*5′- AGAGTTTGATCCTGGCTCAG*, positions 8–27, *Escherichia coli* numbering) and pH* (*5′-AAGGAGGTGATCCAGCCGCA*, positions 1541–1522, *E. coli* numbering) as described previously [10], and subjected to a comparative analysis. The identification of the phylogenetic relatives and calculations of pair-wise 16S rRNA gene sequence similarities were achieved, using the EzTaxon-e server [15]. The 16S rRNA gene sequences of the type strains of all validly published species of the genus *Flavobacterium* were retrieved from GenBank and aligned with the newly determined sequences using the program SeqTools [16]. Phylogenetic trees were constructed according to three different algorithms: neighbour-joining [17], using the programs SeqTools and TREEVIEW [18]; maximum-likelihood, using the PHYML software [19]; and maximum-parsimony, using the software package MEGA (Molecular Evolutionary Genetics Analysis) version 5.0 [20]. Genetic distances for the neighbour-joining and the maximum-likelihood algorithms were calculated by the Kimura two-parameter [21] and close-neighbour-interchange (search level=2, random additions=100) was applied in the maximum-parsimony analysis. The stability of the groupings was estimated by bootstrap analysis (1000 replications).

Genomic DNA G+C Content Determination and DNA-DNA Hybridizations

The G+C content of the genomic DNA of a representative strain (1126-1H-08^T) was determined from the mid-point value (Tm) of the thermal denaturation profile [22], obtained with a Perkin-Elmer UV-Vis Lambda 20 spectrophotometer at 260 nm.

Genomic DNA-DNA hybridizations were carried out between strains 1126-1H-08^T, 986-08, 424-08, 1084B-08 and 51B-09, and between strain 1126-1H-08^T and the type strains of the closest phylogenetically related species. DNA was extracted and purified by the method of Marmur [22]. Hybridization studies were carried out, using the membrane method of Johnson [23], described in detail by Arahal *et al.* [24]. The hybridization experiments were carried out under optimal conditions, at a temperature of 44°C, which is within the limits of validity for the membrane method [25]. The percentages of hybridization were calculated as described by Johnson [26]. Three independent determinations were carried out for each experiment and the results reported as mean values. The type strains of species *F. aquidurense* CCUG 59847^T, *F. araucananum* CCUG 61031^T, *F. hydatis* DSM 2063^T, *F. pectinovorum* CCUG 58916^T, *F. frigidimaris* CCUG 59364^T, *F. chungangense* CCUG 58910T and *F. oncorhynchi* CECT 7678^T were included in this study.

Chemotaxonomic Characteristics

Respiratory quinones of strain 1126-1H-08^T were extracted from 100 mg of freeze-dried cell material, using the two stage method described by Tindall [27], [28], and further separated by thin layer chromatography on silica gel and analyzed, using HPLC, by the identification service of the DSMZ (Braunschweig, Germany).

For cell fatty acid-fatty acid methyl ester (CFA-FAME) analyses, strain 1126-1H-08^T was grown on Columbia II agar base (BBL 4397596) with 5% horse blood, at 30°C for 30–48 h, under aerobic conditions. The CFA-FAME profile was determined using gas chromatography (Hewlett Packard HP 5890) and a standardized protocol similar to that of the MIDI Sherlock MIS system [29], described previously [10]. CFAs were identified and the relative amounts were expressed as percentages of the total fatty acids of the respective strains.

Morphological, Physiological and Biochemical Characteristics

The minimal standards for the description of new taxa in the family *Flavobacteriaceae* [30] were followed for the phenotypic characterization of the strains. Gram-staining was performed as described by Smibert & Krieg [31]. Oxidase activity was determined by monitoring the oxidation of tetramethyl-*p*-phenylenediamine on filter paper and catalase activity was determined, using 3% H_2O_2 solution [31]. Hydrolysis of L-tyrosine (0.5%, w/v), lecithin (5%, w/v) [31], esculin (0.01% esculin and 0.05% ferric citrate, w/v), gelatin (4%; w/v), starch (0.2%, w/v), and casein [50% skimmed milk (Difco), v/v] were tested using nutrient agar as basal medium [30]. DNase test agar (Difco) was used for the DNase assay. Hydrolysis of urea (1%, w/v) was tested as described by Bowman *et al.* [32]. Growth in brain heart infusion broth was assessed at 15, 25, 30, 37 and 42°C, with 3.0, 4.5 and 6.5% added NaCl, and under anaerobic (with 4–10% CO_2) and micro-aerobic (with 5–15% O_2 and 5–12% CO_2) conditions, using GasPak Plus and CampyPak Plus systems (BBL), respectively. Growth was tested on MacConkey (bioMérieux), nutrient (Difco) and trypticase-soy (bioMérieux) agar plates. The presence of gliding motility, using the hanging drop technique, and the production of flexirubin-type pigments and extracellular glycans were assessed, using the KOH and Congo red tests, respectively [1]. The strains were further biochemically characterized using the API 20NE and API ZYM systems (bioMérieux) according to the manufacturer's instructions, except that incubation temperature was 25°C. The type strains of species *F. aquidurense* CCUG 59847^T, *F. araucananum* CCUG 61031^T, *F. hydatis* DSM 2063^T, *F. pectinovorum* CCUG 58916^T, *F. frigidimaris* CCUG 59364^T, *F. chungangense* CCUG 58910^T and *F. oncorhynchi* CECT 7678^T were included in this study as references for the investigation of the phenotypic properties of the trout strains, using the same laboratory conditions.

PGFE Typing

The five strains were characterized by pulsed-field gel electrophoresis (PFGE), after digestion of their genomic DNAs with the restriction enzymes *Bsp120*I and *Xho*I, according to the specifications of Chen *et al.* [33]. DNA fragments were resolved in a 1% agarose gel with a pulse-field gel electrophoresis apparatus, CHEF-DR III (Bio-Rad), at 6V/cm for 40 hours, with switching times ramped from 0.1 to 12 s at 14°C, with an angle of 120°. The gels were stained for 30 min with Syber-Safe and photographed under UV light (Gel-Doc, Bio-Rad). Strains differing in at least one band were considered different.

RESULTS AND DISCUSSION

16S rRNA gene sequences were determined for the five trout strains, displaying 100% 16S rRNA sequence similarity among them. Sequence searches showed that the 16S rRNA gene sequence of the strains were most similar to those of species of the genus *Flavobacterium*, exhibiting the highest levels of similarity with the sequence of the type strains of *Flavobacterium oncorhynchi* CECT 7678^T and *Flavobacterium pectinovorum* DSM 6368^T (98.5% and 97.9% sequence similarity, respectively). In addition, strains exhibited 16S rRNA gene sequence similarities greater than 97.0% with other seventeen other *Flavobacterium* species. It is clear from the phylogenetic analysis (Fig. 1) that the trout strains held a clear affiliation to the genus *Flavobacterium* and represented a distinct sub-lineage clustering with a cluster of four species that included *F. pectinovorum, F. chilense, F. oncorhynchi* and *F. hercynium*. However, their position within this sub-group was not supported by significant bootstrap values. The GenBank accession numbers for the 16S rRNA gene sequences of five strains sequenced in this study are shown in Fig. 1.

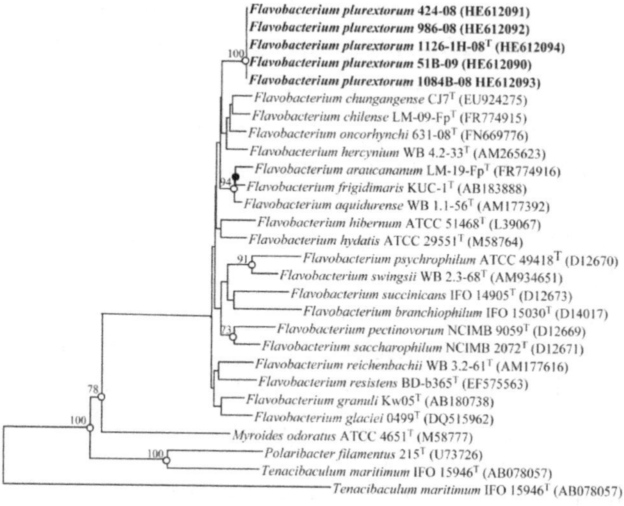

Figure 1: Phylogenetic tree based on 16S rRNA gene sequence comparisons, obtained with the neighbour-joining algorithm, showing the relationships of *Flavobacterium plurextorum* **sp.** nov. with related species.

Flexibacter flexilis ATCC 23079^T was used as an outgroup. Bootstrap values (expressed as a percentage of 1,000 replications) greater than 70% are given at the nodes. Solid circles indicate that the corresponding nodes (groupings) are also obtained on the maximum-likelihood tree. Open circles indicate that the corresponding nodes (groupings) are also obtained on the maximum-likelihood and parsimony trees. Sequence accession numbers are indicated in brackets. Bar, 1% sequence divergence.

Genomic DNA–DNA hybridizations between the trout strains yielded binding values of 87 to 100%. *Flavobacterium* species with 16S rRNA gene sequence similarities to the sequences of the trout strains lower than 98.0% correlated with levels of genomic DNA-DNA relatedness always lower than 70% [9]–[11], [34]–[36]. For that reason, DNA-DNA hybridizations were carried out only between strain 1126-1H-08^T and the type strains of the phylogenetically closest related species; *i.e.*, those species with 16S rRNA gene sequence similarities greater than 97.5%. The levels of DNA-DNA relatedness for strain 1126-1H-08^T with respect to *F. aquidurense* CCUG 59847^T, *F. araucananum* CCUG 61031^T, *F. hydatis* DSM 2063^T, *F. pectinovorum* CCUG 58916^T, *F. frigidimaris* CCUG 59364^T, *F. chungangense* CCUG 58910^T and *F. oncorhynchi* CECT 7678^T ranged between 21 and 48%. These values were below the 70% cut-off point for species delineation [37], [38] and clearly confirmed that the trout strains belong to a distinct genomic species of the genus *Flavobacterium*. The DNA G+C content of strain 1126-1H-08^T was 33.2 mol%, a value consistent with those of the genus *Flavobacterium* [1], [30].

Chemotaxonomic characteristics of strain 1126-1H-08^T were in accordance with those of members of the genus *Flavobacterium* [5], [6]: the major quinone was MK-6 (95%) with minor amounts of MK-5 (5%). The predominant cell fatty acids of strain 1126-1H-08^T were iso-$C_{15:0}$ (19%) and $C_{15:0}$ (15%). Strain 1126-1H-08^T also contained moderate or small amounts of $C_{16:1}$ *ω7c* (10%), $C_{15:1}$ *ω6c* (9%), iso-$C_{15:0}$ 3-OH, $C_{17:1}$ *ω6c*, isoG-$C_{15:1}$ (6%/each), iso-$C_{17:0}$ 3-OH (5%), iso-$C_{17:1}$ *ω9c*, $C_{15:0}$ 3-OH, $C_{16:0}$ 3-OH (3%/each), isoaldehyde-$C_{15:0}$, $C_{16:0}$, iso-$C_{16:0}$ 3-OH, unknown fatty acids with an equivalent chain length of 11.5 (2%/each) and $C_{17:1}$ *ω8c*, iso-C16:0, $C_{12:1}$, aldehyde-$C_{14:0}$, anteiso-$C_{15:0}$ and unknown fatty acids with an equivalent chain lengths of 14.8 and 12.5 (1%/each) (Table 1).

Table 1: Cellular fatty acid compositions of *Flavobacterium plurextorum* **1126-1H-08**^{T} **and its** closest phylogenetic neighbours

Fatty acid	1	2	3	4	5	6	7	8
Saturated								
$C_{12:1}$	1	tr	–	–	–	–	tr	–
$C_{14:0}$	tr	–	tr	1.1	tr	tr	–	tr
$C_{15:0}$	15	13.5	11.9	5.5	5.6	6.9	20.6	15.7
$C_{16:0}$	2	1.6	1.1	2.3	2.8	2.2	tr	2.9
Hydroxy								
$C_{15:0}$ 2OH	–	–	1.1	–	tr	–	–	–
$C_{15:0}$ 3OH	3	3.3	1.9	–	–	–	1.8	–
iso-$C_{15:0}$ 3OH	6	7.8	6.9	7.7	8.6	5.8	7.1	5.8
$C_{16:0}$ 3OH	3	–	1.1	3.5	4.5	1.4	–	2.5
iso-$C_{16:0}$ 3OH	2	1.0	2.1	1.6	1.9	tr	2.1	1.5
iso-$C_{17:0}$ 3OH	5	8.2	7.3	5.9	10.3	5.1	7.0	
Branched								
$C_{14:0}$ aldehyde	1.0	–	–	–	–	–	–	tr
iso-$C_{15:0}$	19	26.1	14.6	28.2	23.5	28.0	24.8	25.5
anteiso-$C_{15:0}$	1.0	1.3	3.0	3.2	tr	4.3	2.5	1.9
iso-$C_{15:0}$ aldehyde	2.0	3.2	1.2	1.3	tr	1.3	2.3	2.0
iso-$C_{15:1}$ G	6.0	2.9	7.4	3.7	5.8	7.2	5.0	5.0
iso-$C_{16:0}$	1	–	1.1	1.0	tr	1.0	–	1.1
iso-$C_{16:1}$ H	tr	–	1.0	1.0	tr	–	–	tr
iso-$C_{17:1}$ ω9c	3	6.0	5.2	4.3	4.1	6.0	1.1	2.9
Unsaturated								

$C_{15:1}$ ω6c	9	12.3	10.1	4.1	2.9	5.5	12.2	7.6
$C_{16:1}$ ω7c	10	3.7	11.2	19.2	15.7	18.1	2.2	9.8
$C_{17:1}$ ω6c	6	5.9	6.4	3.5	2.5	3.2	6.2	2.4
$C_{17:1}$ ω8c	1	1.0	1.5	–	tr	tr	1.3	tr
Summed feature 1 [a]	–	–	–	2.0	1.7	1.4	–	1.7
Unidentified fatty acid [b]								
ECL 11.541	2	1.4	tr	tr	tr	tr	1.2	1.1
ECL 12.555	1	–	tr	–	–	–	1.1	tr
ECL 14.809	1	–	–	–	–	–	–	–
ECL 16.580	–	–	tr	–	1.1	–	–	tr

Taxa: 1, *F. plurextorum* 1126-1H-08ᵀ; 2, *F. pectinovorum* CCUG 58916ᵀ; 3, *F. aquidurense* CCUG 59847ᵀ; 4, *F. frigidimaris* CCUG 59364ᵀ; 5, *F. hydatis* DSM 2063ᵀ; 6, *F. araucananum* CCUG 61031ᵀ; 7, *F. chungangense* CCUG 58910ᵀ; 8, *F. oncorhynchi* CECT 7678ᵀ.

Values are percentages of total fatty acids; fatty acids representing less than 1% in all strains were omitted. tr=trace amount, i.e., <1%. -=not detected.

CFA values for type strains other than *F. plurextorum* 1126-1H-08ᵀ were taken from the CCUG culture collection (http://www.ccug.se/). Strains were cultivated on the same medium and growth conditions.

[a]Summed features represent groups of two or three fatty acids that cannot be separated by GLC with the MIDI system. Summed feature 1 comprised iso-$C_{17:1}$ I/$C_{16:0}$ DMA.

[b]ECL, equivalent chain length.

The trout strains exhibited identical physiological and biochemical characteristics. Cells were Gram-negative rods, 0.7 μm wide and 3 μm long, non-endospore-forming, and non-gliding. Strains grew well under aerobic conditions and grew weakly under micro-aerobic conditions. Strains grew at 15–30°C with optimal growth at approximately 25°C, while no growth was observed at 37°C or 42°C. Growth occurred on trypticase-soy and nutrient agars but not on Marine agar after incubation at 25°C for 72 hours. Colonies were circular, yellow-pigmented, smooth and entire on

TGE agar after 72 hours incubation at 25°C. Colonies are non-hemolytic on Columbia agar after 72 hours incubation at 25°C. Diffusible flexirubin-type pigments were produced and congo red was not absorbed by colonies. Growth did not occur in brain heart infusion broth containing 3, 4.5 and 6.5% NaCl. Catalase and oxidase were produced and nitrate and nitrite were reduced. Starch and tyrosine were degraded but DNA, gelatin, casein or agarose were not. A brown pigment was not produced on tyrosine agar. Aesculin was hydrolyzed but not urea, lecithin and arginine. Indole and H_2S were not produced. Acid was not produced from D-glucose. Arabinose, D-glucose, mannose, N-acetyl-glucosamine, and maltose were used as sole carbon and energy sources but not citrate, mannitol, gluconate, caprate, adipate, and malate. Activities for alkaline phosphatase, leucine arylamidase, N-acetyl-β-glucosaminidase, α-glucosidase, acid phosphatase, and naphthol-AS-BI-phophohydrolase were detected. Esterase C4, valine arylamidase, β-galactosidase, ester lipase C8, lipase C14, cystine arylamidase, α-chymotrypsin, trypsin, α-galactosidase, β-glucuronidase, β-glucosidase, α-mannosidase and α-fucosidase were not detected.

The phenotypic characteristics that differentiated the trout strains from phylogenetically related species are shown in Table 2. The new species also can be also differentiated from the clinically relevant fish pathogens *F. columnare*, *F. psycrophilum* and *F. branchiophilum*, by the inability of these three species to grow in trypticase-soy agar and to hydrolyze aesculin [4]. Other species isolated from diseased fish such as *F. hydatis*, *F. jonshoniae* and *F. succinicans* are motile (gliding), degrade DNA and produce acid from carbohydrates [4], while the new species exhibited opposite results for those tests. Moreover, the new species can be readily differentiated from *F. chilense* and *F. araucananum* because the latter species are motile (gliding), grow in 3% NaCl and assimilate mannitol [9] and from *F. oncorhynchi* which produces β-galactosidase while the new species give opposite results for this test [10].

Table 2: Characteristics that differentiate *Flavobacterium plurextorum* **sp.** nov. from closely related *Flavobacterium* **species based** in the 16S rRNA tree topology

Characteristic	1	2	3	4	5	6	7	8
Growth on Marine agar	–	–	–	+	–	–	–	–
Growth at 30°C	+	+	+	–	+	+	+	+
Hydrolysis of:								

L- tyrosine	+	–	+	–	+	–	–	+
DNA	–	–	–	–	+	–	+	–
Urea	–	–	+	–	–	–	–	–
Nitrate reduction	+	+	–	–	+	+	+	+
Assimilation of:								
Arabinose	+	+	+	+	–	+	+	+
Mannitol	–	–	–	+	–	–	–	–
N-acetyl-glucosamine	+	+	+	+	+	+	-	+
Production of:								
Valine arylamidase	–	+	+	+	+	+	+	–
α-Glucosidase	+	–	–	+	+	+	–	+
β-Glucosidase	–	+	–	+	–	+	+	–
N-Acetyl-β-glucosaminidase	+	–	–	+	+	–	–	+

Taxa: 1, *F. plurextorum* 1126-1H-08^T; 2, *F. pectinovorum* CCUG 58916^T; 3, *F. aquidurense* CCUG 59847^T; 4, *F. frigidimaris* CCUG 59364^T; 5, *F. hydatis* DSM 2063^T; 6, *F. araucananum* CCUG 61031^T; 7, *F. chungangense* CCUG 58910^T; 8, *F. oncorhynchi* CECT 7678^T.

Data are from this study.

+, positive reaction; –, negative reaction.

After PFGE typing, the trout strains were characterized by 3 different restriction profiles with the enzymes *Bsp120*I (Fig. 2) and *Xho*I (not shown). Strains 986-08 and 1084B-08 exhibited indistinguishable restriction profiles with both enzymes and strain 51B-09 could not be characterized because its DNA systematically was autodegraded.

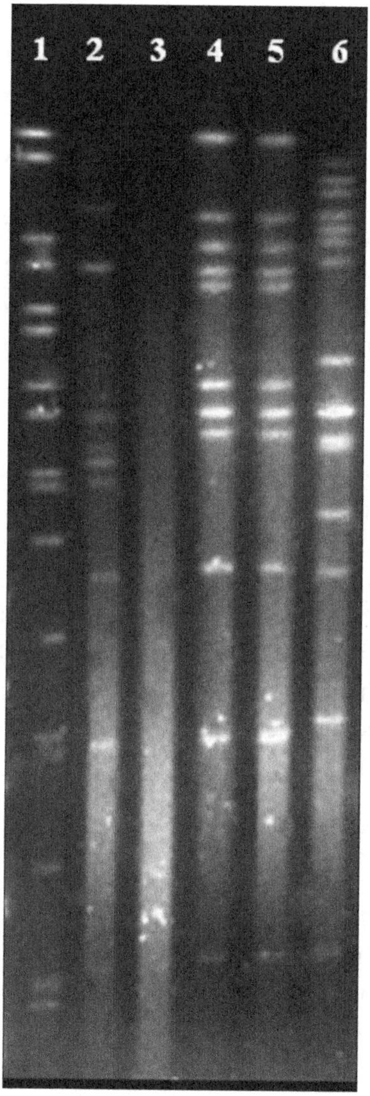

Figure 2: PFGE patterns generated after Bsp120I macrorestriction of *Flavobacterium plurextorum* **sp.** nov.

Lane 1, DNA molecular size marker; Lanes 2 to 6, strains 1126-1H-08^{T}, 51B-09, 986-08, 1084B-08 and 424-08, respectively.

Flavobacteria are known to belong to the microbiota of fish and fish eggs [4], [5]. Therefore, although two strains were isolated from internal organs, the other three were recovered from gills and eggs which suggest that the new species could be saprophytic or commensal and able to colonize fish, and produce disease under stressful conditions or other predisposing circumstances such as coinfections with other bacteria or viruses, poor farming conditions or environmental disorders [4], [39]. This assumption should be confirmed by experimental infection trials. Nevertheless, the formal description of *Flavobacterium plurextorum* and the availability of tests to facilitate its identification from other *Flavobacterium* species associated with fish disease or isolated from diseased fish will aid laboratories in its recognition and identification in the future, and to improve the knowledge of its distribution and possible association with disease.

Conclusion

The phylogenetic, genotypic and phenotypic results of the present polyphasic study demonstrated that the new strains isolated from rainbow trout represented a novel species of the genus *Flavobacterium*, for which the name *Flavobacterium plurextorum* sp. nov. is proposed (plu.rex.to'rum. L. comp. pl. plures, more, several, many; L. pl. n. exta -orum, entrails; N.L. gen. pl. n. plurextorum, of several internal organs). Detailed description of the morphological, physiological and biochemical characteristics of this species were indicated above. The type strain is 1126-1H-08^{T} (=CECT 7844^{T}=CCUG 60112^{T}).

ACKNOWLEDGMENTS

The authors thank Professor J. P. Euzéby of the Ecole Nationale Vétérinaire in Toulouse for advice concerning the Latin species name and A. Casamayor (VISAVET) for technical assistance in PFGE analysis and Kent Molin (CCUG) for the analyses of CFAs.

REFERENCES

1. Bernardet JF, Segers P, Vancanneyt M, Berthe F, Kersters K, et al. (1996) Cutting a Gordian knot: emended classification and description of the genus *Flavobacterium* emended description of the family *Flavobacteriaceae* and proposal of *Flavobacterium hydatis* nom nov (basonym *Cytophaga aquatilis* Strohl and Tait 1978). Int J Syst Bacteriol 46: 128–148.
2. Bernardet JF, Nakagawa Y (2006) An introduction to the family *Flavobacteriaceae.* In: Dworkin M, Falkow S, Rosenberg E, Schleifer KH, Stackebrandt E, editors. The Prokaryotes: a Handbook on the Biology of Bacteria, 3rd ed, vol, 7, New York: Springer. 455–480.
3. Euzéby JP (1997) List of Bacterial Names with Standing in Nomenclature: a folder available on the Internet. Int J Syst Bacteriol 47 590–592. Available: http://wwwbacterionet Accessed (April 13, 2013).
4. Bernardet JF, Bowman JP (2006) The genus *Flavobacterium.* In: Dworkin M, Falkow S, Rosenberg E, Schleifer KH, Stackebrandt E, editors. The Prokaryotes: a Handbook on the Biology of Bacteria, 3rd ed. Vol, 7. New York: Springer. 481–531.
5. Bernardet JF, Bowman JP (2011) Genus I. *Flavobacterium* Bergey, Harrison, Breed, Hammer and Huntoon 1923, 97AL emend. Bernardet, Segers, Vancanneyt, Berthe, Kersters and Vandamme 1996, 139. In: Krieg NR, Staley JT, Brown DR, Hedlund BP, Paster BJ, et al.. editors. Bergey's Manual of Systematic Bacteriology, 2nd ed. Vol, 4. New York; Springer. 112–155.
6. Roberts RJ (2012) The Bacteriology of Teleosts. In Fish Pathology, 4 ed. UK: Wiley-Blackell. 339–382.
7. Starliper CE, Schill WB (2012) Flavobacterial diseases: columniaris disease, coldwater disease and bacterila gill disease. In: Woo TTK, Bruno DW, editors. Fish Diseases and Disorders, 2nd ed, Vol. 3. UK: CAB International. 606–631.
8. Flemming L, Rawlings D, Chenia H (2007) Phenotypic and molecular characterization of fish-borne *Flavobacterium johnsoniae*-like isolates from aquaculture systems in South Africa. Res Microbiol 158: 18–30.
9. Kämpfer P, Lodders N, Martin K, Avendaño-Herrera R (2012) *Flavobacterium chilense* sp nov and *Flavobacterium araucananum* sp. nov. two novel species isolated from farmed salmonid

in Chile. Int J Syst Evol Microbiol 62: 1402–1408.

10. Zamora L, Fernández-Garayzábal JF, Svensson-Stadler LA, Palacios MA, Domínguez L, et al. (2012) *Flavobacterium oncorhynchi* sp. nov. a new species isolated from rainbow trout (*Oncorhynchus mykiss*). Syst Applied Microbiol 35: 86–91.
11. Chen WM, Huang WC, Young CC, Sheu SY (2013) *Flavobacterium tilapiae* sp. nov. isolated from a freshwater pond and emended descriptions of *Flavobacterium defluvii* and *Flavobacterium johnsoniae*. . Int J Syst Evol Microbiol 63: 827–834.
12. Sheu SY, Lin YS, Chen WM. (2012) *Flavobacterium squillarum* sp. nov., isolated from a freshwater shrimp culture pond and emended descriptions of *Flavobacterium haoranii, F. cauense, F. terrae* and *F. aquatile*. Int J Syst Evol Microbiol (in press) doi: 10.1099/ijs.0.046425-0.
13. Sheu SY, Chiu TF, Young CC, Arun AB, Chen WM (2011) *Flavbacterium macrobrachii* sp. nov. isolated from a freshwater shrimp culture pond. Int J Syst Evol Microbiol. 61: 1402–1407.
14. Ilardi P, Avendaño-Herrera R (2008) Isolation of *Flavobacterium*-like bacteria from diseased salmonids cultured in Chile. Bull Eur Assoc Fish Pathol 28: 176–185.
15. Kim OS, Cho YJ, Lee K, Yoon SH, Kim M, et al.. (2012) Introducing EzTaxon-e: a prokaryotic 16S rRNA gene sequence database with phylotypes that represent uncultured species. Int J Syst Evol Microbiol 62: 716–721. Available: http://eztaxon-e.ezbiocloud.net Accessed May 29, 2013.
16. Rasmussen SW (2002) SEQtools a software package for analysis of nucleotide and protein sequences. Available: http://wwwseqtoolsdk Accessed May 27, 2013.
17. Saitou N, Nei M (1987) The neighbour-joining method: a new method for reconstructing phylogenetic trees. Mol Biol Evol 4: 406–425.
18. Page RDM (1996) TREEVIEW: an application to display phylogenetic trees on personal computers Comput Appl Biosci. 12: 357–358.
19. Guindon S, Gascuel O (2003) A simple fast and accurate algorithm to estimate large phylogenies by maximum likelihood. Syst Biol 52: 696–704.
20. Tamura K, Peterson D, Peterson N, Stecher G, Nei M, et al. (2011) MEGA5: Molecular evolutionary genetics analysis using

maximum likelihood, evolutionary distance, and maximum parsimony methods. Mol Biol Evol 28: 2731–2739.

21. Kimura M (1980) A simple method for estimating evolutionary rates of base substitutions through comparative studies of nucleotide sequences. J Mol Evol 16: 111–120.
22. Marmur J (1961) A procedure for the isolation of deoxyribonucleic acid from microoganisms. J Mol Biol 3: 208–219.
23. Johnson JL (1994) Similarity analysis of DNAs. In: Gerhardt P, Murray RGE, Wood WA, Krieg NR, editors. Methods for General and Molecular Bacteriology. Washington DC: American Society for Microbiology. 655–681.
24. Arahal DR, García MT, Vargas C, Canovas D, Nieto JJ, et al. (2001) *Chromohalobacter salexigens* sp. nov. a moderately halophilic species that includes *Halomonas elongata* DSM 3043 and ATCC 33174. Int J Syst Evol Microbiol 51: 1457–1462.
25. De Ley J, Tijtgat R (1970) Evaluation of membrane filter methods for DNA-DNA hybridization. A Van Leeuw J Microb 36: 461–474.
26. Johnson JL (1994) Similarity analysis of DNAs. In: Gerhardt P, Murray RGE, Wood WA, Krieg NR, editors. Methods for General and Molecular Bacteriology, Washington DC: American Society for Microbiology. 655–681.
27. Tindall BJ (1990) A comparative study of the lipid composition of *Halobacterium saccharovorum* from various sources. Syst Appl Microbiol 13: 128–130.
28. Tindall BJ (1990) Lipid composition *of Halobacterium lacusprofundi.* . FEMS Microbiol Letts 66: 199–202.
29. Sasser M (2001) Identification of bacteria by gas chromatography of cellular fatty acids MIDI. Available: http://wwwmicrobialidcom/PDF/TechNote_101pdf Accessed May 27, 2013.
30. Bernardet JF, Nakagawa Y, Holmes B (2002) Proposed minimal standards for describing new taxa of the family *Flavobacteriaceae* and emended description of the family. Int J Syst Evol Microbiol 52: 1049–1070.
31. Smibert RM, Krieg NR (1994) Phenotypic Characterization. In: Gerhardt P, Murray RGE, Wood WA, Krieg NR, editors. Methods for General and Molecular Bacteriology, Washington DC: American Society for Microbiology. 607–653.

32. Bowman JP, Cavanagh J, Austin JJ, Sanderson K (1996) Novel *Psychrobacter* species from Antarctic ornithogenic soils. Int J Syst Bacteriol 46: 841–848.
33. Chen YC, Davis MA, Lapatra SE, Cain KD, Snekvik KR, et al. (2008) Genetic diversity of *Flavobacterium psychrophilum* recovered from commercially raised rainbow trout *Oncorhynchus mykiss* (Walbaum) and spawning coho salmon *O. kisutch* (Walbaum). J Fish Dis 31: 765–73.
34. Lim CS, Oh YS, Lee JK, Park AR, Yoo JS, et al. (2011) *Flavobacterium chungbukense* sp. nov. isolated from soil. Int J Syst Evol Microbiol 61: 2734–2739.
35. Xu M, Xin Y, Tian J, Dong K, Yu Y, et al. (2011) *Flavobacterium sinopsychrotolerans* sp. nov. isolated from a glacier. Int J Syst Evol Microbiol 61: 20–24.
36. Yoon JH, Park S, Kang SJ, Oh SJ, Myung SC, et al. (2011) *Flavobacterium ponti* sp. nov.isolated from seawater. Int J Syst Evol Microbiol 61: 81–85.
37. Wayne LG, Brenner DJ, Colwell RR, Grimont PAD, Kandler O, et al. (1987) International Committee on Systematic Bacteriology Report of the ad hoc Committee on Reconciliation of Approaches to Bacterial Systematics. Int J Syst Bacteriol 37: 463–464.
38. Stackkebrandt E, Goebel BM (1994) Taxonomic note: a place for DNA-DNA reassociation and 16S rRNA sequence analysis in the present species definition in bacteriology. Int J Syst Bacteriol 44: 846–849.
39. Georgiadis MP, Gardner IA, Hedrick RP (2001) The role of epidemiology in the prevention, diagnosis, and control of infectious diseases of fish. Prev Vet Med 48: 287–302.

SECTION 6: AVIAN

CHAPTER 19

Outbreak of Listeria Monocytogenes in an Urban Poultry Flock

Rocio Crespo[1], Michael M Garner[2] , Sharon G Hopkins[3] and Devendra H Shah[4,5]

[1] Avian Health and Food Safety Laboratory, Washington Animal Disease Diagnostic Laboratory, Washington State University, 2607 West Pioneer, Puyallup, WA 98371, USA

[2] Northwest ZooPath, 654 W. Main St., Monroe, WA 98272, USA

[3] Public Health–Seattle & King County, 401 5th Avenue, Ste 1100, Seattle, WA 98104, USA

[4] Department of Veterinary Microbiology, Washington State University, Pullman, WA 99164-7040, USA

[5] Pathology and Paul Allen School for Global Animal Health, Washington State University, Pullman, WA 99164-7040, USA

Citation: Crespo, R., Garner, M.M., Hopkins, S.G. et al. Outbreak of Listeria monocytogenes in an urban poultry flock. BMC Vet Res 9, 204 (2013). https://doi.org/10.1186/1746-6148-9-204.

ABSTRACT

Background

Listeria monocytogenes infection is most commonly recognized in ruminants, including cattle, sheep, and goats; but it is rarely diagnosed in poultry. This report describes an outbreak of *L. monocytogenes* in a backyard poultry flock. Also, it points out the importance of collaboration between veterinarians and public health departments and the possible implications of zoonotic diseases.

Case Presentation

Depression, lack of appetite, labored breathing, and increased mortality were noted for 5 months in several affected birds within the flock. The pathologic changes in the internal organs of infected birds included severe myocarditis, pericarditis, pneumonia, hepatitis, and splenitis. No lesions were noted in the brain. Gram-positive organisms were seen in histologic sections of the heart and spleen. *Listeria monocytogenes* was detected by real time PCR from formalin fixed heart and spleen, and was isolated from fresh lung, spleen, and liver. This isolate was identified as *L. monocytogenes* serotype 4b by 16S rDNA sequencing and by PCR-based serotyping assay.

Conclusions

This is the first report describing outbreak of *L. monocytogenes* in backyard poultry flock in Washington State and use of molecular methods to confirm *L. monocytogenes* infection from formalin fixed tissues.

BACKGROUND

Listeriosis is caused by the bacterium *Listeria monocytogenes* and has a worldwide distribution. Although many species of birds are susceptible to infection, clinical disease in birds is rare. Young birds are most susceptible. *L. monocytogenes* infections in chickens occur in two forms: encephalitic or septicemic form. The encephalitic form is characterized by neurologic signs such as depression, incoordination and torticollis; whereas the septicemic form is characterized by diarrhea and emaciation.

L. monocytogenes has been isolated from poultry, poultry products, and ready-to-eat poultry meat [1-3]. One study showed that *L. monocytogenes* was isolated from >40% of the organic as well as non-organic commercial chicken meat in Maryland [4]. The vast majority of serotypes included 1/2a, 1/2b and 4b. Interestingly, out of more than 14 reported serotypes of *L. monocytogenes,* these three serotypes cause most of the clinical cases [5]. Several studies have reported that handling and consumption of contaminated raw broiler meat is an important risk factor for human infection. Because of the epidemiological importance of certain serotypes of *L. monocytogenes* to human health and potential transmission of the pathogen from poultry to humans, accurate detection of *L. monocytogenes* followed by subtyping methods to identify the specific serotype or genotype involved in outbreaks is essential. This report describes an unusual presentation of listeriosis in adult chickens and the systematic use of molecular tests performed to diagnose listeriosis in this backyard poultry flock. Also, it stresses the importance of being aware of potentially zoonotic diseases.

CASE PRESENTATION

Clinical History

Several chickens from a backyard flock (Seattle, WA) were presented to a private veterinarian. The affected flock consisted of 20 chickens that were approximately 8 months of age at the time of presentation. During the course of the previous five months, a total of 7 birds had died and other 5 had shown clinical signs of illness. Most birds stopped producing eggs. The clinical signs included depression, anorexia and panting. The owner medicated the flock with sulfadimethoxine sodium in the water (1 fl oz of 12.5% Sulmet per gallon of drinking water; Fort Dodge Animal Health, Fort Dodge, IA) for 5 days. Despite this treatment, no improvement in clinical signs or egg production was noticed. The owner provided commercial layer feed (Payback®, CHS Inc., Sioux Falls, SD) purchased from the local feed store in Issaquah, WA. The owner also reported occasionally feeding some fresh garden produce grown on the premises, but no processed foods (e.g., cold cuts, hot dogs, cheese) were offered to birds.

Pathology

The owner submitted a dead chicken (the 6th one that died) to a local veterinarian for necropsy. The veterinarian collected several tissues (including heart, lung, air sac, intestine, pancreas, and liver) and submitted them for histologic examination to a private veterinary diagnostic laboratory. Two weeks later, another bird, an 8-month-old female chicken, died (the 7th death in the flock) and was submitted to the Avian Health and Food Safety Laboratory (AHFSL) for necropsy. The bird submitted to AHFSL had small comb fecal pasting around the cloaca. The bird was in good body condition and weighed 2,065 grams. The gross findings were 1-2 cc serous fibrinous pericardial effusion, thickening and opacity of pericardium with multiple pale foci throughout the myocardium. Pulmonary congestion, fibrinous exudate on the pleura, and thickening of air sacs were noted. The liver and spleen were enlarged and mottled.

Portions of the heart, liver, lung, kidney, gizzard, proventriculus, intestine, pancreas, ovary, oviduct, thyroid, parathyroid skeletal muscle, peripheral nerve, and brain were collected for histopathology. Tissues were fixed in 10% buffered neutral formalin, routinely processed, embedded in paraffin, sectioned at 4 μm, stained with hematoxylin and eosin, and examined by light microscopy. Additionally, sections of heart, liver and spleen were stained with modified Brown-Hopps tissue Gram stain [6].

Histologic findings were similar in the two birds submitted separately. The most prominent lesion was extensive inflammation of the heart (Figure 1), consisted of lymphocytes, macrophages, numerous multinucleated giant cells, and occasional heterophils. The lumen of larger blood vessels contained numerous large lymphocytes and scattered heterophils. In the liver there were random, small necrotic foci and accumulation of heterophils and macrophages (Figure 2). Lung sections revealed interstitial pneumonia, with infiltration of heterophils in the air capillaries and edema accumulating in the interparabronchial spaces. Lesions in the pancreas consisted of mild lymphocytic inflammation around the ducts. Spleen sections revealed severe lymphoid depletion, random accumulation of fibrin in the white pulp, and several macrophages contained intracytoplasmic cellular debris. Rod-shaped, Gram-positive bacteria were noted in the heart (Figure 3) and spleen.

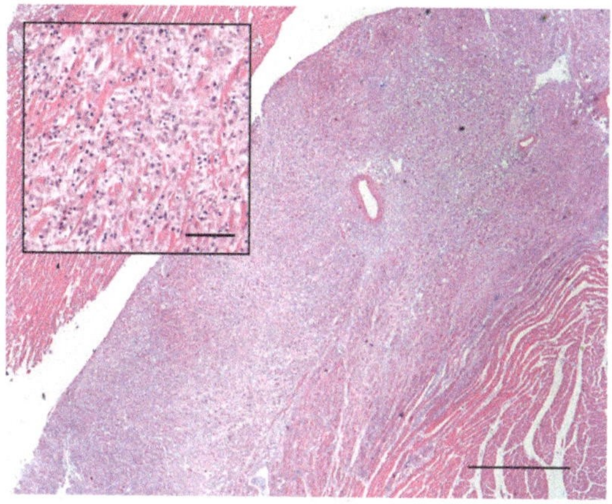

Figure 1: Extensive chronic-active myocardial inflammation. Histologic section of the heart showing extensive inflammation. H&E. Bar = 1 mm. Insert: higher magnification of the heart showing the inflammatory cell infiltration and disruption of myofibers. Inflammatory cells consisted mostly of lymphocytes and macrophages with scattered heterophils. H&E. Bar = 50 μm.

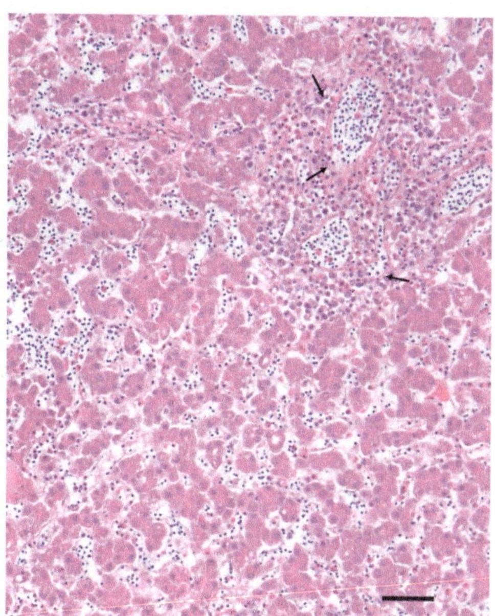

Figure 2: Periportal hepatic inflammation. Histologic section of the liver showing accumulation of heterophils and macrophages mostly around portal areas (arrows). H&E. Bar = 50 μm.

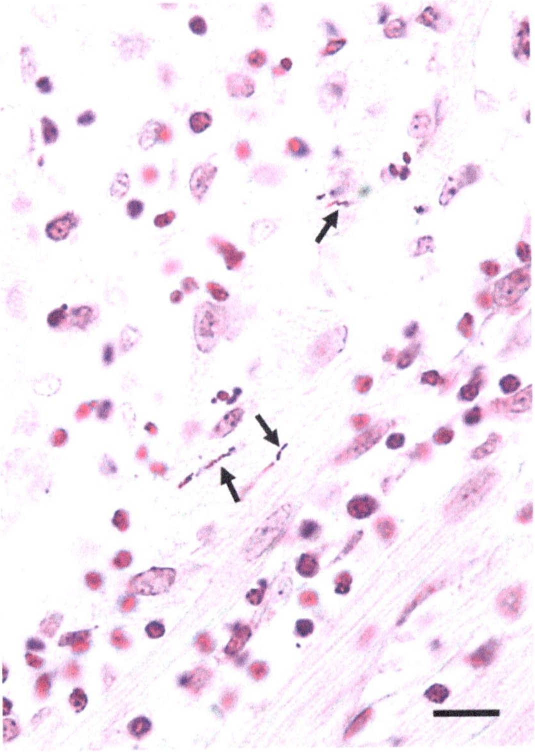

Figure 3: Gram positive bacteria in the heart. Histologic section of the heart showing inflammation and presence of Gram-positive rod-shaped bacteria (arrows) between the myofibers. Modified Brown-Hopps gram stain. Bar = 10 μm.

Bacteriology

Heart, liver, spleen and lung specimens were inoculated on blood and McConkey's agar (Remel, Lenaxa, KS) plates and incubated aerobically at 37°C for a minimum of up to 48 h. Cultures were examined for bacterial growth at approximately 24 and 48 h post-inoculation. Additionally, multiple sections of intestine were collected and cultured for *Salmonella spp*. The intestinal contents were selectively enriched in tetrathionate (TT) Hajna (BD Worlwide, Franklin Lakes, NJ) broth at 40°C for 20-24 hr. After enrichment, TT broth was plated on brilliant green with novobiocin and xylose-lysine-tergitol agar 4 agars (Remel) and the plates were incubated at 35°C for an additional 24 h.

Identical small round colonies with a narrow zone of β-hemolysis on the blood agar were isolated in pure form from liver, spleen, and lung. The bacteria were identified as Gram-positive rods, catalase-positive and oxidase-negative. The liver isolate was biochemically characterized as *L. monocytogenes* by testing with the API Coryne kit (bioMérieux, Inc., Hazelwood, MO) and Christie–Atkins–Munch–Peterson (CAMP) test.

Molecular Characterization

The isolate was confirmed to be *L. monocytogenes* by the real-time polymerase chain reaction (RT-PCR) using iQ-Check Listeria monocytogenes II kit (BioRad, Hercules, CA) according to the manufacturer's instructions. In addition, the genomic DNA was extracted from paraffin sections of heart from both birds, and the liver and the spleen from the bird submitted at AHFSL using a QIAamp® DNA Mini Kit (Qiagen, Valencia, CA) and tested by RT-PCR. *L. monocytogenes* was detected from both fixed sections of heart (Ct = 28.35 and 38.45) and the single fixed section of spleen (Ct = 32.48). No bacteria were isolated from the heart. None of the samples were positive for *Salmonella* sp.

A complete ORF of 16S rDNA (~1500 bp) was PCR amplified, cloned in pGEM-T Easy vector system (Promega, Madison, WI) and sequenced using protocols described previously [7]. The sequences were edited, aligned and phyologenetic analysis (data not shown) was performed as described previously [7]. The 16S rRNA gene sequence from *L. monocytogenes* strain showed 99.9% similarity with several *L. monocytogenes* strains in the GenBank database, but serotype of this isolate could not be determined based on the 16S rDNA sequence similarity. Consequently, PCR based serotyping was performed following the procedures described previously using following sets of primers that amplify variable regions of *L. monocytogenes* genome: D1, D2, FlaA, GLT and MAMA-C [8]. These primer pairs can accurately classify *L. monocytogenes* into specific serotype. The *L. monocytogenes* isolate in this study tested PCR positive only by one primer set (D1), but negative for all the other primer pairs tested (Figure 4). These results suggested that the isolate belonged to *L. monocytogenes* serotype 4b.

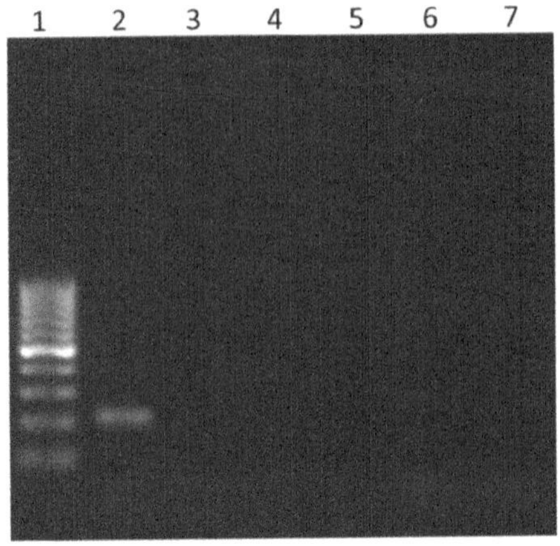

Figure 4: Molecular serotyping of L. monocytogenes. Results of PCR-based serotyping of *L. monocytogenes*. Lane 1: 100-bp ladder; lane 2: D1 primers; lane 3: D2 primers; lane 4: FlaA primers; lane 5: GLT primers; lane 6: MAMA-C primers; lane 7: no template control. Positive amplification with D1 primer pair only. This suggests that the *L. monocytogenes* isolate is serotype 4b [8].

DISCUSSION

In animals and humans the most common *Listeria* infections are caused by three serotypes: 1/2a, 1/2b, and 4b. PCR-based serotyping of the *L. monocytogenes* isolate in the current study identified this as serotype 4b, one of the most pathogenic serovars that is known to cause majority of human infections [9]. Experimental studies have shown that serotype 4b strains may have increased virulence compared with other serotypes [9] and are often responsible for invasive disease outbreaks in mammals. This serotype is also found in poultry meat [4]. In addition, serotype 4b has been isolated from a chicken flock with encephalitis [10]. However, no signs of central nervous system disease were observed in the current outbreak.

In the current case, *L. monocytogens* were isolated from the lung, liver and spleen, but not from the heart. On the other hand the bacteria were detected by RT-PCR from fixed heart and spleen. Molecular techniques have been used to detect *L. monocytogenes* in enriched samples or pure cultures. The limit of detection for the commercial real time PCR kit used in this study is 1,000 coliform forming units. The Ct value for qPCR for this bird was 38.45 which indicated that this bird likely had low numbers of *L. monocytogenes* in heart and that could have led to failure of isolation. It is possible that cold-enrichment might have helped recovering the bacteria from the heart. However cold-enrichment was not conducted because all other samples were positive.

The exact source of infection for this flock is unknown. Possible sources of the *Listeria* infection include feces, soil, decaying plant material, feed, and water. Neither the environment where the chickens were housed nor the commercial ration or drinking water were collected or tested. Furthermore, various species of mammals and birds may be infected with *Listeria* and may serve as asymptomatic carriers [11].

The first clinical signs in the flock appeared when investigation from a multistate outbreak of listeriosis linked to whole cantaloupes was ongoing [12]. The number of outbreak-associated deaths in the mentioned outbreak was 33 people. According to the history and confirmed by the King's County Public Health investigation, the chicken flock was fed a commercial layer's ration and occasional fresh produce grown in the owner's garden. The birds did not have access to compost or kitchen leftovers that may be contaminated with listeria (e.g. like cold cuts, hot dogs, soft cheeses, or cantaloupes). Conjunctivitis due to *L. monocytogenes* has been reported in individuals handling apparently healthy but infected chickens [13]. Human infections have also resulted from the consumption of contaminated poultry or ready-to-eat poultry products. Although the family continued to consume the eggs produced by the flock during the outbreak, no family member was ever sick during the outbreak or diagnosed with Listeriosis.

The sick birds were treated intramuscularly with 25 mg/kg of enrofloxacin (Bayer Corp., Pittsburgh, PA) for 5 days. All the birds recovered. Enrofloxacin is a fluoroquinolone antibiotic used for the treatment of a wide variety of bacterial infections in companion animals and horses. In 2005, the Food and Drug Adminstraton withdrew approval of enrofloxacin for use in water to treat flocks of poultry, as this practice was noted to promote the evolution of fluoroquinolone-resistant strains of the bacterium Campylobacter, a human pathogen [14]. Unlike commercial poultry, these birds or their eggs were not sold for food. The veterinarian educated the clients regarding the potential of contamination of eggs with the antibiotic and recommended not to consume the eggs laid by treated hens.

CONCLUSION

This is the first report of outbreak of listeriosis in the backyard poultry flock in Washington State. In addition, this paper describes the use of real time PCR, for the first time, to confirm *L. monocytogenes* infection from formalin fixed tissues. Because most of human infections are caused by three major serotypes (serotyping must be performed to identify the specific serotypes of *Listeria* (1/2a, 1/2b and 4b), serotyping must be undertaken to determine potential public health risk associated with exposure to infected birds. It is important to note that most reported outbreaks of listeriosis in poultry involve relatively very young birds. Systemic listeriosis in birds older than 5 months of age described in this report is relatively rare, but the current outbreak suggests that veterinarians and public health officials must be vigilant and should raise suspicion for listeriosis irrespective of age of the chickens. Finally, because antibiotics may remain in meat and other animal products, it is essential to educate the owners of urban poultry on these potential risks.

AUTHORS' CONTRIBUTIONS

RC performed the necropsy, the histologic examination, isolation, and PCR from fixed tissues, reviewed the literature and prepared the manuscript. MG performed the histologic examination and prepared the photos. SH performed the field investigation and interviewed the owner. DS performed the molecular characterization and serotype-specific identification of *Listeria* isolate. All authors read and approved the final manuscript.

ACKNOWLEDGEMENTS

The authors gratefully acknowledge Dennis Schaberg, Sylvia Weber, and Gail Deckert for their technical support. The authors also thank Elaine Lively and Chandra Miller of Phoenix Central Laboratory, Mukilteo, WA, for data retrieval. This work was partially funded by the Agricultural Animal Health Program, College of Veterinary Medicine, Washington State University.

REFERENCES

1. Berrang ME, Meinersmann RJ, Frank JF, Smith DP, Genzlinger LL. Distribution of Listeria monocytogenes subtypes within a poultry further processing plant. J Food Prot. 2005;68:980–985.
2. Berrang ME, Meinersmann RJ, Frank JF, Ladely SR. Colonization of a newly constructed commercial chicken further processing plant with *Listeria monocytogenes*. J Food Prot. 2010;73:286–291.
3. Loura CAC, Almeida RCC, Almeida PF. The incidence and level of Listeria spp. and *Listeria monocytogenes* contamination in processed poultry at a poultry processing plant. J Food Prot. 2005;25:19–29.
4. Yeh ET. Characterization of Listeria monocytogenes isolated from retail organic chicken. University of Maryland, MD, USA: Food Science; 2004.
5. Tappero JW, Schuchat A, Deaver KA, Mascola L, Wenger JD. Reduction in the incidence of human listeriosis in the United States: effectiveness of prevention efforts? JAMA. 1995;273:1118–1122. doi: 10.1001/jama.1995.03520380054035.
6. Carson FL, Hladik C. Histotechnology: a self instructional text. 3. Hong Kong: American Society for Clinical Pathology Press; 2009. Brown-Hopps modification of the gram strain; pp. 231–233.
7. Burbank DR, Shah DH, LaPatra SE, Fornshell G, Cain KD. Enhanced resistance to coldwater disease following feeding of probiotic bacterial strains to rainbow trout (*Oncorhynchus mykiss*). Aquaculture. 2011;321:185–190. doi: 10.1016/j.aquaculture.2011.09.004.
8. Borucki MK, Call DR. Listeria monocytogenes serotype identification by PCR. J Clin Microbiol. 2003;41:5537–5540. doi: 10.1128/JCM.41.12.5537-5540.2003. [PMC free article]
9. Czuprynski CJ, Faith NG, Steinberg H. Ability of the *Listeria monocytogenes* strain Scot A to cause systemic infection in mice infected by the intragastric route. Appl Environ Microbiol. 2002;68:2893–2900. doi: 10.1128/AEM.68.6.2893-2900.2002.
10. Kurazono M, Nakamura K, Yamada M, Yonemaru T, Sakoda T. Pathology of listerial encephalitis in chickens in Japan. Avian Dis. 2003;47:1496–1502. doi: 10.1637/7066.
11. Barnes HJ. In: Diseases of poultry. 12. Saif YM, Fadly AA, Glisson JR, McDougald LR, Nolan L, Swayne DE, editor. Ames, IA: ISU Press;

2008. Miscellaneous and sporadic bacterial infections; pp. 952–970.

12. Centers for Disease Control and Prevention. Multistate outbreak of listeriosis linked to whole cantaloupes from Jensen Farms, Colorado. 2012. http://www.cdc.gov/listeria/outbreaks/cantaloupes-jensen-farms/index.html; (accessed on March 29, 2013)
13. Morishita TY. In: The merck veterinary manual. 10. Kahn CM, Allen DG, Constable PD, Davies PR, Quesenberry KE, Reeves PT, Sharma JM, Smith RKW, Treadwell T, editor. Kendallville, IN: Merck & Co., Inc; 2010. Listeriosis; p. 2441.
14. U. S. Food and Drug Administration. Enrofloxacin for poultry. 2012. http://www.fda.gov/AnimalVeterinary/SafetyHealth/RecallsWithdrawals/ucm042004.htm; (accessed on March 29, 2013)

CHAPTER 20

Development of an Allele-Specific PCR Assay for Simultaneous Sero-Typing of Avian Pathogenic Escherichia coli Predominant O1, O2, O18 and O78 Strains

Shaohui Wang[1]., Qingmei Meng[1,2]., Jianjun Dai[2] , Xiangan Han[1] , Yue Han[1] , Chan Ding[1] , Haiwen Liu[1] , Shengqing Yu[1]

[1] Shanghai Veterinary Research Institute, Chinese Academy of Agricultural Sciences, Shanghai, China

[2] Key Lab of Animal Bacteriology, Ministry of Agriculture, Nanjing Agricultural University, Nanjing, Jiangsu, China

ABSTRACT

Systemic infections by avian pathogenic *Escherichia coli* (APEC) are economically devastating to poultry industries worldwide. *E. coli* strains belonging to serotypes O1, O2, O18 and O78 are preferentially associated

Citation: Wang S, Meng Q, Dai J, Han X, Han Y, Ding C, et al. (2014) Development of an Allele-Specific PCR Assay for Simultaneous Sero-Typing of Avian Pathogenic Escherichia coli Predominant O1, O2, O18 and O78 Strains. PLoS ONE 9(5): e96904. https://doi.org/10.1371/journal.pone.0096904.

with avian colibacillosis. The *rfb* gene cluster controlling O antigen synthesis is usually various among different *E. coli* serotypes. In present study, the *rfb* gene clusters of *E. coli* serotypes O1, O2, O18 and O78 were characterized and compared. Based on the serotype-specific genes in *rfb* gene cluster, an allele-specific polymerase chain reaction (PCR) assay was developed. This PCR assay was highly specific and reliable for sero-typing of APEC O1, O2, O18 and O78 strains. The sensitivity of the assay was determined as 10 pg DNA or 10 colony forming units (CFUs) bacteria for serotypes O2 and O18 strains, and 500 pg DNA or 1,000 CFUs bacteria for serotypes O1 and O78 strains. Using this PCR system, APEC isolates and the infected tissue samples were categorized successfully. Furthermore, it was able to differentiate the serotypes for the samples with multi-agglutination in the traditional serum agglutination assay. Therefore, the allele-specific PCR is more simple, rapid and accurate assay for APEC diagnosis, epidemiologic study and vaccine development.

INTRODUCTION

Escherichia coli typically colonize the mammalian and avian gastrointestinal tract and other mucosal surfaces. Although many *E. coli* strains are commensal, certain pathogenic strains can cause a wide variety of intestinal and extraintestinal diseases [1]–[2]. *E. coli* could be sero-typed by somatic (O), capsular (K), and flagellar (H) antigens [3], and a close connection exists among specific O-antigen serotypes and certain pathogenicity of pathogens. Avian pathogenic *E. coli* (APEC) are economically devastating to poultry industries worldwide. Previous studies indicated that varied serotypes including O1, O2, O18 and O78 are preferentially associated with APEC outbreaks, which accounted for more than 50% of the APEC issues [4]–[6]. Our previous epidemiology study showed that more than 85% APEC were O1, O2, O18 and O78 in the farms of Eastern China [7]–[8]. Moreover, there is less cross-reaction among serotypes. Thus, sero-typing of APEC bacteria isolated or in infected tissues would be a crucial modality for disease diagnosis, epidemiologic study and vaccine development.

APEC isolates are generally sero-typed by serum agglutination assay using specific O-antigen antiserum. This traditional assay needs isolated bacterial colony and specific antiserum for the sero-typing. Therefore, it is complex, costly, and time consuming. Moreover, cross-reactivity of the antisera with multiple O-antigen strains occurs occasionally. Recently, PCR-based method has been used as a rapid analytical technique for detection of a variety of bacterial strains [9]. Genes controlling O-antigen synthesis are in the *rfb* gene cluster, ranging from 4.2 to 20 kb, which is generally bordered by the *gnd* and *galF* genes in *E. coli*. Sequence analysis shows that the number and arrange of genes in the *rfb* gene cluster are various for different serotypes of *E. coli* [10]–[12]. Thus, PCR assays based on O-antigen gene clusters have been developed to determine predominant O serotypes of several pathogenic *E. coli* [13]–[21]. However, no rapid and sensitive PCR is available for sero-typing of APEC predominant O1, O2, O18 and O78 strains yet.

This study attempted to analyze the *rfb* gene clusters in APEC predominant serotypes O1, O2, O18 and O78 strains and develop an allele-specific PCR method for sero-typing of the O-antigens. The allele-specific PCR method was evaluated for its specificity, sensitivity, and application for APEC diagnosis.

MATERIALS AND METHODS

Bacterial Strains, Growth Conditions and DNA Preparation

The bacterial strains used in this study are listed in Table 1. The *E. coli*, *Salmonella enterica* (*S. enterica*), *Riemerella anatipestifer* (*R. anatipestifer*) and *Pasteurella multocida* (*P. multocida*) strains were grown in appropriate medium at 37°C with aeration. Other reference strains from the Chinese Veterinary Culture Collection Center (CVCC, Beijing, China) or the American Type Culture Collection (ATCC, Manassas, VA, USA) were cultured in recommended conditions. Sixty-five APEC serotype O1, O2, O18 and O78 strains were isolated from chickens or ducks between 2010 and 2012, and kept in Shanghai Veterinary Research Institute, Chinese Academy of Agricultural Sciences (CAAS) [7]–[8]. Bacterial genomic DNA was prepared using TIANamp Bacterial DNA Kit (Tiangen, Beijing, China) according to the manufacturer's guidelines.

Table 1: Bacterial strains used in this study

Strains	Description	Source or reference
APEC O1	APEC strain, serotype O1	[22]
DE47	APEC strain, serotype O1	[7]–[8]
DE14	APEC strain, serotype O2	[7]–[8]

DE17	APEC strain, serotype O2	[7]–[8]
RS218	NMEC strain, serotype O18	[28]
CE66	APEC strain, serotype O18	[7]–[8]
DE48	APEC strain, serotype O78	[7]–[8]
DE65	APEC strain, serotype O78	[7]–[8]
CFT 073	UPEC strain, serotype O6	[29]
MG1655	*E. coli* strain, serotype O16	[30]
CVCC 1543	*E. coli* strain, serotype O38	CVCC [a]
CVCC 1547	APEC strain, serotype O73	CVCC
O131	*E. coli* strain, serotype O131	[31]
ATCC 43889	EHEC O157:H7	ATCC [b]
O138	*E. coli* strain, serotype O138	[31]
CVCC 3384	*S. enterica serovar typhimurium*	CVCC
CVCC 1805	*S. enterica serovar enteritidis*	CVCC
CVCC 519	*S. enterica serovar pullorum*	CVCC
CAU 0118	*S. enterica serovar anatum*	CVCC
CH3	*R. anatipestifer* strain, serotype 1	[27]
Th4	*R. anatipestifer* strain, serotype 2	[27]
HXb2	*R. anatipestifer* strain, serotype 10	[27]
CVCC 493	*Pasteurella multocida*	CVCC
CVCC 1651	*Mycoplasma gallisepticum*	CVCC
CVCC 274	*Mycoplasma avium*	CVCC
CVCC 543	*Staphylococcus aureus*	CVCC
IPDH 591-77	*Bordetella avium*	ATCC

[a]CVCC, Chinese Veterinary Culture Collection Center, China.

[b]ATCC, American Type Culture Collection, USA.

Primer Design and Development of Allele-specific PCR Assay

The *rfb* gene cluster is generally bordered by the *gnd* and *galF* genes, which controls *E. coli* O-antigen synthesis and shows serotype-dependent differences in its gene sets and organization. To design the suitable primers, the *rfb* gene clusters of eight *E. coli* strains, including serotype O1 strains (strain APEC O1 [Acc No. CP000468.1] and G1632 [Acc No. GU299791.1]), serotype O2 strains (strain G1674 [Acc No. GU299792.1] and *E. coli* O2 [Acc No. EU549863.1]), serotype O18 strains (strains IHE3034 [Acc No. CP001969.1] and G1630 [Acc No. GU299793.1]), and serotype O78 strains (strain APEC O78 [Acc No. CP004009.1] and *E. coli* O78 [Acc No. FJ940775.1]) [21]–[26], were analyzed using software Vector NTI (Invitrogen, Carlsbad, CA, USA). The putative functions of the identified genes were determined based on their sequence similarity to genes of known function from the available databases, in which they were named according to the bacterial polysaccharide gene nomenclature (BPGN) system (www.microbio.usyd.edu.au/BPGD/default.htm). Then, the universal forward primer and specific reverse primers were designed based on the sequence of *gnd* gene and specific *rfb* genes to develop the allele-specific PCR assays (Table 2 and Fig. 1). Furthermore, the absence of amplification by our primers was tested *in silico* on the other available *rfb* gene cluster sequences.

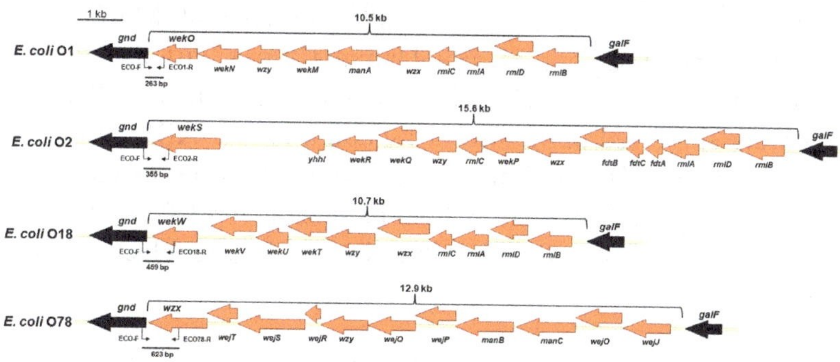

Figure 1: The *rfb* **gene clusters of** *E. coli* **serotypes** O1, O2, O18 and O78 strains

The black arrows correspond to *gnd* and *galF* genes. Grey arrows correspond to *rfb* gene cluster and the gene names are *italic* indicated. The length of *rfb* gene cluster was also shown. In the PCR reaction system, the universal forward primer was used for all the sero-typing amplification with specific reverse primers. The bold lines below the *gnd* gene indicate the size of the PCR products for different *E. coli* serotype strains, which allow the differentiation of the O types. Primers and their locations were also indicated.

Table 2: Primers used in this study

Primer	Sequence (5′ to 3′)[a]	Target gene	Size of PCR product
ECO-F	*CGATGTTGAGCGCAAGGTTG*	*gnd*	
ECO1-R	*CATTAGGTGTCTCTGGCACG*	*rfbO1*	263 bp
ECO2-R	*GATAAGGAATGCACATCGCC*	*rfbO2*	355 bp
ECO18-R	*AGAAGCATTGAGCTGTGGAC*	*rfbO18*	459 bp
ECO78-R	*TAGGTATTCCTGTTGCGGAG*	*rfbO78*	623 bp

[a]The primers were designed based on the gene sequences of *wekO*, *wekS*, *wekW* and *wzx* in the *rfb* gene cluster of respective serotypes of *E. coli* strains.

The *E. coli* reference strains were used for the allele-specific PCR development. Briefly, 1 μL template DNA was added to the reaction mixture (25 μL) containing 2.5 μL 10× PCR buffer with $MgCl_2$ (25 mM), 1.5 U *Taq* DNA polymerase (TaKaRa, Dalian, China), 2 μL dNTPs (2.5 mM for each dNTP), and 0.5 μL (10 μM) of each primer pair. The PCR reaction mixtures were subjected to the following conditions in ABI thermal cycler: pre-denaturation at 95°C for 5 min, followed by 30 cycles of 95°C for 35 s, 57°C for 30 s, 72°C for 40 s and a final extension at 72°C for 10 min. The PCR products were observed under ultraviolet light after electrophoresis on a 2% agarose gel.

Specificity of the Allele-specific PCR

For the specificity analysis of the allele-specific PCR, *E. coli* reference strains, *S. enterica*, *R. anatipestifer*, and other species of bacteria strains were used as the templates (Table 1). The bacterial DNAs from *E. coli* serotypes O1, O2, O18 and O78 strains were used as positive controls. The negative control contained sterile distilled water in place of template DNA. Under optimized condition, each serotype-specific gene fragment was amplified with respective primer pairs. The PCR products were then cloned into pMD18-T vector (TaKaRa, Dalian, China) and DNA sequencing was performed on an Applied Biosystems DNA sequencer ABI PRISM 377. The reliability and specificity of the assay were also verified by comparing the sero-typing results of 65 APEC isolates to traditional serum agglutination assay. The traditional agglutination assay was carried out with rabbit anti-*E. coli* immune serum produced against *E. coli* O1, O2, O18 or O78 (Statens Serum Institut, Copenhagen, Denmark) according to the manufacturer's guidelines. Briefly, the *E. coli* bacterial culture was boiled for 1 h. Then, the boiled culture was mixed with equal amount of O antiserum in glass tubes, which was incubated in a humid atmosphere at 50–52°C overnight. The reaction was read against artificial light with a black background. Physiological saline was used as a negative control and must be negative. If the negative control was positive, the strain was auto-agglutinating.

Sensitivity of the Allele-specific PCR

The sensitivity was determined using diluted DNA templates, ranged from 100 ng to 1 pg. On the other hand, 10-fold serial dilutions of bacterial broth culture, ranged from 1×10^7 to 1 colony forming units (CFUs), were also used as templates for the allele-specific PCR. The PCR assays were performed under the optimized condition.

Sero-typing of APEC from Experimental and Clinical Infected Tissue Samples

APEC experimental infection was performed as described previously [27]. Briefly, 20 Cherry Valley ducks (7-day-old) were randomly divided into 5 groups. The ducks in groups 1–4 were infected intratracheally with a bacterial suspension containing APEC serotypes O1, O2, O18 or O78 strains at 10^8 CFUs, respectively. The ducks in group 5 were used as controls. At 24 h post-infection, ducks were dissected which livers were aseptically collected. The genomic DNAs were extracted and eluted with 100 μL of distilled water from each 100 mg of duck liver tissue using QIAamp DNA mini kit (Qiagen, Santa Clarita, CA, USA) according to the manufacturer's instructions. The allele-specific PCR was carried out as described above. Genomic DNA of *E. coli* serotypes O1, O2, O18 or O78 strains were used as positive controls. In addition, 20 clinical diseased ducks with colibacillosis from APEC infected farms and 20 healthy ducks were obtained and subjected to APEC sero-typing using allele-specific PCR. To evaluate the results obtained from the PCR assays, all the samples were also subjected to an extended bacteriological examination and traditional serum agglutination assay.

Ethics Statement

All procedures were carried out in accordance with guidelines of the Association for Assessment and Accreditation of Laboratory Animal Care

International (AAALAC). The animal study protocol was approved by the Animal Care and Use Committee of Shanghai Veterinary Research Institute, CAAS, China.

RESULTS

Bioinformatics Analysis of the APEC *rfb* Gene Clusters

E. coli rfb gene cluster mainly encodes glycosyltransferase, acetyltransferase, polysaccharide polymerase and flippase. Bioinformatics analysis showed that the gene numbers and arrange in the cluster were completely different among APEC serotypes. The cluster length between genes *gnd* and *galF* was 10.5 kb (10 ORFs), 15.6 kb (14 ORFs), 10.7 kb (10 ORFs), and 12.9 kb (11 ORFs) for serotypes O1, O2, O18 and O78 strains, respectively (Fig. 1). The putative functions of the identified genes were determined based on their sequence similarity to genes of known function from the available databases. As shown in Figure 1, the genes near to *gnd* in the cluster of serotype O1, O2, and O18 strains were glycosyltransferase encoding genes *wekO*, *wekS* and *wekW*, respectively. The gene near to *gnd* in the cluster of serotype O78 strains was O antigen flippase encoding gene *wzx*.

Development of the Allele-specific PCR Assay for APEC Serotyping

The universal forward primer was designed based on the sequence of *gnd* gene. The specific reverse primers were designed based on the glycosyltransferase encoding genes (APEC O1, O2 and O18 strains) and the flippase encoding gene (APEC O78 strain), respectively (Fig. 1). The allele-specific PCR was optimized by adjustment of different parameters, and the resulting optimal condition was described in the Method section. As a result, the serotypes of the reference strains were specifically differentiated by the PCR assays (Fig. 2).

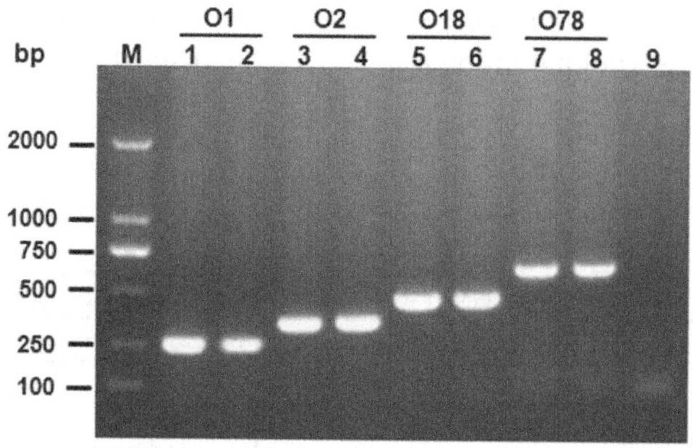

Figure 2: The product profiles of *E. coli* **serotypes O1, O2, O18 and O78 strains amplified using the allele-specific PCR.**

Lane M: DL2000 DNA Marker; O1, O2, O18 and O78 represent PCR products for O1, O2, O18 and O78 strains respectively. Lane 1: APEC O1 strain; Lane 2: APEC strain DE47; Lane 3: APEC strain DE14; Lane 4: APEC strain DE17; Lane 5: APEC strain RS218; Lane 6: APEC strain CE66; Lane 7: APEC strain DE48; Lane 8: APEC strain DE65; Lane 9: Negative control.

Determination of the Specificity for the PCR Assays

In order to evaluate the specificity of the primers used in this study, PCR were performed using different bacterial template listed in Table 1. The results showed that the serotype-specific fragments were amplified using respective primers from *E. coli* serotypes O1, O2, O18 and O78 reference strains. No fragment was amplified from other serotypes of *E. coli* reference strains or other species of bacteria tested in the allele-specific PCR assays (data not shown). The PCR products were further confirmed for the specificity by sequencing analysis. The results showed that all primers were specific and compatible in PCR reactions.

The reliability and specificity of the PCR assay was also verified by comparing the sero-typing results of 65 APEC O1, O2, O18 and O78 isolates with serum agglutination assays. As shown in Table 3, 65 isolates were distinctly sero-typed using the PCR assay. However, serum agglutination identified 60 of them, which were completely matched to those of PCR sero-typing. While the other 5 PCR sero-typed isolates, 3 for serotype O78, 1 for serotype O1 and 1 for serotype O2, were shown in a multi-agglutination pattern in the serum agglutination assay. This result further confirmed the specificity of the assay.

Table 3: Comparison of PCR and serum agglutination assays for differentiating the serotypes of APEC isolates and clinical infected samples

Serotypes	APEC isolates (n=65)		Clinical infected samples (n=20)	
	PCR	Serum agglutination	PCR	Serum agglutination
O1	9	8	2	2
O2	11	10	3	3
O18	6	6	2	2
O78	39	36	10	9
O1/O2/O18/O78	0	5	0	1

Assessment of the Sensitivity for the PCR Assays

The sensitivity of the assay was determined using a series of diluted chromosomal DNA or bacterial culture of *E. coli* serotypes O1, O2, O18 and O78 strains. The results showed that 10 pg DNA or 10 CFUs of O2 and O18 bacteria were sufficient for the amplification of serotype-specific fragments. However, 500 pg DNA or 1,000 CFUs of O1 and O78 bacteria were needed for the amplification of serotype-specific fragments (Fig. 3).

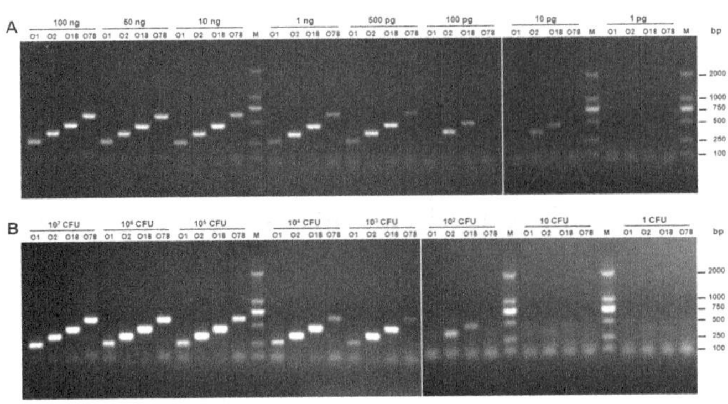

Figure 3: Sensitivity analysis of the allele-specific PCR assay.

(A) Sensitivity analysis using the bacterial genomic DNA. The detection limit was determined as 10 pg of bacterial DNA for *E. coli* serotypes O2 and O18 strains, and 500 pg of bacterial DNA for *E. coli* serotypes O1 and O78 strains, respectively. (B) Sensitivity analysis using the bacterial culture. The detection limit was determined as 10 CFUs of *E. coli* serotypes O2 and O18 strains, and 1,000 CFUs of *E. coli* serotypes O1 and O78 strains, respectively. Lane M: DL2000 Marker.

Sero-typing of the APEC from Experimental or Clinical Infected Samples

The results indicated that 16 isolates from experimentally infected ducks showed corresponding PCR bands at respective size, which was in accordance with bacteriological examination and traditional serum agglutination assay. Four samples from normal control ducks showed no any PCR band. Sero-typing results for 20 clinical samples with colibacillosis showed that 17 of them were serotypes O1, O2, O18 and O78 (Table 3), which was further confirmed by bacterial isolation and regular serum agglutination assay. In order to get more positive samples to verify our allele-specific PCR assay, we collected the samples from APEC positive farms. Therefore, a high positive rate (85%, 17/20) of the predominant serotype strains was obtained, which was similar to that in our previous prevalence studies [7]–[8]. Moreover, one clinical sample, which shows multi-agglutination with O2 and O78 antiserum in a traditional serum agglutination assay, was identified as O78 using this PCR assay. The other 3 clinical samples gave negative results of serotypes O1, O2, O18 and O78 in both PCR and serum agglutination assays, which were belonged to other serotypes (Table 3). The 20 healthy duck samples were negative for serotypes O1, O2, O18 and O78 by bacteriological examination and PCR assay. These results revealed that the developed PCR assay in this study was more sensitive and specific than traditional serum agglutination assay, and it achieved the requirement for the detection of clinical samples.

DISCUSSION

Colibacillosis is one of the principal causes of morbidity and mortality in poultry worldwide. APEC serotypes O1, O2, O18 and O78 strains are responsible for most of the poultry colibacillosis. Various strains with different serotypes were occasionally found at one APEC outbreak. In addition, the cross-protection among different APEC serotypes is poor [1]–[2], [4]–[7]. Therefore, a rapid and accurate sero-typing method is very important for the APEC control. Serum agglutination assay is a traditional method for APEC sero-typing, however, isolated bacteria and a panel of high quality antisera against different O-antigens are needed for getting the results. Serum agglutination can not simultaneously differentiate APEC serotypes at one test. In addition, occasionally one strain may react with multiple APEC antisera [3]. Thus, an allele-specific PCR assay for sero-typing APEC predominant serotypes was developed in this study.

To design effective primers for the sero-typing, the *rfb* gene cluster and flanking sequence in APEC serotypes O1, O2, O18 and O78 strains were subjected to bioinformatics analysis. The results showed that O-antigen associated genes were highly specific to individual serotypes, suggesting it could be typically used as the target in PCR based typing methods. By taking advantage of these features, a rapid and simple allele-specific PCR assay was developed. The PCR was designed to amplify the specific O-antigen sequences between the *rfb* locus and *gnd* gene at different size. The PCR assay could be used for sero-typing of APEC O1, O2, O18 and O78 strains in bacterial culture and in infected tissue samples, and showed no reaction with other serotypes of *E. coli* reference strains and other species of bacteria, demonstrating a very good specificity of the assay. Moreover, the multi-agglutination isolates could be distinctly sero-typed by this method, suggesting the PCR assay was specific and reliable.

The sensitivity assay of the PCR indicated that a quantity of 10 pg DNA or 10 CFUs bacteria was sufficient for detection of *E. coli* serotypes O2 and O18 strains, and 500 pg DNA or 1,000 CFUs bacteria was needed for detection of *E. coli* serotypes O1 and O78 strains (Fig. 3). Other single PCR system reveals that 1 pg DNA is good for detection of *E. coli* serotypes O15, O174 and O177 [14]–[15]. However, 500 pg DNA are needed to amplify the predominant serotypes of uropathogenic *E. coli* in multiplex PCR assays [21]. The lower sensitivity of multiplex PCR assays may because of possible interference among the primers in the system. The PCR assay was successful for typing of the APEC strains in both experimental and clinical infected tissue samples, suggesting it could be used for clinical and laboratory detection. Moreover, it was able to differentiate the serotypes of the samples with multi-agglutination in the regular bacteriological examination. Thus, this PCR assay was more specific and sensitive than the traditional serum agglutination assay. It achieved the requirements for APEC clinical diagnosis and epidemiology studies with reduced workload and shorted the time.

In summary, an allele-specific PCR assay was developed in this study, which was able to differentiate APEC predominate serotypes of O1, O2, O18 and O78 strains with high specificity and sensitivity. This PCR assay was an efficient and convenient strategy for sero-typing of APEC predominant strains, avoiding the disadvantage of traditional serologic assays. Thus, development of this PCR assay benefits for clinical diagnostics, epidemiology studies, and disease control.

REFERENCES

1. Kaper JB (2005) Pathogenic *Escherichia coli* . Int J Med Microbiol 295: 355–356.
2. Croxen MA, Finlay BB (2010) Molecular mechanisms of *Escherichia coli* pathogenicity. Nat Rev Microbiol 8: 26–38.
3. Kauffmann F (1947) The serology of the *coli* group. J Immunol 57: 71–100.
4. Ewers C, Janssen T, Wieler LH (2003) Avian pathogenic *Escherichia coli* (APEC). Berl Munch Tierarztl Wochenschr 116: 381–395.
5. Ewers C, Li G, Wilking H, Kiessling S, Alt K, et al. (2007) Avian pathogenic, uropathogenic, and newborn meningitis-causing *Escherichia coli*: how closely related are they? Int J Med Microbiol 297: 163–176.
6. Ewers C, Janssen T, Kiessling S, Philipp HC, Wieler LH (2004) Molecular epidemiology of avian pathogenic *Escherichia coli* (APEC) isolated from colisepticemia in poultry. Vet Microbiol 104: 91–101.
7. Bai H, Ji H, Han X, Gong J, Dong H, et al. (2013) Characterization of Avian Pathogenic *Escherichia coli* isolated from Jiangsu and Anhui provinces. Microbiol Chin 40: 1315–1322.
8. Chen W, Han X, He L, Hu Q, Yu S (2010) Characterization of duck pathogenic *Escherichia coli* . Chin J Anim Infect Dis 18: 34–40.
9. Wright PA, Wynford-Thomas D (1990) The polymerase chain reaction: miracle or mirage? A critical review of its uses and limitations in diagnosis and research. J Pathol 162: 99–117.
10. Samuel G, Reeves P (2003) Biosynthesis of O-antigens: genes and pathways involved in nucleotide sugar precursor synthesis and O-antigen assembly. Carbohydr Res 338: 2503–2519.
11. Shepherd JG, Wang L, Reeves PR (2000) Comparison of O-antigen gene clusters of *Escherichia coli* (*Shigella*) sonnei and *Plesiomonas shigelloides* O17: sonnei gained its current plasmid-borne O-antigen genes from *P. shigelloides* in a recent event. Infect Immun 68: 6056–6061.
12. Bastin DA, Reeves PR (1995) Sequence and analysis of the O antigen gene (*rfb*) cluster of *Escherichia coli* O111. Gene 164: 17–23.
13. DebRoy C, Fratamico PM, Roberts E, Davis MA, Liu Y (2005) Development of PCR assays targeting genes in O-antigen gene

clusters for detection and identification of *Escherichia coli* O45 and O55 serogroups. Appl Environ Microbiol 71: 4919–4924.

14. Beutin L, Tao J, Feng L, Krause G, Zimmermann S, et al. (2005) Sequence analysis of the *Escherichia coli* O15 antigen gene cluster and development of a PCR assay for rapid detection of intestinal and extraintestinal pathogenic *E. coli* O15 strains. J Clin Microbiol 43: 703–710.
15. Beutin L, Kong Q, Feng L, Wang Q, Krause G, et al. (2005) Development of PCR assays targeting the genes involved in synthesis and assembly of the new *Escherichia coli* O 174 and O 177 O antigens. J Clin Microbiol 43: 5143–5149.
16. Feng L, Wang W, Tao J, Guo H, Krause G, et al. (2004) Identification of *Escherichia coli* O114 O-antigen gene cluster and development of an O114 serogroup-specific PCR assay. J Clin Microbiol 42: 3799–3804.
17. Fratamico PM, Briggs CE, Needle D, Chen CY, DebRoy C (2003) Sequence of the *Escherichia coli* O121 O-antigen gene cluster and detection of enterohemorrhagic *E. coli* O121 by PCR amplification of the *wzx* and *wzy* genes. J Clin Microbiol 41: 3379–3383.
18. Paton AW, Paton JC (1999) Direct detection of Shiga toxigenic *Escherichia coli* strains belonging to serogroups O111, O157, and O113 by multiplex PCR. J Clin Microbiol 37: 3362–3365.
19. Perelle S, Dilasser F, Grout J, Fach P (2002) Identification of the O-antigen biosynthesis genes of *Escherichia coli* O91 and development of a O91 PCR serotyping test. J Appl Microbiol 93: 758–764.
20. Clermont O, Johnson JR, Menard M, Denamur E (2007) Determination of *Escherichia coli* O types by allele-specific polymerase chain reaction: application to the O types involved in human septicemia. Diagn Microbiol Infect Dis 57: 129–136.
21. Li D, Liu B, Chen M, Guo D, Guo X, et al. (2010) A multiplex PCR method to detect 14 *Escherichia coli* serogroups associated with urinary tract infections. J Microbiol Methods 82: 71–77.
22. Johnson TJ, Kariyawasam S, Wannemuehler Y, Mangiamele P, Johnson SJ, et al. (2007) The genome sequence of avian pathogenic *Escherichia coli* strain O1:K1:H7 shares strong similarities with human extraintestinal pathogenic *E. coli* genomes. J Bacteriol 189: 3228–3236.
23. Fratamico PM, Yan X, Liu Y, DebRoy C, Byrne B, et al.

(2010) Escherichia coli serogroup O2 and O28ac O-antigen gene cluster sequences and detection of pathogenic *E. coli* O2 and O28ac by PCR. Can J Microbiol 56: 308–316.

24. Moriel DG, Bertoldi I, Spagnuolo A, Marchi S, Rosini R, et al. (2010) Identification of protective and broadly conserved vaccine antigens from the genome of extraintestinal pathogenic *Escherichia coli* . Proc Natl Acad Sci U S A 107: 9072–9077.
25. Mangiamele P, Nicholson B, Wannemuehler Y, Seemann T, Logue CM, et al. (2013) Complete genome sequence of the avian pathogenic *Escherichia coli* strain APEC O78. Genome Announc 1: e0002613.
26. Liu B, Wu F, Li D, Beutin L, Chen M, et al. (2010) Development of a serogroup-specific DNA microarray for identification of *Escherichia coli* strains associated with bovine septicemia and diarrhea. Vet Microbiol 142: 373–378.
27. Hu Q, Tu J, Han X, Zhu Y, Ding C, et al. (2011) Development of multiplex PCR assay for rapid detection of *Riemerella anatipestifer*, *Escherichia coli*, and *Salmonella enterica* simultaneously from ducks. J Microbiol Methods 87: 64–69.
28. Xie Y, Kolisnychenko V, Paul-Satyaseela M, Elliott S, Parthasarathy G, et al. (2006) Identification and characterization of *Escherichia coli* RS218-derived islands in the pathogenesis of *E. coli* meningitis. J Infect Dis 194: 358–364.
29. Welch RA, Burland V, Plunkett G 3rd, Redford P, Roesch P, et al. (2002) Extensive mosaic structure revealed by the complete genome sequence of uropathogenic *Escherichia coli* . Proc Natl Acad Sci U S A 99: 17020–17024.
30. Blattner FR, Plunkett G 3rd, Bloch CA, Perna NT, Burland V, et al. (1997) The complete genome sequence of *Escherichia coli* K-12. Science 277: 1453–1462.
31. Ma X, Wang S, Liu Y, Dai J (2010) Expression and biological characterizations of virulence-related gene fragment E9 of avian pathogenic *Escherichia coli* IMT5155. Anim Husb Vet Med 42: 20–24.

INDEX

Printed in the United States
By Bookmasters